COMPREHENSIVE BIOCHEMISTRY

COMPREHENSIVE BIOCHEMISTRY

SECTION I (VOLUMES 1-4)
PHYSICO-CHEMICAL AND ORGANIC ASPECTS OF BIOCHEMISTRY

SECTION II (VOLUMES 5-11)
CHEMISTRY OF BIOLOGICAL COMPOUNDS

SECTION III (VOLUMES 12-16)
BIOCHEMICAL REACTION MECHANISMS

SECTION IV (VOLUMES 17-21)
METABOLISM

SECTION V (VOLUMES 22-29)
CHEMICAL BIOLOGY

SECTION VI (VOLUMES 30-36)
A HISTORY OF BIOCHEMISTRY

COMPREHENSIVE BIOCHEMISTRY

Series Editors:
ALBERT NEUBERGER
Chairman of Governing Body, The Lister Institute of Preventive Medicine, University of London, London (U.K.)

LAURENS L.M. VAN DEENEN
Professor of Biochemistry, Biochemical Laboratory, Utrecht (The Netherlands)

VOLUME 37

SELECTED TOPICS IN THE HISTORY OF BIOCHEMISTRY PERSONAL RECOLLECTIONS. III.

Volume Editors:
GIORGIO SEMENZA
Laboratorium für Biochemie, ETH-Zentrum Zurich (Switzerland)

RAINER JAENICKE
Institut für Biophysik und Physikalische Biochemie Universität Regensburg Regensburg (F.R.G.)

ELSEVIER
AMSTERDAM · NEW YORK · OXFORD
1990

ISBN 0-444-81216-4 (volume) ISBN 0-444-80151-0 (series)

With 14 plates, 7 figures and 3 tables. This book is printed on acid-free paper.

Published by:
Elsevier Science Publishers B.V.
(Biomedical Division)
P.O. Box 211
1000 AE Amsterdam
The Netherlands

Sole distributors for the USA and Canada:
Elsevier Science Publishing Company, Inc.
655 Avenue of the Americas
New York, NY 10010
USA

Library of Congress Cataloging-in-Publication Data

Selected topics in the history of biochemistry : personal recollections / volume editors, Giorgio Semenza, Rainer Jaenicke.
p. cm. -- (Comprehensive biochemisty ; v. 37. Section VI, A history of biochemistry.)
Includes bibliographical references.
Includes index.
ISBN 0-444-81216-4 (alk. paper)
1. Biochemists--Biography. 2. Biochemistry--Research--History.
I. Semenza, G., 1928- . II. Jaenicke, R. (Rainer), 1930- .
III. Series: Comprehensive biochemistry ; v. 37.
[DNLM: 1. Biochemistry--history--personal narratives. QU 4 C743 v. 37]
QD415.F54 vol. 37
[QP511.7]
574.19'2 s--dc20
[574.19'2'0922]
[B]
DNLM/DLC
for Library of Congress 90-13926
CIP

Printed in The Netherlands

GENERAL PREFACE

The Editors are keenly aware that the literature of Biochemistry is already very large, in fact so widespread that it is increasingly difficult to assemble the most pertinent material in a given area. Beyond the ordinary textbook the subject matter of the rapidly expanding knowledge of biochemistry is spread among innumerable journals, monographs, and series of reviews. The Editors believe that there is a real place for an advanced treatise in biochemistry which assembles the principal areas of the subject in a single set of books.

It would be ideal if an individual or a small group of biochemists could produce such an advanced treatise, and within the time to keep reasonably abreast of rapid advances, but this is at least difficult if not impossible. Instead, the Editors with the advice of the Advisory Board, have assembled what they consider the best possible sequence of chapters written by competent authors; they must take the responsibility for inevitable gaps of subject matter and duplication which may result from this procedure.

Most evident to the modern biochemist, apart from the body of knowledge of the chemistry and metabolism of biological substances, is the extent to which we must draw from recent concepts of physical and organic chemistry, and in turn project into the vast field of biology. Thus in the organization of Comprehensive Biochemistry, sections II, III and IV, Chemistry of Biological Compounds, Biochemical Reaction Mechanisms, and Metabolism may be considered classical biochemistry, while the first and fifth sections provide selected material on the origins and projections of the subject.

It is hoped that sub-division of the sections into bound volumes will not only be convenient, but will find favour among students concerned with specialized areas, and will permit easier future revisions of the individual volumes. Towards the latter end particularly, the Editors will welcome all comments in their effort to produce a useful and efficient source of biochemical knowledge.

M. Florkin†
E.H. Stotz

Liège/Rochester

There is a history in all men's lives.
W. Shapespeare, Henry IV, Pt. 2

History is the essence of innumerable biographies.
T. Carlyle, On History

PREFACE TO VOLUME 37

Perhaps one of the most exciting developments in the biological sciences of our times has been their merging with chemistry and physics with the resulting appearance of biochemistry, biophysics, molecular biology, and related sciences. The nearly explosive development of these 'newcomers' has led to the almost unique situation that these new biological sciences have come of age at a time when their founding fathers, or their scientific sons, are alive and active.

It was therefore an almost obvious idea to ask them to write, for the benefit of both students and senior scientists, personal accounts of their scientific lives. With this idea in mind one of us (G.S.) had already edited two volumes for John Wiley & Sons, which had, however, a somewhat different format.

As in Vols. 35 and 36, the chapters in this volume are meant to complement, with personal recollections, the History of Biochemistry in the Comprehensive Biochemistry series (Vols. 30-33, by M. Florkin and Vol. 34A, by P. Laszlo). It is hoped that the biographical or autobiographical chapters will convey to the reader lively, albeit at times subjective, views on the scientific scene as well as the social environment in which the authors have operated and brought about new concepts and pieces of knowledge. The Editors considered it presumptuous to give the authors narrow guidelines or to suggest changes in the chapters they received; they think that directness and straightforwardness should be given priority over uniformity. The contributions assembled in this volume will convey the flavour of each author's particular personality; whatever the optical distortion of one chapter, it will be compensated for by the views in another.

The development of today's life sciences was acted upon by serious and often tragic historical events. The Editors hope that this message also will reach the readers, especially the young ones.

It proved an impossible task to group the contributions in a strictly logical manner, whether according to subject matter, geographical area, or time. In fact, most contributions cross each of these borders. Nevertheless, the Editors hope that the reader will find these contributions as interesting as they did.

The Editors want to express their gratitude to all individuals who made this series possible; first of all to the authors themselves, who not only wrote the texts, but also willingly collaborated in suggesting further potential contributors, thereby acting as a kind of 'Editorial Board at Large'. Thanks are due to Ms. U. Zilian who typed most of the correspondence.

Swiss Institute of Technology
Zurich, 1989

Giorgio Semenza

University of Regensburg
Regensburg, 1989

Rainer Jaenicke

CONTRIBUTORS TO THIS VOLUME

J.J. ARAGÓN (Introduction for A. Sols)
Instituto de Investigaciones Biomédicas C.S.I.C. and Departamento de Bioquímica, Facultad de Medicina U.A.M., Arzobispo Morcillo 4, 28029 Madrid (Spain)

H. CHANTRENNE
82, Chaussée de Tervuren, 1160 Brussels (Belgium)

H. EISENBERG
Polymer Department, The Weizmann Institute of Science, Rehovot 76100 (Israel)

C. GANCEDO (Introduction for A. Sols)
Instituto de Investigaciones Biomédicas C.S.I.C. and Departamento de Bioquímica, Facultad de Medicina U.A.M., Arzobispo Morcillo 4, 28029 Madrid (Spain)

H.M. KALCKAR
Department of Chemistry, Metcalf Center for Science and Engineering, 590 Commonwealth Ave., Boston, MA 02215 (U.S.A.)

L.L. KISSELEV
Engelhardt Institute of Molecular Biology, The U.S.S.R. Academy of Sciences, 32 Vavilov Street, Moscow 117984 (U.S.S.R.)

A. NEUBERGER
Charing Cross Hospital Medical School, Reynolds Building, St. Dunstans Road, London W6 8RP (U.K.)

M. PERUTZ
MRC Laboratory of Molecular Biology, Hills Road, Cambridge CB2 2QH (U.K.)

A. ROTHSTEIN
Research Institute, The Hospital for Sick Children, Toronto (Canada)

A. SOLS †
Instituto de Investigaciones Biomédicas C.S.I.C. and Departamento de Bioquímica, Facultad de Medicina U.A.M., Arzobispo Morcillo 4, 28029 Madrid (Spain)

C.-L. TSOU
National Laboratory of Biomacromolecules, Institute of Biophysics, Academia Sinica, Beijing, 100080 (China)

LIST OF PLATES

(Photographs reproduced with permission of authors, publishers, and/or owners)

Section VI

A HISTORY OF BIOCHEMISTRY

CONTENTS

VOLUME 37

A HISTORY OF BIOCHEMISTRY

Selected Topics in the History of Biochemistry
Personal Recollections. III.

G. Semenza and R. Jaenicke (Eds.) Selected Topics in the History of Biochemistry: Personal Recollections, III. (Comprehensive Biochemistry Vol. 37) © 1990 Elsevier Science Publishers BV (Biomedical Division)

Chapter 1

Physics and the Riddle of Life

M. PERUTZ

MRC Laboratory of Molecular Biology, Hills Road, Cambridge CB2 2QH (U.K.)

In the early 1940s Erwin Schrödinger, the discoverer of wave mechanics, worked at the Institute for Advanced Studies in Dublin. One day he met P.P. Ewald, another German theoretician who was then professor at the University of Belfast. Ewald, who had been a student in Göttingen before the First World War, gave Schrödinger a paper that had been published in the *Nachrichten aus der Biologie der Gesellschaft der Wissenschaften* in Göttingen in 1935. It was by N.W. Timoféeff-Ressovsky, K.G. Zimmer, and Max Delbrück and was titled 'The Nature of Genetic Mutations and the Structure of the Gene' [1]. Apparently Schrödinger had been interested in that subject for some time, but the paper fascinated him so much that he made it the basis of a series of lectures at Trinity College, Dublin, in February 1943; they were published by Cambridge University Press in the following year, under the title *What Is Life? The Physical Aspect of the Living Cell.*

The book is written in an engaging, lively, almost poetic style ('The probable life time of a radioactive atom is less predictable than that of a healthy sparrow'). It aroused wide interest, especially among young physicists. Up to 1948 it drew 65 reviews, and it has probably by now sold 100 000 copies. It has since become a classic that has provided a nourishing habitat for historians, sociologists, and philosophers of science who have commented on it, on the comments on it, or on the comments on the comments on it. A Ph.D. thesis published on the subject in 1979 contains over 120 references, excluding the 65 reviews [2]. François Jacob has explained the reasons for the book's impact best:

> 'After the war, many young physicists were disgusted by the military use that had been made of atomic energy. Moreover, some of them had wearied of the turn experimental physics had taken . . . of the complexity imposed by the use of big machines. They saw in it the end of a science and looked around for other activities. Some looked to biology with a mixture of diffidence and hope. Diffidence because they had about living beings only the vague notions of the zoology and botany they remembered from school. Hope, because the most famous of their elders had painted biology as full of promise. Niels Bohr saw it as the source of new laws of physics. So did Schrödinger, who foretold revival and exaltation to those entering biology, especially the domain of genetics. To hear one of the fathers of quantum mechanics ask himself: 'What is Life?' and to describe heredity in terms of molecular structure, of interatomic bonds, of thermodynamic stability, sufficed to draw towards biology the enthusiasm of young physicists and *to confer on them a certain legitimacy* [The emphasis is mine]. Their ambitions and their interests were confined to a single problem: the physical nature of the genetic information.' [3]

Ilya Prigogine found that Schrödinger's book was an inspiration for his work on nonequilibrium thermodynamics; Seymour Benzer, Maurice Wilkins, and Gunther Stent have said that the book was decisive in drawing them from physics into biology; Francis Crick told me that he found it interesting but would have switched to biology anyway. I was already in the thick of trying to solve the structure of proteins when it was published, and I may have been encouraged by its quotation of C.D. Darlington's view than genes are made of protein. Crick wrote in 1965:

> 'On those who came into the subject just after the 1939–1945 war Schrödinger's little book . . . seems to have been peculiarly influential. Its main point – that biology needs the stability of chemical bonds and that only quantum mechanics can explain this – was one that only a physicist would feel it necessary to make, but the book was extremely well written and conveyed in an exciting way the idea that, in biology, molecular explanations would not only be extremely important but also that they were just around the corner. This had been said before, but Schrödinger's book was very timely and attracted people who might otherwise not have entered biology at all.' [4]

However, in 1971 Crick added:

> 'I cannot recall any occasion when Jim Watson and I discussed the limitations of Schrödinger's book. I think the main reason for this is that we were strongly influenced by Pauling, who had essentially the correct set of ideas. We therefore never wasted any time discussing whether we should think in the way Schrödinger did or the way Pauling did. It seemed quite obvious to us we should follow Pauling.' [5]

Neither can I recall Crick, Watson, John Kendrew, and me ever discussing the bearing of Schrödinger's book on structural molecular biology during our years together at the Cavendish Laboratory. Stanley Cohen wrote that few of the many scientists participating in Delbrück's phage course at Cold Spring Harbor in 1944 had read Schrödinger, 'and in all the social and intellectual activities of these [postwar] summers I do not recall any mention of Schrödinger.' [2]. The participants included such pioneers of molecular genetics and biochemistry as Salvador Luria, Alfred Hershey, André Lwoff, Jacques Monod, and Jean Brachet. Hence the book does not appear to have had much impact on the people already working in the field.

Schrödinger's book is written for the layman and begins with the chapter 'The Classical Physicist's Approach to the Subject.' He asks how events in space and time taking place in a living organism can be accounted for by physics and chemistry.

> 'Enough is known about the material structure of life to tell exactly why present-day physics cannot account for life. That difference lies in the statistical point of view. It is well-nigh unthinkable that the laws and regulations thus discovered [i.e., by physics] should apply immediately to the behaviour of systems which do not exhibit the structure on which these laws and regularities are based.'

Schrödinger jumps to this conclusion after reading that genes are specific molecules of which each cell generally contains no more than two copies. He had entered Vienna University in 1906, the year that Ludwig Boltzmann died, and had been taught physics by Boltzmann's pupils. He remained deeply influenced by Boltzmann's thoughts throughout his life. According to Boltzmann's statistical thermodynamics, the behaviour of single molecules is unpredictable; only the behaviour of large numbers is predictable. In genetics, therefore, Schrödinger concludes: 'We are faced with a mechanism entirely different from the probabilistic ones of physics'. This difference forms the guiding theme of his book.

In the first chapter, Schrödinger illustrates the meaning of statistical thermodynamics by the examples of Curie's law, of Brownian motion and diffusion, and of the $\sqrt{n}$ rule. His next two chapters, on the hereditary mechanisms and on mutations, give brief popular introductions to textbook knowledge on these subjects available at the time. They reveal one vital misconception in Schrödinger's mind: 'Chromosomes,' he writes, 'are both the law code and the executive power of the living cell.' In fact, biochemists had shown that the executive power resides in enzyme catalysts, and in 1941 G.W. Beadle and E.I. Tatum discovered that single genes determine single enzymatic activities [6]. That discovery led to the one-gene, one-enzyme hypothesis, an idea that had already been foreshadowed by the Cambridge biochemist and geneticist J.B.S. Haldane [7] and that has become central to an understanding of biology. Schrödinger does not appear to have heard of this.

The next two chapters form the backbone of his book; they are called 'The Quantum-Mechanical Evidence' and 'Delbrück's Model Discussed and Tested' and, as C.H. Waddington first spotted, they are largely paraphrased versions of the paper by Timoféeff, Zimmer, and Delbrück [8]. That paper covers 55 pages and is divided into four sections. The first section, by Timoféeff, describes the mutagenic effects of x-rays and γ-rays on the fruit fly *Drosophila melanogaster*. He shows that the spontaneous mutation rate of the fly is low and that it is raised about fivefold by a rise in temperature of 10°C. Ionizing radiations increase that rate as a linear function of the dose, independent of its time distribution, of the wavelength, and of the temperature during irradiation.

The second section of the paper is by Zimmer and applies the target theory to Timoféeff's results. The number of mutations $x = a(1 - e^{-kD})$, where a and k are constants and D is the dose. Zimmer next asks whether the mutations had arisen by the direct absorption of quanta, by the passage of secondary electrons through a sensitive volume, or by the generation of ion pairs. If the dose is measured in röntgens, the number of quanta required to produce a given dose diminishes with diminishing wavelength. Thus, direct absorption of of quanta is inconsistent with the linear dependence of the mutation rate on the dose. The same applies to secondary electrons. Only the number of ion pairs is proportional to the dose, obviously, since that is how the dose is measured. Zimmer, therefore, concludes that a single hit suffices for the production of one mutation and that this hit consists of either the formation of an ion pair or a transition to higher energy.

The third section of the paper is by Delbrück and bears the title 'Atomphysikalisches Modell der Mutation' ('A Model of Genetic Mutation Based on Atomic Physics'). Delbrück reminds us that the concept of the gene began as an abstract one, independent of physics and chemistry, until it was linked to chromosomes and later to parts of chromosomes, which were estimated to be of molecular size. Since he and his colleagues had no means of discovering the chemical nature of genes directly, they attacked the problem indirectly by studying the nature and the limits of their stability and by asking if these were consistent with the knowledge that atomic theory has provided about the behaviour of well-defined assemblies of atoms.

Such assemblies can undergo discrete and spontaneous transitions of vibrational and electronic states. Vibrational transitions are very frequent and involve no chemical changes. From electronic transitions the assemblies may either revert to the ground state or reach a new equilibrium state after undergoing an atomic rearrangement, for example to a tautomeric form. The fivefold rise in spontaneous mutation frequency for a 10°C rise in temperature leads Delbrück to derive an activation energy of about 1.5 eV and an average lifetime of a few years, when half the molecules composing the gene will have undergone an electronic transition.

Delbrück then describes how on the average x-rays lose energy to secondary electrons in portions of 30 eV ionization, which is 1000 × kT and 20 times the energy of activation of 1.5 eV needed for a spontaneous mutation. However, to produce as much as 1.5 eV, the ionization must not occur too far away from its target. We knew too little about the ways in which the energy of photoelectrons is dissipated to determine the absolute value of the dose needed to induce a mutation with a probability of unity, but that dose, expressed as the number of ionizations per unit volume, was likely to be about ten to 100 times smaller than the number of atoms of the gene per unit volume. Delbrück now calculates that dose as follows.

A frequently observed x-ray mutation (eosin) occurs with a dose of 6000 röntgens once in 7000 gametes. Hence a probability of unity of its occurrence needs a dose of 42×10^6 röntgens. One röntgen produces about 2×10^{12} ion pairs in 1 ml of water, whence 42×10^6 röntgens produce about 10^{20} ion pairs. Since 1 ml of water contains about 10^{23} atoms, this means that at least 1 in 1000 atoms becomes ionized. However, Delbrück cautiously refrains from concluding that a gene is likely to consist of a thousand atoms.

Schrödinger used Delbrück's result to point out 'that there is a fair chance of producing a mutation when an ionization occurs not more than about ten atoms away from a particular spot on the chromosome,' but research published while Schrödinger was writing his book showed such calculations to be meaningless. In a paper that appeared in *Nature* in June 1944, Joseph Weiss pointed out that the biological effects of ionizing radiation are caused principally by the generation of hydroxyl radicals and hydrogen atoms in the surrounding water [9]. E. Collinson, F.S. Dainton, D.R. Smith, and S. Tazuke [10] and, independently, G. Czapski and H.A. Schwartz [11], later discovered that the supposed hydrogen atoms were in fact hydrated electrons [12]. Hydroxyl ions and hydrated electrons have half-lives of about 1.0 ms (assuming a concentration of 1 μmol H_2O_2) and of 0.5 ms, respectively, in which times they can diffuse to their targets even if they are generated more than a thousand atomic diameters away from them.

Delbrück concludes that it is premature to make the description of the gene any more concrete than the following:

> 'We leave open the question whether the single gene is a polymeric entity that arises by the repetition of identical atomic structures or whether such periodicity is absent; and whether individual genes are separate atomic assemblies or largely autonomous parts of a large structure, i.e., whether a chromosome contains a row of separate genes like a string of pearls, or a physicochemical continuum.'

I found the Timoféeff, Zimmer, and Delbrück paper, and especially Delbrück's part, most impressive. Delbrück was a theoretical physicist whose interest in biology had been aroused by Niels Bohr's lecture 'Light and Life', delivered in Copenhagen in 1932. In that lecture Bohr had said:

> 'The existence of life must be considered as an elementary fact that cannot be explained, but must be taken as a starting point in biology, in a similar way as the quantum of action, which appears as an irrational element from the point of view of classical mechanical physics, taken together with the existence of elementary particles, forms the foundation of atomic physics. The asserted impossibility of a physical or chemical explanation of the function peculiar to life would be . . . analogous to the insufficiency of the mechanical analysis for the understanding of the stability of atoms.' [13]

The search for Bohr's elementary fact of life had fired Delbrück's imagination. He was only 29 years old, working as assistant to Otto Hahn and Lise Meitner in the Kaiser Wilhelm Institut für Chemie in Berlin and doing his biological work as a sideline, but his paper shows the maturity, judgement, and breadth of knowledge of someone who had been in the field for years. It is imaginative and sober, and its carefully worded predictions have stood the test of time. The paper won him a Rockefeller fellowship to Pasadena to work with the *Drosophila* geneticist T.H. Morgan. There he met Linus Pauling, with whom he published an important paper in 1940. That paper was an attack on the German theoretician Pascual Jordan, who had advanced the idea that there exists a quantum-mechanical stabilizing interaction, operating preferentially between identical or near-identical molecules, which is important in biological processes such as the reproduction of genes. Pauling and Delbrück pointed out that interactions between molecules were now rather well understood and give stability to two molecules of *complementary* structure in juxtaposition, rather than to two molecules with necessarily *identical* structures. Complementariness should be given primary consideration in the discussion of the specific attraction between molecules and their enzymatic synthesis [14]. In 1937 the Cambridge geneticist and biochemist J.B.S. Haldane had made a similar suggestion:

> 'We could conceive of a [copying] process [of the gene] analogous to the copying of a grammophone record by the intermediation of a negative, perhaps related to the original as an antibody is to an antigen.' [15]

Schrödinger mentions neither of these important ideas.

Schrödinger's last two chapters do contain his own thoughts on the nature of life. In 'Order, Disorder, and Entropy' he argues that

> 'the living organism seems to be a macroscopic system which in part of its behaviour approaches to that purely mechanical (as contrasted with thermodynamical) conduct to which all systems tend, as the temperature approaches the absolute zero and the molecular disorder is removed.'

He comes to this strange conclusion on the ground that living systems do not come to thermodynamic equilibrium, defined as the state of maximum entropy. They avoid doing so, according to Schrödinger, by feeding on negative entropy. I suspect that Schrödinger got that idea from a lecture by Ludwig Boltzmann on the second law, delivered before the Imperial Austrian Academy of Sciences in 1886:

> 'Hence the general battle for existence of living organisms is not one for the basic substances – these substances are abundant in the air, in water and on the ground – also not for energy that every body contains abundantly, though unfortunately in a non-available form, but for entropy which becomes available by the transition of energy from the hot sun to the cold earth.' [16]

Franz (later Sir Francis) Simon, then at Oxford, pointed out to Schrödinger that we do not live on $-T\Delta S$ alone but on free energy [2] Schrödinger deals with that objection in the second edition of his book; he writes that he had realized the importance of free energy but had regarded it as too difficult a term for his lay audience; to me this seems a strange argument, since the meaning of entropy is surely harder to grasp. Schrödinger's postscript did not satisfy Simon, who pointed out to him in a letter that

> 'the reactions in the living body are only partly reversible and consequently heat is developed of which we have to get rid to the surroundings. With this irreversibly produced heat also flow small amounts (either + or –) of reversibly produced heat ($T\Delta S$), but they are quite insignificant and therefore cannot have the important effects on life processes which you assign to them.' [2]

In fact, it was known when Schrödinger wrote his book that the primary currency of chemical energy in the cells is ATP, and that the free energy stored in ATP is predominantly enthalpic. Prigogine disagrees with Simon's and my objections and explains that organisms in a steady state liberate as much positive entropy as the negative entropy they absorb. I find this argument hard to follow, because plants absorb free energy only in the form of radiation, which they use to create order from disorder. In other words, they convert enthalpy into negative entropy.

The final chapter, 'Is Life Based on the Laws of Physics?', reiterates and amplifies the central argument already stated at the beginning of the book. According to Delbrück, Schrödinger writes, the gene is a molecule, but the bond energies in molecules are of the same order as the energy between atoms in solids, for example, in crystals, where the same pattern is repeated periodically in three dimensions and where there exists a continuity of chemical bonds extending over large distances. This leads him to the famous hypothesis that the gene is a linear one-dimensional crystal, simply lacking a periodic repeat: an aperiodic crystal. A single such crystal, or a pair of them, directs the orderly process of life. Yet, according to Boltzmann's laws, their behaviour must be unpredictably erratic. Schrödinger concludes that

> 'we are faced with a mechanism entirely different from the probabilistic one of physics, one that cannot be reduced to the ordinary laws of physics, not on the ground that there is any 'new force' directing the behaviour of single atoms within an organism, but because the construction is different from any yet tested in the physical laboratory.'

I wonder why Schrödinger did not adhere to Delbrück's much better formulation of 'a polymeric entity that arises by the repetition of identical atomic structures'. One could argue over the distinction between aperiodic and identical, but Delbrück could not have meant structures that are *completely* identical, since these could contain no information. Schrödinger does suggest that the genetic information might take the form of a linear code, analogous to the Morse code.

He argues that the nature of the gene allows only one general conclusion:

> 'Living matter, while not eluding the laws of physics as established to date, is likely to involve other laws of physics hitherto unknown which, however, once they have been revealed, will form as integral a part of this science as the former.'

Schrödinger is thus drawn to the same conclusion that Niels Bohr had been, apparently unknown to Schrödinger, 12 years earlier, and one that young physicists found equally inspiring.

Schrödinger next refers to a paper by Max Planck, 'Dynamical and statistical laws.' Dynamical laws control large-scale events such as the motions of the planets or of clocks. Clockworks function dynamically, because they are made of solids kept in shape by London-Heitler forces, strong enough to elude disorderly heat motions at ordinary temperatures. An organism is like a clockwork in that it also hinges on a solid: the aperiodic crystal forming the hereditary substance, largely withdrawn from the disorder of heat motion. The single cog of this clockwork is not of coarse human make but is the finest piece ever achieved along the lines of the Lord's quantum mechanics. C.D. Darlington at Oxford had advised Schrödinger that genes are likely to be protein molecules, as was then generally believed; Schrödinger quotes that information but does not mention that proteins are long chain polymers made up of some 20 different links that might form the kind of aperiodic patterns or linear code he had in mind. He must also have been unaware that the true chemical nature of that 'finest piece' was actually published while he was writing his book. In January 1944 there appeared in the *Journal of Experimental Medicine* a paper by O.T. Avery, C.M. McLeod and Maclyn McCarty reporting conclusive evidence that genes are made not of protein but of DNA [17]. In the fullness of time, that discovery has led the majority of scientists to the recognition that life can be explained on the basis of the existing laws of physics.

The apparent contradictions between life and the statistical laws of physics can be resolved by invoking a science largely ignored by Schrödinger. That science is chemistry. When Schrödinger wrote

> 'The regular course of events, governed by the laws of physics, is never the consequence of one well-ordered configuration of atoms, not unless that configuration repeats itself many times,'

he failed to realize that this is exactly how chemical catalysts work. Given a source of free energy, a well-ordered configuration of atoms in a single molecule of an enzyme catalyst can direct the formation of an ordered stereospecific compound at a rate of 10^3 to 10^5 molecules a second, thus creating order from disorder at the ultimate expense of solar energy. Haldane pointed this out in 1945, in his review of Schrödinger's book [18].

Chemists could also have told him that there is no problem in explaining the stability of polymers that living matter is made of, because their bond energies range from 3 *eV* upward, corresponding to a half-life for each bond of at least 10^{30} years at room temperature. The difficulty resides in explaining how their aperiodic patterns are accurately reproduced in each generation. There is no mention of this central problem in Schrödinger's book.

Research has cleared away the apparent contradiction between the randomness of single molecular events and the orderliness of life that exercised Schrödinger. The orderliness depends on fidelity of reproduction of the genetic message every time a cell divides, and the fidelity of protein synthesis. The genetic message is encoded in a sequence of nucleotides along a chain of DNA. That chain is paired to another carrying a complementary sequence of bases. The two chains are coiled around each other in a double helix in which each adenine (A) forms two hydrogen bonds with a thymine (T) and each guanine (G) forms three hydrogen bonds with a cytidine (C). At cell division the two strands of the parent double helix separate, and each forms a template for the formation of a new complementary strand, resulting in two daughter double helices with the same base sequence as the parent double helix. The necessary monomers are supplied as nucleoside triphosphates, which carry the energy for the formation of the growing chain in the form of an energy-rich phosphorus-oxygen bond; the synthesis of new chain links is catalyzed by an enzyme or system of enzymes that attach themselves to the end of the double helix, unwind it, hold each parent strand rigidly in the conformation needed to catalyze the formation of a new chain link, move forward one step, catalyze the formation of the next link, and so on. Arthur Kornberg and his colleagues at Stanford University have worked out how these enzymes function in *Escherichia coli* [19].

How do they ensure that at each step of elongation only the nucleotide complementary to that on the parent strand is linked to the daughter strand? On the one hand, chemical kinetics tell us that the four alternative trinucleotides must be bombarding the active site of the enzyme at the diffusion rate of about 10^9 molecules a second. On the other hand, their rates of dissociation from the active site vary, depending on their ability to form complementary hydrogen bonds with the base of the parent strand. Only if the incoming nucleoside triphosphate is oriented correctly in the active site of the enzyme and if the hydrogen-bonding groups of its base are complementary to those of the parent base will the new nucleotide remain in the active site long enough for a new chain link to be formed.

As Delbrück foresaw, the main source of spontaneous mutations is not the cleavage of covalent bonds in the parent strand. One source was believed to be the existence of tautomeric forms of the bases, which differ in their arrangement of hydrogen-bond donor and acceptor groups. Such changes allow a G to pair with a T, or a C with an A, or a G with an A, but in fact such mispairing probably occurs by another mechanism at apparently less cost in free energy. X-ray analysis of synthetic oligodeoxyribonucleotides has shown that mismatched bases can form hydrogen bonds with each other and be incorporated in the double helix with only slight distortions of the bond angles in the phosphate ester chain. Finally, a G-A pair can form, again with only minor distortions of the double helix, if either of the two bases is inverted about its bond to the ribose [20]. Judging by the frequency of these mistakes, the error rate in the reproduction of the genetic message should be 10^{-4} to 10^{-5} per nucleotide; in fact, the measured error rate in *E. coli* is 10^{-8} to 10^{-10}, at least three orders of magnitude less than the theoretically expected one.

How does Nature defeat statistical thermodynamics? One of its tricks is a proofreading and editing mechanism unraveled initially by Kornberg and others at Stanford and subsequently by A.R. Fersht and his colleagues, first at Stanford and later in London at Imperial College [19]. In *E. coli* the enzyme that catalyzes the elongation of the DNA chain has a 'second look' at the base pair just joined to the daughter double helix; it excises wrongly paired bases, then incorporates the correct ones. Proofreading and editing, however, are also subject to the errors imposed by chemical kinetics, which means that once in a thousand times, say, the correct base is excised and must then be reincorporated in the growing chain. This costs energy. If proofreading is too rigorous, too much energy is wasted in excising and reincorporating correctly paired bases; if it is not rigorous enough, too many copying errors are left uncorrected.

Using bacterial mutants that are either exceptionally error-prone or error-free, Fersht has measured the cost of fidelity and shown how Nature achieves the best compromise by increasing fidelity by two orders of magnitude to about 5×10^{-7}, but still not enough to account for the observed mutation frequency of only 10^{-8} to 10^{-10} in the replication of *E. coli.* [21] B.W. Glickman and Miroslav Radman have discovered a second proofreading mechanism that can distinguish the parent strand from the daughter strand by virtue of the fact that some of the parent strand's bases have become methylated, while those of the daughter strand are still bare. When the mechanism finds a mismatched base pair in the daughter strand, it excises it and replaces it with the correct one, thus reducing the error rate by the missing one or two orders of magnitude [22]. The error rate of a viral RNA by an enzyme that is incapable of repairing mismatches was found to be about 10^{-4} per doubling, a quite unacceptably high rate even for a bacterium [23].

The genetic message's function is to code for the sequence of amino acids along protein chains, but DNA does not code for proteins directly. Instead, the genetic message is first transcribed into messenger RNA and then translated into a sequence of amino acids in a protein chain. If enzymes are to work effectively, mistakes in the sequence must be infrequent. Transcription of DNA into RNA is not subject to proofreading and excision repair, perhaps because messenger RNAs are rarely longer than 10^4 base pairs, so that greater error rates are acceptable in them than in DNA replication. However, translation of RNA into protein presents problems that were first pointed out by Linus Pauling, characteristically, before the enzymatic machinery for protein synthesis was unraveled [24].

Certain pairs of amino acids differ by only one methyl group. It is easy to imagine an active site of an enzyme that efficiently rejects an amino acid that is a misfit because it is too large by one methyl group, but it is hard to see how an active site could discriminate well against an amino acid that merely leaves a hole, because it is too short by one methyl group. The ratio of the reaction rates V of two amino acids A and B whose side chains differ in length by a single methyl group, one just fitting the active site and the other being too short, is given by the equation.

$$V_A/V_B = \frac{[A]}{[B]} e^{-\Delta G_b/RT}$$

where ΔG_b is the difference in Gibbs binding energy resulting from the contribution of the side chain. ΔG_b is not likely to be more than 3 *k*cal/mol. If [A] = [B] this means that $V_A/V_B < 200$, implying an error rate greater than 0.5% [21]. Yet when R.B. Loftfield and D. Vanderjagt tried to measure the error rate in such a situation, they found it to be only 3 parts in 10 000 and concluded

> 'that the precision . . . of peptide assembly is very great, much greater than can be deduced from the study of non-biological chemical reactions.' [25]

Fersht showed how this low error rate is achieved without making Boltzmann turn in his grave [21]. Nature makes use of the fact that the selection of the correct amino acid into the growing protein chain proceeds in two stages, both catalyzed by the same enzyme. In the first stage the amino acid is coupled to a phosphate to give it an energy-rich bond; in the second stage it is transferred to an adapter, a molecule of RNA that carries the anticodon triplet of nucleotide bases complementary to the coding triplet for that particular amino acid. At the first stage of the reaction, the enzyme rejects amino acids with side chains that are misfits in the active site because they are too long, but reacts with amino acids whose side chains are too short with the large error rates predicted by Pauling. The second stage of the reaction takes place at a different active site of the same enzyme. It is constructed so as to fit those amino acids that were too short for the previous active site. It cleaves them from the adapter RNA and sets them free some hundreds of times faster than the correct amino acid. This second stage can thus reduce the error by a further two orders of magnitude, giving a total error date of only 10^{-4}. Fersht calls this a double sieve mechanism: the first sieve rejects the amino acids that are too large; the second rejects the ones that are too small. A further stage of editing may reduce possible errors in the recognition by the coding triplet on the messenger RNA of the anticodon triplet on the adaptor RNA, thus ensuring incorporation of the correct amino acid into the growing protein chain [26–28].

We can see that life has resolved the apparent conflict between the unpredictable behaviour of single molecules and the need for order by making enzymes sufficiently large to stabilize them in unique structures, capable of immobilizing other molecules in their active sites and bringing them into juxtaposition so that they can react at high rates. Yet enzymes are long-chain polymers. What makes their chains fold up to form unique and largely rigid structures when entropy drives them to form random coils? X-ray analysis has shown the interior of proteins to be closely packed jigsaws of amino acids with hydrocarbon side chains that adhere to each other. They do so partly by dispersion forces, which are enthalpic, and partly thanks to the entropy that is gained by the exclusion of water from the protein interior. When the protein chain achieves maximum entropy by unfolding to a random coil, both polar and nonpolar groups are exposed to water, which adheres to them and becomes immobilized so that its entropy is diminished. When the chain folds up to its unique structure, the polar groups on the main chain form hydrogen bonds with each other, the nonpolar side chains pack together, and the bound water molecules are set free. The resulting gain in translational and rotational entropy of the water more than compensates for the loss of rotational entropy of the protein chain. Thus, it is the water molecules' anarchic distaste for the orderly regimentation imposed on them by the unfolded protein chain that provides a major part of the stabilizing free energy of the folded one and keeps it in its unique, enzymatically active structure.

I feel that I should not close this story without telling what became of the scientists whose paper Schrödinger popularized.

Delbrück, whose Rockefeller fellowship had taken him to Pasadena, stayed there, with short interruptions, for the remainder of his life. In the early 1940s he founded bacteriophage genetics and later, with Salvador Luria, bacterial genetics; he became the leader of an enthusiastic band of young people who developed these new fields of research. In 1969 he, Luria, and A.D. Hershey received the Nobel Prize in physiology or medicine 'for their discoveries concerning the replication mechanism and the genetic structure of viruses'. Delbrück died in Pasadena in 1981.

While Delbrück's life was happy, Timoféeff's life seems tragic to me, even though I am told that he did not regard it as such. He began his research on *Drosophila* in Moscow in the early 1920s. According to Zhores Medvedev,

> 'In 1924, the Soviet Government made a special exchange agreement with Germany. The famous Kaiser Wilhelm Institute for Brain Research in Berlin-Buch was invited to help organize in Moscow a laboratory for brain research, specially designed to study the brain of Lenin, who had died in January 1924. (At the time of his death, Lenin was considered to be the greatest of geniuses, and his brain was expected to be unique.)' [29]

In a lecture held in the new Moscow Laboratory, Oskar Vogt, the director of the Kaiser Wilhelm Institute, said that deep in the third layer of Lenin's cerebral cortex he found pyramidal cells larger and more numerous than he had ever observed before, and he linked them to Lenin's exceptional powers of associative thought, just as an athlete's strength is linked to his powerfully developed muscles [30]. Modern brain research discounts these conclusions, but at the time they inspired an enthusiastic, popular article in one of the big Berlin dailies by Arthur Koestler, the later novelist, who had embraced the Communist faith.

In exchange for Vogt's services, the Soviet Academy of Sciences promised to help set up in Vogt's institute in Berlin a laboratory of experimental genetics. Vogt owned a large collection of bumble bees. He was convinced that the different species of bumble bees had arisen by Lamarckian inheritance of acquired characters rather than by mutation and natural selection, and he needed a geneticist to prove his theory. Among the young scientists who were recommended to start work in Berlin was Timoféeff. He left for Germany in 1926 and founded a laboratory for the study of *Drosophila* in

Vogt's institute. He never proved Vogt's Lamarckian ideas, but instead became one of the world's leading Mendelian geneticists. Contemporaries describe him as a physical and intellectual giant; in Russia his nickname had been the Wild Boar.

In the 1930s Timoféeff thought of returning to Russia, but his friends advised him that it would be unsafe because Stalin's persecution of Mendelian geneticists had already begun. His younger brothers were arrested, and one was executed. In Germany, during the war, his 17-year-old son joined an underground anti-Fascist group. He was caught by the Gestapo and disappeared. After the end of the war, when the Russians occupied Berlin, German colleagues advised him to flee to the West, but he decided to stay with his precious collection of flies. In August 1945, Timoféeff was arrested by the Soviet secret police, sentenced to ten years' hard labour, and sent to a prison camp in North Kazakhstan. Later he shared a prison cell with Alexander Solzhenitsyn and 22 other prisoners in Bytyrsky. Solzhenitsyn described in *The Gulag Archipelago* how Timoféeff's irrepressible enthusiasm for science made him organize scientific seminars even in that prison cell. Apparently Solzhenitsyn also used him as a model for the scientist in *The First Circle.*

In 1947 the physicist Frédéric Joliot-Curie wrote to L.P. Beria, the head of the Russian secret police, on behalf of the French Academy of Sciences and asked for Timoféeff's release on the grounds that he was a valuable scientist and should be given a chance to do research. Joliot-Curie's intervention saved Timoféeff's life. He had been at death's door, but after several months of recuperation in a Moscow hospital, he was sufficiently restored to set up a new secret prison research institute on radiation biology east of the Urals.

Previously, in September 1945, the Russians had also arrested Karl G. Zimmer with two of his colleagues and taken them to the Lubljanka Prison in Moscow for interrogation. After some time they were sent to work in a uranium factory not far from the city. When Timoféeff set up his new institute, he asked that Zimmer and his colleagues, as well as his own wife, should be allowed to join him there. Timoféeff's eyesight had been ruined by starvation, and his wife read the scientific literature to him. After Stalin's death they were released from prison but continued to work in Sverdlovsk. In 1964 Timoféeff was asked to organize the Department of Genetics and Radiobiology at the new Institute of Medical Radiology in Obninsk, where Medvedev, the geneticist and author of the famous book *The Rise and Fall of Lysenko*, joined him. Medvedev described Timoféeff as a great man and a brilliant scientist; his mastery of many fields of genetics and biology, his dynamism and personal magnetism stimulated the work of the entire laboratory. Peter Kapitsa became his close friend.

Timoféeff retired in 1970 on a pension so meager that it left him almost destitute. He died in 1981, in the same year as his friend Delbrück, who actually came to visit him in Obninsk after having been given the Nobel Prize in Stockholm. But for Schrödinger's book, the name of Timoféeff would have remained unknown outside the circles of genetics and radiation biology.

Zimmer returned to West Germany in 1955. He became one of the first to recognize the importance of electron spin resonance for radiation biology and to prove that ionizing radiations generate free radicals in biological molecules. In 1957 he was offered a chair in Heidelberg combined with the directorship of a new Department of Radiation Biology at the Institute for Nuclear Research in Karlsruhe. There he worked on the effects of ionizing radiations on DNA and other biologically important molecules, and his laboratory became a successful center for fundamental and applied radiobiology. He also published a book on that subject [31]. He died in Karlsruhe in 1988.

As a final irony H. Traut, working in Zimmer's laboratory, found Timoféeff's linear dose–response curve to have been an artifact. He showed that the mutation rate of *Drosophila* germ cells varies widely at different stages of their development. If males are irradiated and then mated, the frequency of mutation among the offspring varies with the time that elapsed between the two events, because the sperm that fertilizes a female five days after irradiation was at an earlier stage of development when it was irradiated than the sperm that fertilizes a female one day after irradiation. At all stages the dose–response curves are nonlinear. Traut demonstrated that a linear response curve, similar to those observed by Timoféeff, is

obtained by summing the different dose–response curves produced by matings during the first four days of irradiation [32].
by summing the different dose–response curves produced by matings during the first four days of irradiation [32].

Zimmer comments:

> 'This result removes one of the foundation stones of the Green Pamphlet [as the Timoféeff, Zimmer, and Delbrück paper became known]. Strangely enough, that does not seem to matter any more, for two reasons: (i) the concept of the gene and modern trends in genetic research as well as in radiation biology have changed considerably during thirty years, and (ii) the Green Pamphlet has served a useful purpose by helping to initiate these modern trends.' [33]

Timoféeff's observation of a fivefold rise in the mutation rate accompanying a 10°C rise in temperature formed the basis for Delbrück's estimate of the energy needed for spontaneous mutations. That observation is now known to have no general validity, because other mutation rates have been found to be independent of temperature or even to drop with rising temperature [34]. These findings remove the other cornerstone of the Green Pamphlet, confirming Karl Popper's dictum that even wrong experimental results may sometimes help the progress of science.

REFERENCES

1 N.W. Timoféeff-Ressovsky, K.G. Zimmer and Max Delbrück, Nachrichten aus der Biologie der Gesellschaft der Wissenschaften Göttingen, 1 (1935) 189–245.
2 E.J. Yoxen, History of Science, 17 (1979) 17–52.
3 F. Jacob, The Logic of Living Systems, Allen Lane, London, 1974.
4 F.H.C. Crick, Brit. Med. Bull., 21 (1965) 183–186.
5 F.H.C. Crick, quoted by R.C. Olby, J. Hist. Biol., 4 (1971) 119–148.
6 G.W. Beadle and E.L. Tatum, Proc. Natl. Acad. Sci. 27 (1941) 499–506.
7 J.B.S. Haldane, in J. Needham and D.E. Green (Eds.), The Biochemistry of the Individual in Perspectives of Biochemistry, Cambridge University Press, Cambridge, 1937, pp. 1–10.
8 C.H. Waddington, Nature 221 (1969) 318–321.
9 J. Weiss, Nature 153 (1944) 748–750.
10 E. Collinson, F.S. Dainton, D.R. Smith, and S. Tazuke, Proc. Chem. Soc. (1962) 140–144.
11 G. Czapski and H.A. Schwartz, J. Phys. Chem. 66 (1962) 471–479.
12 F.S. Dainton, Chem. Soc. Rev. 4 (1975) 323–362.

13 N. Bohr, Nature 131 (1933) 458–460.
14 L. Pauling and M. Delbrück, Science 92 (1940) 77–79.
15 J.B.C. Haldane, The Biochemistry of the Individual, in Perspectives in Biochemistry. Cambridge University Press, Cambridge, 1937, pp. 1–10.
16 L. Boltzmann, Der zweite Hauptsatz der mechanischen Wärmetheorie, Sitzungsberichte der Kaiserlichen Akademie der Wissenschaften, Vienna, 1886.
17 O.T. Avery, C.M. McLeod, and Maclyn McCarty, J. Exp. Med. 79 (1944) 137–158.
18 J.B.S. Haldane, Nature 155 (1945) 375–376.
19 A. Kornberg, DNA Replication, Freeman, San Francisco, 1980; and A. Kornberg, Supplement to DNA Replication, Freeman, San Francisco, 1982.
20 O. Kennard, Structural Studies of Base Pair Mismatches, in Structure and Expression, DNA and Drug Complexes, R.H. Sarma and M.H. Sarma (Eds.), Adenine Press, New York, 1988, pp. 1-25.
21 A.R. Fersht, Proc. Roy. Soc. B 212 (1981) 351–379.
22 B.W. Glickman and M. Radman, Proc. Nat. Acad. Sci. USA 77 (1980) 1063–1067. For a review of mismatch repair in *E. coli* see M. Radman and R. Wagner, Annu. Rev. Genet. 20 (1986) 523-538; Sci. Am. (1988) 24.
23 E. Batschelet, E. Domingo and C. Weissman, Gene 1 (1976) 27–33.
24 L. Pauling, Festschrift Prof. Dr. Arthur Stoll Siebzigsten Geburtstag, 1958, pp. 597-602.
25 R.B. Loftfield and D. Vanderjagt, Biochem. J. 128 (1972) 1353–1356.
26 J.J. Hopfield, Proc. Nat. Acad. Sci. USA 77 (1974) 4135–4139.
27 R.C. Thompson and P.J. Stone, Proc. Natl. Acad. Sci. USA 74 (1977) 198–202.
28 J.L. Yates, J. Biol. Chem. 254 (1979) 1150–1154.
29 Z.A. Medvedev, Genetics 100 (1982) 1–5.
30 O. Vogt, J. Psychol. Neurol. 40 (1929) 108.
31 K.G. Zimmer, Quantitative Radiation Biology, Oliver and Boyd, Edinburgh, 1961.
32 H. Traut, Dose-Dependence of the Frequency of Radiation-induced Recessive Sex-linked Lethals in *Drosophila melanogaster,* with Special Consideration of the Stage Sensitivity of the Irradiated Germ Cells, in F.H. Sobels (Ed.), Repair from Genetic Radiation Damage, Pergamon, London, 1963, p. 359.
33 K.G. Zimmer, The target theory, in J. Cairns, G.S. Steng and J.D. Watson (Eds.), Phage and the Origins of Molecular Biology, Cold Spring Harbor Laboratory, Cold Spring Harbor, NY, 1966, pp. 33–42.
34 B.L. Sheldon and J.S.F. Barker, The effect of temperature on mutation in *Drosophila melanogaster,* Mutation Res. (1964) 310–317.

G. Semenza and R. Jaenicke (Eds.) Selected Topics in the History of Biochemistry: Personal Recollections, III.
(Comprehensive Biochemistry Vol. 37) © 1990 Elsevier Science Publishers BV (Biomedical Division)

Chapter 2

An Octogenarian Looks Back

ALBERT NEUBERGER

Charing Cross Hospital Medical School, St. Dunstans Road, London W6 8RP (U.K.)

Family background and early education

I was born in 1908 in a small town in northern Bavaria, a region where my family had been settled for several generations. When I visited the cemetery where members of my family were buried I discovered gravestones of my ancestors going back for more than 120 years. My father was a businessman who was more interested in a great variety of intellectual activities than in his business and, having an unusual combination of qualities, had the greatest influence on the early part of my life. My mother was an excellent housewife and devoted to her children. I had a very secure and typically middle-class family background, and the values which I assimilated and fully accepted for practically my entire life were old-fashioned but combined with a fair degree of tolerance. My parents and grandparents were religious Jews who, however, were fully involved in the intellectual and cultural activities of the wider community.

I learned how to read and write before I entered the elementary school in my home town at the age of six. Apart from writing German I learned Hebrew at about the age of five and became a reasonably good Hebrew scholar in my early youth. When I was nine years old I was sent to the secondary school in Würzburg, where I

Plate 1. Albert Neuberger.

boarded with a family. However, at the end of the First World War conditions in Germany were very difficult, particularly in the bigger towns, and my father decided that for the next few years I should be educated privately. I returned to Hassfurt, my birthplace, and for the next four or five years I had a tutor who lived at our house and, in addition, I had tuition from various teachers or clergymen who were available. One of my teachers, for instance, was a local Protestant pastor who was quite a good scholar and who taught me Latin. After the age of 11 or 12 I also learnt Greek and had fairly good teaching in mathematics and history. My family invited, especially at weekends, various academics from the nearby University of Würzburg to stay at our house and this provided an interesting and enjoyable intellectual atmosphere. Looking back now, it seems very surprising to me that I was completely unaffected by the political turmoil which existed in Germany up to the end of 1923. Living in the country we always had a good supply of food, and even the monster inflation of 1923 did not impinge on the security of our life in any way. During this time, apart from my education in the classics, I received a remarkably good Jewish education which had a permanent impact on my life. At the end of 1923 it was felt that I should return to a more conventional education and I entered the Neue Gymnasium in Würzburg. I became interested in mediaeval and baroque architecture, enjoyed the sculpture of Riemenschneider, and was greatly impressed by Greek prose, particularly by Plato's Dialogues. I also enjoyed Latin grammar and thus my interests were largely confined to the Classics, but I had good teaching in mathematics and physics but no biology or chemistry.

During my years in Würzburg I came under the influence of a distant relative, Dr. J. Neubauer, who combined great scholarship in Roman law, a general interest in legal history, and a wide range of Jewish learning. I found it very diffucult to make up my mind, when I entered university, what career to follow. My interests were entirely in the field of the humanities, and I combined courses in law, particularly Roman law, with history and philosophy. After a while I felt the need to enter a career which would bring me into closer contact with ordinary human beings and I also felt the need to combine academic studies with life in the community at large. After a period of uncertainty I decided to change to medicine, and I was allowed for one university term to combine a course in human anatomy with seminars on the history of law. Following this short period I decided to devote my further studies entirely to medicine. The variety of interests which filled my early life had a permanent effect on me. I had read Kant, Ranke and Mommsen with great enjoyment, also Russian novelists like Dostoievski and Tolstoy. Although I have forgotten most of my Latin and Greek I have retained an interest in the humanities, particularly history.

My medical career

My pre-clinical studies were mainly done at the University of Würzburg, which I thoroughly enjoyed. At that time I had a particularly good memory and I found it easy to absorb a large amount of factual information which forms such a large part of the teaching of anatomy for instance. This subject was taught by a Prof. Petersen, who had what now seems to me a very modern approach to his subject. The professor of physiology was Max von Frey, who made a number of important contributions to certain branches of neurology, but he was a poor teacher. Somehow, in a way which would be difficult these days, I managed to interpolate classes in chemistry during my medical course, passing two examinations in organic and inorganic chemistry. The head of the chemistry department was Prof. Dimroth, who, together with members of his staff, made the subject extraordinarily interesting.

After passing the appropriate examination, I transferred at the age of 20 to the University of Berlin. In 1928 this city was a very stimulating and exciting place with remarkably varied cultural and artistic activities. During this period, and again in 1929 when I spent another term in Berlin, I worked in my spare time during term and also through the long university vacations, in the laboratory of Prof. Peter Rona, who acquainted me with the modern aspects of biochemistry. Rona had worked in the past with Leonor Michaelis, who introduced physicochemical concepts, and particularly enzymology, into biochemistry. It is interesting what a large number of German biochemists, who later did important research in different countries, were stimulated and influenced by Rona. It was in his laboratory that I first met Ernst Chain, with whom I maintained friendship throughout his life. I remained in contact with Rona for some years after I left Germany and I shall always be grateful to him for his support and encouragement during a critical point in my career. He perished in the holocaust in 1944 in circumstances which are still not fully known. The remarkable influence which Rona exercised on many chemists has been described amongst others by Blaschko [1] and in the autobiography of Hans Krebs [2].

Clinical teachers in Berlin whom I remember particularly are the surgeon Ferdinand Sauerbruch, an extrovert whose mode of instruction had certain theatrical features which I did not find altogether attractive; the physician von Bergmann, who again was an interesting teacher but did not seem to me altogether sound scientifically, and the psychiatrist Bonhöffer, a member of a distinguished family; a close relative of his was executed following the unsuccessful attempt on Hitler's life in 1944.

After my Berlin interlude I returned to Würzburg where I completed my medical studies and where I was influenced by the pathologist M.B. Schmidt, a truly remarkable person. He was a sound pathologist, but he gave almost equal weight to the descriptive and functional aspects of his subject, and made pathology appear a really important field of medical research. I formed a very close personal relationship with H. Reinwein who was a young lecturer in the department of medicine and later became professor of medicine at the University of Kiel. He allowed me to spend time in his wards doing extra-curricular work with patients and made clinical medicine appear one of the most stimulating fields of modern science. Reinwein encouraged me to work on a topic of my own choice, intracellular proteolysis, and this ultimately resulted in two papers which were published in 1931. I do not think these papers were of permanent value, but they were submitted as part of my M.D. thesis and to my great surprise the M.D. given to me was designated 'summa cum laude'.

After completion of my medical studies and passing my final examination in medicine I took up appointments as House Officer, one at the university hospital in Frankfurt and the other at the university hospital in Würzburg, which gave me quite extensive clinical experience. In the last months of 1932 I was faced with a difficult decision: I had enjoyed the 18 months doing clinical work, but I was at least equally attracted to experimental work in biochemistry, and I found it hard to make up my mind which career I should follow. I decided to increase my knowledge and experience of biochemistry, with the possible intention of going back to clinical work, and returned to Rona's department in October 1932. However, the situation in Germany was becoming more uncertain and I decided to explore the possibility of transferring my activities at least for a time either to The Netherlands or Great Britain. A few months before Hitler came to power I visited both Amsterdam and London to find out whether there were any prospects of getting scientific employment. My visit to England convinced me that I would be quite happy living in London, where I already had a few friends, but I returned to Berlin without having made any final decision. As soon as Hitler came to power in January 1933 I left Berlin and came to London.

University College Hospital Medical School (UCH), 1933–1939

An introduction from Prof. Rona enabled me to start work in London in the laboratory of Prof. C.R. Harington (later Sir Charles). At that time UCH was one of the outstanding centres of scientific medicine in the United Kingdom. It was next door to University College, where the chemistry department was headed by Prof. F.G. Donnan, one of the leading physical chemists of his time. Another distinguished member of the chemistry department was C.K. Ingold, who might be considered one of the leaders of modern physical organic chemistry. I was allowed to work on my own, but I was influenced greatly by Donnan and Harington. The latter, who was trained as a chemist, taught me the importance of using sound chemical methods and his integrity, both personal and intellectual, greatly impressed me. I therefore started work, somewhat influenced by members of Donnan's department, on the electrochemistry of amino acids and proteins.

At first I lived rather precariously on my small savings, but after a period of work in the laboratory I was supported by a grant from the Academic Assistance Council, which tided me over one of the most difficult periods of my career.

After obtaining my Ph.D. I was awarded a Beit Memorial Research Fellowship, which gave me the opportunity of planning research with some degree of security. I chose as my major problem the question of whether sugars are true components of ordinary proteins. During the work, which I will describe later in detail, I constructed a primitive apparatus for the extraction of acetyl amino acids which was seen by R.L.M. Synge, who had started work on the separation and quantitative estimation of amino acids, and with whom I was at that time in relatively frequent contact. Synge invited me to join him in work on making this approach into a quantitative method for the estimation of amino acids, but I was not sufficiently interested in a purely analytical problem and I declined his invitation. Reference is made to this in Synge's Nobel Prize lecture.

In my first few years in London I spent a lot of my spare time reading extensively, particularly the English classics, and I acquired a special interest in Jane Austen and the Brontës. My reading also covered English history and I became attracted by the periods of the Stuarts and the Georgians. I spent a fair amount of my time at weekends or during holidays visiting various parts of the English, Welsh, and Scottish countryside, and I formed to an increasing extent a strong attachment to the country in all its aspects.

My Cambridge period

Just about the outbreak of war in 1939 I tried to join the armed forces, but as I was in what was called a 'reserved occupation' my offer was not accepted. I was told that I could be more useful carrying on with research for the time being, particularly if I could direct my activities into fields which might be of direct value to the nation in time of war. Most of the colleges of the University of London were transferred to other centres, and just before the outbreak of war I was invited by Sir Frederick Gowland Hopkins to join his department in Cambridge [3]. The Cambridge department of biochemistry was at that time one of the world centres of biomedical research, and I was fortunate enough to be given facilities in a laboratory close to that of Hopkins, or 'Hoppie' as he was called by all members of his staff. Whilst I never worked with Hoppie myself, I had frequent conversations with him and was greatly influenced by his imagination and his breadth of outlook. It was usual amongst members of the staff of the biochemistry department to attend the advanced lectures of their colleagues. I particularly enjoyed listening to Malcolm Dixon's lectures on enzymology and to Robin Hill who dealt in a most stimulating manner with plant biochemistry. Another person who impressed me was Marjorie Stephenson, who might be considered one of the founders of modern bacterial chemistry. I saw quite a lot of David Keilin [4] and T.R. Mann, and got to know Sir Charles Martin, who had for many years been an outstanding director of the Lister Institute and offered the use of his house and its outbuildings to the nutrition workers of the Lister Institute who had been exiled to Cambridge, amongst whom was Dame Harriette Chick.

The 'Tea Clubs', even during war-time, gave an opportunity of discussing a great variety of biochemical problems in a most interesting manner. I also appreciated the fact that Cambridge provided the opportunity of meeting people from other faculties: thus I had friends who were professional historians, specialists in various languages, and a few philosophers. I also enjoyed walking in the Fens and punting on the Cam river to Grantchester. I still remember vividly the day that Paris fell to the Germans. I was in a punt moving slowly up the river and the lovely English countryside with its calm beauty seemed worlds away from the momentous events a few hundred miles away.

I also became involved in teaching biochemistry and was made responsible for instruction in protein chemistry for the Part II Course of the Biochemistry Tripos (Cambridge name for the Honours examination). I still remember my embarrassment when I found Davíd Keilin to be among my audience when I gave a series of lectures on the structure and biochemistry of proteins. I was also responsible for a practical course on the biochemistry of amino acids and proteins which fully occupied me for five weeks each year.

Apart from doing some research for the Royal Air Force which was connected with adhesives, I decided, probably under the influence of Sir Charles Martin, to work on the nutritional value of the potato. Whilst the protein content of the potato tuber is relatively low, there was some evidence that the nutritional value of the protein of the potato was unexpectedly high. At that time, i.e. 1940, Fred Sanger, who had just passed his Part II examination with first-class honours, chose to join me as a Ph.D. student. We isolated a number of free amino acids and other nitrogenous substances from the potato, but I do not feel that the results of this investigation are of great permanent interest. However, we did some other work on lysine metabolism which I shall mention later.

While Fred Sanger and I were mainly involved in preparative and nutritional work which at the time was considered to have some bearing on national problems of practical application, we were both interested in some more fundamental problems of protein structure. I had been intrigued for some time by the report of Jensen and Evans made in 1935 that treatment of insulin with phenylisocyanate produced relatively large amounts of phenylhydantoin of phenylalanine. The amounts were much in excess of what one might have expected on the basis of the molecular weight of insulin, which at that time was believed to be about 40 000. I carried out some experiments with benzene sulphonyl chloride which confirmed the findings of Jensen and Evans. Later work of Fred Sanger, which was based on the introduction of fluorodinitrobenzene as an end group reagent, was carried out without any participation on my part. I realized at the time that Fred Sanger was an unusually gifted and persistent investigator, but I did not visualize his outstanding achievements made in the ensuing 20 or 30 years.

Move to the National Institute for Medical Research

In the middle of 1942 I was invited by Harington to join his staff at the National Institute for Medical Research (NIMR), then located in an old building, a former hospital, in Hampstead. Harington had taken over the directorship of the NIMR on 1 October 1942, succeeding Sir Henry Dale, and I joined his staff in the following January. At Cambridge there had been no immediate prospect of a permanent appointment and the attraction of having an appointment of unlimited tenure in one of the best centres of medical research in Great Britain led me to accept the invitation. As I expected, when I started work at Hampstead I missed the stimulating atmosphere of Cambridge and my contacts with nonscientists, but this was largely compensated for by the excellent facilities which included the provision of good technical assistance and also by the high calibre of my colleagues, especially Christopher Andrewes, Lindor Brown, Harold King, O. Rosenheim, A. Parkes, R.B. Bourdillon, and the protozoologist, C. Dobell.

My work at the Institute at first followed some of the lines I had pursued at Cambridge. In continuation of my work on protein nutrition I felt there was the need to have a well-defined diet, and in particular to provide protein in the form of pure amino acids. Only a few of these compounds were available commercially, and some of the amino acid preparations commercially available were not pure. I therefore set out to synthesize a large number of amino acids by established methods and to obtain the optically active components, either by resolution with optically active bases or by enzymatic digestion. This involved the preparation of aliphatic α-bromo acids, which had a very strong and persistent smell and interfered to some extent with my social life. In spite of the fact that I changed my shirt and trousers before leaving the laboratory I discovered I practically emptied a pub when I sat down for a meal after leaving the laboratory. However, I managed to accumulate reasonably large quantities of amino acids either by synthesis or by isolation from protein hydrolysates. This work became of some practical importance as I became involved with the Protein Requirements Committee of the Medical Research Council, of which I became secretary, succeeding

Dr. D.P. Cuthbertson. This committee dealt with protein or amino acid requirements of the armed forces and of the civilian population in countries involved in the War, particularly in relation to starvation and injuries such as wounds and burns.

Thus, I became involved with the problem of preparing protein hydrolysates for intravenous administration in starvation, and the problem of protein deficiency became increasingly important towards the end of the War when there was extensive malnutrition in countries such as The Netherlands, Denmark and Germany. This also led to close contacts with manufacturers, who were either involved in the preparation of protein hydrolysates which were free of side effects, or the large-scale synthesis of amino acids by purely synthetic methods. The use of protein hydrolysates in the field, i.e., to a starving population, did not come fully up to expectation. A detailed report of the work of the Protein Requirements Committee during the War is given in the Report of the Medical Research Council entitled 'Medical Research in War. Report for the Years 1939–1945'.

I might describe here briefly my involvement with the structure of penicillin. By about 1940, it was known that penicillin was a labile acid of relatively low molecular weight which was fairly quickly destroyed in either alkaline or acid solutions. During the following few years more information was obtained about the breakdown products of penicillin, but at that time the importance of this new antibiotic in war medicine was recognized and an official ban on the publication of the results of further chemical work was imposed early in 1943. At that point the structure of penicillamine and of penilloic aldehyde had been established as degradation products, but the structure of the penicillin molecule was still the subject of considerable discussion. The Oxford Group, particularly Sir Robert Robinson, suggested the 'thiazolidine-oxazoline' structure, whilst others favoured the 'β-lactam' structure. As I had some experience in electrometric titration of amino acids I was asked to measure the ionisation behaviour of penicillin. Since only small amounts of the material were available, a special apparatus for the micro-electrometric titration of penicillin was designed jointly by Dr. Macfarlane and Dr. Schuster, and I carried out a number of titrations at different temperatures and in different solvents. The results of this work, which were finally published in the book entitled *The Chemistry of Penicillin,* clearly favoured the β-lactam structure which is now generally accepted.

The work was carried out under conditions of great secrecy. I remember that any piece of paper on which possible structures and other findings were discussed had to be torn up, or alternatively put in a safe, and it was not until the end of the War that the chemistry of penicillin could be openly discussed.

In 1944 I was asked to go to India to act as consultant in nutrition to the Indian Army, which so far as I remember had about 2 million men. In order to make this assignment effective I was given the equivalent rank of brigadier, a status which I found somewhat embarrassing but it had many pleasant aspects. The only other military experience I had had was during my Cambridge period when I served in the Home Guard and was promoted to lance-corporal. During my stay in India I visited a large number of cities and saw a fair amount of village life, and I became involved in the medical and scientific problems of India. This visit created a link with that country which I have maintained up to the present time. I might mention here that soon after my arrival in New Delhi I became aware of the magnitude of the population problem, and in my reports to the authorities I suggested that the marked increase in population was likely to become the major medical and economic problem for the Indian sub-continent. At that time there was little awareness of the importance of demographic factors.

In 1950 the NIMR moved to Mill Hill, which is about ten miles from the City of London. This move was associated with a large increase in staff and a fairly complete change in the whole atmosphere of the institute. However, one of the most important features was retained, and this was the good scientific co-operation across departmental boundaries. Amongst those with whom I enjoyed either frequent discussions or actual scientific collaboration were J.W. Cornforth, George Popjak, A.J.P. Martin, Rosalind Pitt-Rivers, and Henry Arnstein. I had a relatively light administrative burden as head of the biochemistry division and could devote most of my time to research. The director gave me complete freedom on the scientific side, but had a fairly complete hold over all other matters affecting the institute. There was little delegation on non-scientific matters, even to senior members of staff.

Throughout my career I have been greatly helped by the advice and continuous support of Charles Harington, and it might be appropriate if I quote verbatim from an obituary which I wrote after his death in 1972:

> 'Harington was essentially a very shy person, who appeared to many who did not penetrate his reserve as being cold and somewhat forbidding. In fact he was a man of strong emotion and capable of great affection and generosity. He was particularly kind to younger research workers, and there are many scientists, including the present writer, whom he helped and advised at critical stages in their careers.
>
> Harington was possessed of a strong belief in justice and of an unusually marked sense of duty, which at times seemed almost puritanical. He had a feeling for quality and a great loyalty for the causes and institutions with which he was associated throughout his life.' [Biochem. J. 129 (1972) 801–804]

Move from the NIMR to St. Mary's Hospital Medical School

In 1954 I was asked by George Pickering (later Sir George) for advice concerning the appointment of a professor of chemical pathology at St. Mary's. I suggested various names, but later the Dean of St. Mary's asked me whether I would consider accepting the post. I had for some time felt that I wanted to return to a more clinical environment and I also liked the prospect of being able to do at least some teaching. St. Mary's Hospital Medical School's administration was very generous and gave me all I wanted in terms of staff appointments and equipment, and I accepted the invitation. My own teaching load was relatively light, and the administrative burden not too heavy, thus I had a fair amount of time for continuing my research, which proceeded in several directions. I maintained my interest in porphyrin biochemistry, which I had started in 1946, but I also took up problems of glycoprotein structure, a field which I had neglected since the outbreak of the War.

The 18 years I spent at St. Mary's were most enjoyable and, I believe, scientifically productive. Apart from teaching biochemistry to students in their clinical years, I was also in charge of the routine biochemical services of the hospital, but it was possible to delegate most of the detailed work to my colleagues. Somewhat infrequently, I attended the weekly joint staff rounds and maintained fairly close personal relations with many of my clinical colleagues. At the same time I was free to devote most of my time to research in general biochemistry. There was, I believe, a happy relationship within the department of chemical pathology and I tried to keep a proper balance between maintenance of high scientific standards and tolerance and freedom for senior members of the department to engage in their own line of work. It would be inappropriate to enumerate all the people who worked with me, but I might perhaps mention particularly my cooperation with Robin Marshall on glycoproteins, and work with J.J. Scott, K. Gibson, G.H. Tait and A. Gorschein on porphyrins. Among the many visitors to the department over the years I might mention especially Nathan Sharon and Sidney Udenfriend.

In 1973, having reached the age limit of 65, I had to retire from St. Mary's, but I wanted to continue my research in the fields of lectin, glycoprotein and lectin, as well as porphyrin. Brenda Ryman, who was head of the biochemistry department at Charing Cross Hospital Medical School, kindly offered me facilities which I gratefully accepted. I was able to accommodate some of my colleagues, Richard Davies, John Sandy and particularly Tony Allen, who was offered a staff appointment in the department. I also had responsibilities for the preceding 15 years as chairman of the Lister Institute of Preventive Medicine, and I was able to combine the various activities in one location. I am most grateful to all concerned for making my retirement such a pleasant and productive period.

Research activities

Unlike most of my contemporaries I did not concentrate my research on one or two specific fields, but covered a large section of the entire subject. In other words, I 'dabbled' somewhat in many problems of biochemistry. However, I will try to be selective in this account and concentrate on those aspects of my work, extending over a period of more than 50 years, which may have some permanent significance.

Electrochemistry of amino acids and proteins

My Ph.D. work was concerned with the examination of dissociation constants of amino acids, particularly those of glutamic acid and its esters using thermodynamically acceptable techniques [5]. I used the iodination of proteins, especially of zein and insulin which affected only the phenolic residues of tyrosine at least of these proteins and led to a shift in pK values of about 3.5. In a joint paper with Harington I showed that this iodinated insulin was almost biologically inactive [6]. However, it was possible to remove, at least partially, the halogen from the iodinated insulin and restore about 60% of its activity. The restoration of activity appeared to be proportional to the amount of de-iodination of insulin.

The use of spectroscopic methods led John Crammer and myself to investigate the ultraviolet spectroscopy of proteins, and it occurred to us to use the changes of the ultraviolet spectrum of proteins on ionisation for a specific estimation of tyrosine, and particularly of its dissociation constant. This was considered helpful in the general interpretation of the interaction of proteins with protons and hydroxyl ions. At that time modern spectrometers were not available and measurements were made with a Hilger spectrograph and a Spekker photometer. One could obtain just two spectrograms of a protein in a day. We found that with insulin the pK of the phenolic groups of tyrosine showed only a slight shift compared with that of the free amino acid; but with egg albumin there was no change in the absorption spectrum with increasing pH until it was greater than about 13.0. We concluded that the phenolic groups of egg albumin were probably buried inside the molecule, or stabilized by hydrogen bonds formed between the phenolic hydroxyl group and a negatively charged carboxylate ion [7]. We postulated that both protons and hydroxyl ions would destroy this hydrogen bond. Denaturation was also associated with spectral changes. Moreover, we suggested that such hydrogen bonds might be largely responsible for the stability of native proteins. This, I believe, was the first time that ultraviolet spectroscopy had been used in the investigation of protein denaturation.

General aspects of protein structure

During the 1930s I was also interested in the general structure of proteins and published short notes showing that the cyclol hypothesis of Dorothy Wrinch, which had quite a lot of support from Langmuir, was wrong and the subsequent hypothesis of Bergmann and Niemann was unlikely to be correct. This point is discussed in some detail in the account by Laszlo in Volume 34A of Comprehensive Biochemistry.

Glycoproteins

It had been recognized of course for some time that mucins and mucoproteins contain sugars which are in some way covalently bound to peptides, but the majority of biochemists up to the mid-1950s believed that, with other proteins, carbohydrates were essentially present as impurities. I noticed, however, that many proteins give a Molisch reaction, which indicates the presence of carbohydrate, and I suspected therefore that a large proportion of proteins contain sugars as part of their structure.

To investigate this problem I chose egg albumin, a protein which was available in quantity and which could be crystallized. In 1936 when the work was started chromatography of proteins and amino acids had not yet been developed, and identification depended on elementary analysis, preparation of crystalline derivatives, and mixed melting-point determination. I first showed that the sugar content of this protein did not change on repeated crystallization, ultrafiltration, or denaturation. Hydrolysis by acid and alkali appeared unsuitable and I therefore chose exhaustive proteolytic digestion, which yielded a mixture of amino acids and the carbohydrate moiety.

The next problem was to separate these two components. I chose acetylation with ketene as a first step. This led me into an investigation of the action of ketene on amino acids in general and in related compounds. During this work I was taken ill with hepatitis, and the possibility arose that this was caused by a toxic action of this reactive gas on the liver. While I was in hospital I discussed this possibility with Dr. Roy Cameron (later Sir Roy) and we decided to do some animal experiments to test the toxicity of this substance. We found that ketene had a marked toxic effect on the lungs associated with a high mortality in various animal species at any concentration which could be measured, but there was no specific effect on the liver. As ketene in aqueous solution did not react with aliphatic hydroxyl groups, I assumed that the carbohydrate moiety was unaffected by this treatment. It was certain now that many of the amino acids could be removed from aqueous solution by extraction with organic solvents at an acidic pH.

The next problem was to separate from the carbohydrate the free acetyl amino acids not extracted by organic solvents. This was achieved by carrying out an acetylation of the dried material with pyridine and acetic anhydride. This treatment rendered the carbohydrate moiety soluble in chloroform and separated it cleanly from the amino acid fraction. The chloroform-soluble material was then treated carefully with lithium hydroxide, which gave the carbohydrate moiety in a reasonably pure form. I found that the molecular weight of the carbohydrate was 1200; that it was composed almost entirely of four molecules of mannose and two of *N*-acetylglucosamine; its nitrogen content indicated that there was another component present which accounted for two additional nitrogen atoms. These experiments established that the sugar moiety was covalently bound to the protein, and that it was present mainly as a hexasaccharide in the egg albumin molecule [8]. The linkage of the carbohydrate to the sugar was not established at that time. I did show, however, that the linking group was an amino acid containing two nitrogen atoms, one of which yielded ammonia on acid hydrolysis. I suspected the presence of glutamine or asparagine, but with the minute amounts of material available at the end of these experiments I could not establish the linkage.

The outbreak of war put an end to this work, which was not taken up again until the mid 1950s.

Work on amino sugars

In connection with my efforts in the glycoprotein field I became interested in the chemistry of glucosamine, and I prepared the methyl glycosides of glucosamine and its *N*-acetyl derivatives. A methyl derivative of glucosamine had already been prepared in 1911 by the great carbohydrate chemist Irvine, who did not consider this compound to have the normal glycosidic structure because of its unusually strong resistance to hydrolysis by acid. Together with R.C.G. Moggridge I proposed that this methyl derivative did indeed have a normal glycosidic structure, but that its resistance to acid hydrolysis was caused by the fact that, unlike similar derivatives of non-nitrogenous sugars, it contained a positively charged amino group in acid solution, and this increased resistance to acid was caused by an electrostatic effect [9]. We carried out kinetic experiments in which it was shown that the rate of hydrolysis was about 100 times lower than that of similar, uncharged glycosides. These findings had immediate relevance to the estimation of amino sugars in glycoproteins and other compounds in which glucosamine is generally present in the *N*-acetylated form. The α-anomer of *N*-acetyl-methylglucoside was prepared and its hydrolysis by acid was studied. It was shown that this involved two competing reactions: the removal of the acetyl group, and the splitting of the glycosidic bond. On the basis of comparison of optical rotations of glucosamine and glucose it had been assumed for some time that D-glucosamine had the same configuration on carbon atom-2 as glucose, but no direct and compelling chemical evidence was available for this assumption. In 1940, I published a paper [10] in which conclusive proof was obtained that D-glucosamine was really 2-amino-2-deoxyglucose and that its glucosides were indeed pyranosides. *O*-Methylation of an *N*-acetylglucosamine glycoside

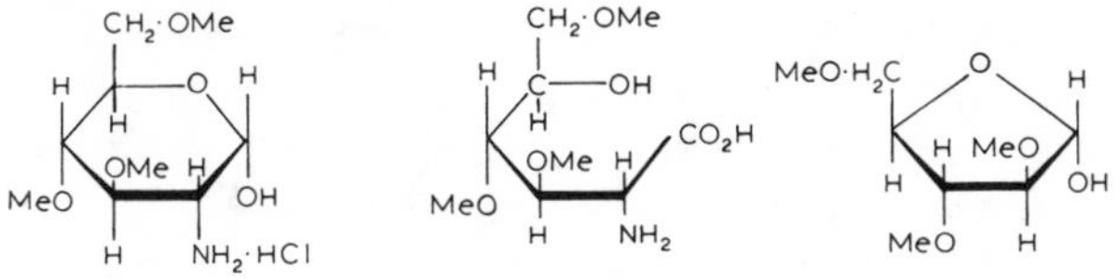

I II III

Fig. 1. *N*-Acetyl 3:4:6-trimethyl-α-methylglucosaminide was de-acetylated to yield 3:4:6-trimethyl glucosamine hydrochloride (I). This compound was oxidized with naphthalene-1-sulphonechloroamide to yield 2:3:5-arabofuranose (III) with (II) as a probable intermediate.

gave, after hydrolysis by acid on oxidation by chloramine-T, a 2,3,5-D-arabinofuranose which could only have been derived from the pyranose structure of glucosamine (Fig. 1). These various glycosides of glucosamine were also used in a variety of enzymatic experiments carried out with Mrs. Pitt-Rivers, which demonstrated the specificity of β-acetylglucosaminidase from emulsin.

Continuation of glycoprotein work

In 1955 I took up again the problem of glycoproteins and started afresh with egg albumin, as chromatographic methods were then available, which made the isolation of a glycopeptide easier than was possible 20 years earlier. This work was done jointly with Dr. R.D. Marshall and Miss Johansen [11]. We soon established that the linkage between the carbohydrate and peptide moieties was one between the reducing carbon atom of *N*-acetylglucosamine and the amide group of an asparagine residue [12]. Similar work was done at the same time by Cunningham in the United States. This work was finally put on a sound basis by preparative work carried out together with G.S. Marks, now professor of pharmacology, Queen's University, Kingston, Ontario, which also proved that the *N*-glycosidic linkage had the β-conformation (Fig. 2). Synthesis of the linkage compound established the structure mentioned above beyond any doubt [13].

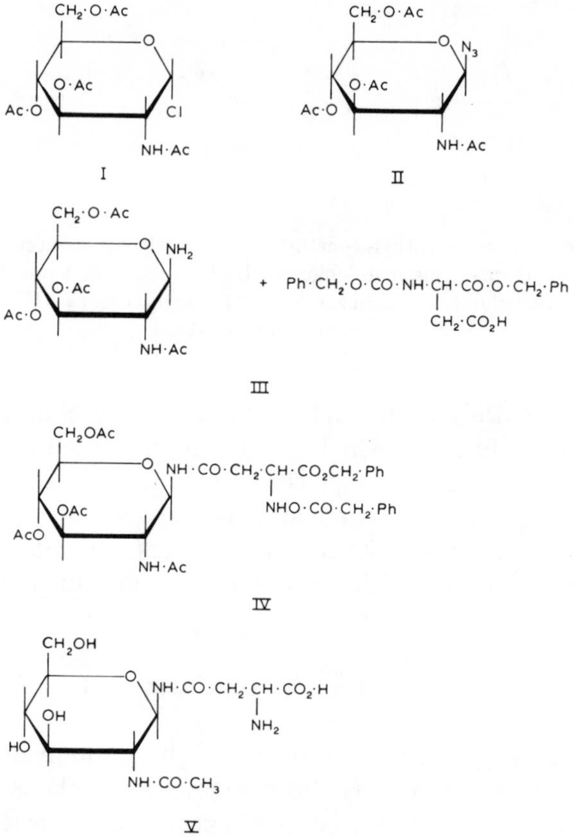

Fig. 2. Acetochloroglucosamine (I) was converted by a Walden inversion to the 2-acetamido-3,4,6-tri-*O*-acetyl-1-β-azido-1,2-dideoxy-D-glucose(II). This was reduced by catalytic reduction to the β-glycosylamine (III). The latter was condensed in the presence of dicyclohexylcarbodi-imide with an α-*N*-benzyloxycarbonyl-L-aspartate (IV). The product (V) was isolated and the protecting groups were removed, yielding the desired product.

At that time in the late 1950s, it became clear that the majority of proteins in living tissue were probably glycoproteins, and the β-linkage between *N*-acetylglucosamine and the amide group of asparagine was a feature characteristic of probably the majority of glycoproteins. Other linkages have, of course, been described involving, for instance, the hydroxy groups of serine and threonine, and the hydroxy groups of hydroxylysine and hydroxyproline. The work I have mentioned above on egg albumin was done at a time when the general importance of glycoproteins was being realized to an increasing extent.

As a model of the linkage of the carbohydrate in collagen which involves the hydroxylysine moiety, 2-aminoethyl-β-D-glucopyranoside was prepared in collaboration with E.R.B. Graham, and its properties were compared with that of the relevant linkage in collagen.

Peptide sequence and glycosylation

In 1965/66 Robin Marshall and I tried to establish which particular factors in the amino sequence of peptide were responsible for glycosylation. We concentrated initially on such glycoproteins as were *N*-glycosylated, i.e., the asparagine moiety linked to acetylglucosamine. This general problem was discussed by Robin Marshall at an international congress of biochemists in Tokyo in 1967, but considered in more detail in a paper I presented at a colloquium in Oregon in 1968 [13a]. It was then suggested that the necessary sequence was Asn.X.Thr (or Ser), and this was further considered in a review paper I gave in 1971 [14], where it was shown that a few sequences which had been reported in the literature as not complying with this rule were erroneous. The sequence Asn.X.Thr (or Ser) is necessary but not a sufficient condition for *N*-glycosylation, as this sequence occurs in proteins which do not contain carbohydrate. Our collaboration was so close at that time that I cannot remember whether it was Dr. Marshall or myself who spotted this fact. This rule is now considered to apply generally, and so far as I know no exception has been reported. No such simple regularity has been discovered for *O*-glycosylation.

Further chemical and physicochemical work on amino sugars

Jointly with A.P. Fletcher and D.M.L. Morgan I carried out some work on measuring the ionization constants of the 2-amino-2-deoxy-D-glucopyranoside. It was found that the basicity of the two anomeric forms differs significantly [15]. By measuring the pH changes of partly neutralized solutions of glucosamine hydrochloride it was possible to determine with considerable accuracy the rate of mutarotation at different temperatures.

We also measured the splitting of the glycosidic bond of 2-*O*-methyl-*N*-acetyl-α-D-neuraminic acid. These experiments indicated that the rate constant was not linearly proportional to the acid concentration, and the experimental results fitted closely the assumption that hydrolysis involves an intermediate existing probably in a half-chair conformation having a protonated ketosidic oxygen atom and a negatively charged carbohydrate ion. In other words, the postulated intermediate is a 'Zwitterion' [16].

David Morgan and I published a paper describing the isolation and characterization of the four anomeric furanosides of glucosamine and *N*-acetylglucosamine. The behaviour of these compounds on acid hydrolysis were investigated and found to be of the expected magnitude. So far as I know, such structures have not been described in natural compounds.

Tamm–Horsfall (T–H) urinary glycoprotein

In 1950 Tamm and Horsfall described a protein which is present in normal urine and which contains moderately large amounts of a variety of sugar residues. We carried out measurements on the molecular weight and composition of this protein and it was shown that it contained 31% carbohydrate and it had a molecular weight of 85 000: we also examined its immunological behaviour. Some aspects of this work were further developed by Dr. R.D. Marshall. The biological function of this protein is still, so far as I know, obscure.

This urinary glycoprotein contains approximately 5% of sialic acid, which is only slowly released on mild acid hydrolysis or treatment by neuraminidase. It was shown that this resistance is associated with the fact that almost all the sialic acid residues contain two *O*-acetyl groups, and the removal of the acetyl groups by moderate alkaline treatment normalizes the subsequent release of the sialic acid by acid and by enzymic hydrolysis. Together with Miss Ratcliffe I undertook some kinetic studies on the acid hydrolysis of the methyl ketoside of unsubstituted neuraminic acid. It was found first of all that the methyl ketoside of tetra-*O*-acetyl-α-D-neuraminic acid resembles in its slow rate of hydrolysis very closely that of the T–H glycoprotein. Again, removal of the *O*-acetyl groups restored a fast rate of hydrolysis. This parallelism was also observed in experiments with neuraminidase. Various electronic mechanisms responsible for the decrease in the rate of hydrolysis were considered.

Experiments on the enzymatic activity of lysozyme

In 1965 D.C. Phillips and his co-workers carried out their classical experiments on the crystalline structure of lysozyme and defined accurately the conformation of the active centre. They suggested a mechanism of action which is now generally accepted. Having worked on the chemistry and physical chemistry of amino sugars, we used our experience to increase our understanding of the enzymic mechanism involved. It was found that *N*-formylglucosamine, which could form the same hydrogen bonds as *N*-acetylglucosamine, was completely inactive as an inhibitor; and the same applied to the *N*-propionyl derivative. This clearly indicated that the methyl moiety of the acetamido group is specifically involved in a hydrophobic interaction and the size of this alkyl group is crucial for the binding.

This work was continued with R.C. Davies. In work on lysozyme two types of substrates are generally used: one group consists of chitin derivatives, such as tetramer of *N*-acetylglucosamine, and the second group of substrates are cell suspensions of *Micrococcus lysodeikticus*. In the latter type of experiment a glucosamine-like linkage employing *N*-acetyl muramic acid is involved. We showed that in the interaction of lysozyme binding of the positively charged enzyme with the negative charges of the cell wall must precede enzymatic splitting [17]. This factor does not affect the action on soluble substrates of low molecular weight. Acetylation of lysine residues of lysozyme abolished reaction with the cell walls, but did not affect cleavage of the tetramer of *N*-acetylglucosamine. The same observations were made when the arginine residues were modified. These findings were in accordance with the accepted concept of the mechanism of lysozyme [18].

Chemistry and biochemistry of amino acids

The complete mixtures of synthetic and natural amino acids were used in a variety of investigations, to which I referred in an earlier part of this essay. In 1945, together with L.E. Glynn and H.P. Himsworth (later Sir Harold), it was shown that in the absence of both methionine and cystine from the diet of the rat, hepatic necrosis developed, and this effect was abolished by cystine. Massive hepatic necrosis also developed if small amounts of methionine were supplied but no cystine. It was later found, however, that the hepatic necrosis could be prevented by adding vitamin E to the diet. It appeared from this work that hepatic necrosis was produced by a combination of deficiency of sulphur-containing amino acids and tocopherol. It was also shown in work together with G. Leaf that a diet low in sulphur-containing amino acids produced a very marked effect on the glutathione content of the liver. Even after one day on a low protein diet, there was a marked reduction in the glutathione content, and after about two weeks the glutathione content of the liver fell to a value of about 25% of normal. It thus appeared that some of the pathological findings with diets deficient in cystine were due to a reduction in liver glutathione.

Osborne and Mendel had shown in 1916 that, in rats, lysine was required for growth, but later work suggested that it was not necessary for the maintenance of body weight in adult animals. In work with T.A. Webster it was shown that the adult animal also requires lysine, but the amount needed for maintenance of body weight was only about one-sixth of that needed by the growing animal [19].

From this we drew the conclusion that in general the amino acid requirements for the maintenance of body weight were likely to be different from those needed for growth. In addition, we also showed that histidine was needed not only for growth but, contrary to earlier findings, also in the adult rat.

Special nutritional position of threonine

Somewhat later, when isotopic nitrogen became available, I carried out an investigation with D.F. Elliott on the metabolism of threonine both in the rabbit and the rat. When glycine labelled with ^{15}N was given to animals, the amino acids isolated from the tissues contained labelled nitrogen with the exception of lysine and threonine [20]. The exceptional position of lysine was already known from the work of Schoenheimer, but it was quite important to show that threonine shares this property with lysine and that, unlike almost all other essential amino acids, its deamination is irreversible.

Metabolism of glycine and serine

In the early 1950s Henry Arnstein and I carried out a number of experiments in which radioactively labelled glycine and serine, and compounds related to these amino acids, were fed to growing rats over long periods. The total amounts of labelled and unlabelled compounds were known and at least approximate conclusions could be made about the quantitative relationships of the metabolism of these amino acids. It was deduced that in rats the endogenous synthesis of serine per day amounted to about 3.5 mmol, and that of glycine to 2.5 mmol. It was also deduced that most of the glycine originated from serine with the formation of a 1-carbon compound, which presumably was linked to tetrahydrofolic acid. It was also shown that the amount of endogenous formation of labile methyl groups was very considerable, and these were mainly derived from the β-carbon atom of serine: in fact, it appeared from our measurements, using a variety of compounds which might give rise to 1-carbon moieties, that the β-carbon atom of serine was the most important endogenous source quantitatively of such labile methyl groups. However, it appeared that the demand for such labile methyl groups is so large that endogenous serine synthesis is not sufficient for optimum growth even in the presence of cobalamin, which greatly improved the utilization of the β-carbon of serine as a methyl precursor [21].

Work on tryptophan metabolism

Also in the 1950s I became interested in tryptophan metabolism and considered the possibility that oxindolylalanine might be an intermediate in the conversion of tryptophan to kynurenine. I joined forces with Dr. J.W. Cornforth and Dr. Rita Cornforth and we worked out a satisfactory synthesis of this amino acid, which had earlier been prepared in small amounts by Witkop. Oxindolylalanine had not been found in mammalian tissues, but Wieland and Witkop had obtained it by hydrolysis of phalloidin, a toxic peptide present in the fungus *Amanita phalloides.* Comparison of the spectra of oxindolylalanine and phalloidin indicated that the oxindole structure was not present in phalloidin itself, and it was suggested that the structure in the peptide contained sulphur instead of oxygen in the β-position of the indole ring, and that hydrolysis removed the substituted sulphur atom giving rise to the oxindole structure. This prediction was verified by later experimental work. Work with animals proved my prediction that oxindolylalanine was the first product on oxidation of tryptophan to be wrong.

Stereochemistry of hydroxyproline

In 1945 I carried out some work bearing on the stereochemistry of hydroxyproline (Fig. 3). I showed that a derivative of the natural compound could be converted to D-(+)methoxysuccinic acid, which

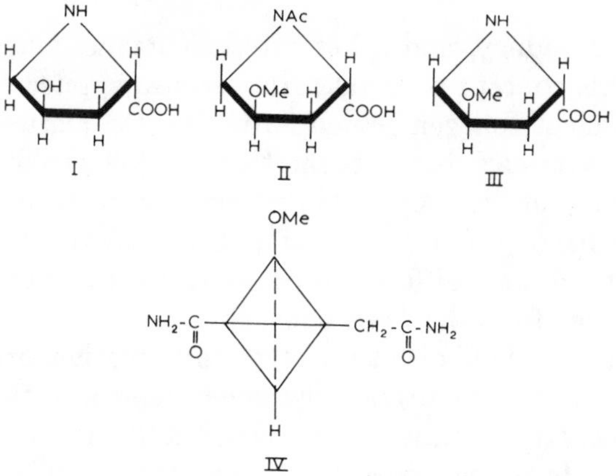

Fig. 3. Hydroxyproline (I) was first converted into its methyl ester and acetylated. The product (II) was further methylated and saponified to yield *O*-methoxyproline (III). This was then oxidized with permanganate and the resulting D-methoxy succinic acid was isolated as a diamide (IV).

indicated that the configuration of the C_4 of hydroxyproline is D. Later on I corresponded with C.S. Hudson, the great carbohydrate chemist in the United States, and discussed in detail the consequence of this finding in relation to the stereochemistry of hydroxyproline, *allo*-hydroxyproline, and various derivatives of these amino acids. This resulted in a joint paper with Hudson [23] which emphasized the fact that the proline which occurs in proteins has the *trans*-configuration. This work led me to review the stereochemistry of amino acids in a paper which was published in *Advances in Protein Chemistry,* Vol. IV (pp. 297–383).

Protein metabolism

The work of Schönheimer had shown that a large part of the protein of experimental animals and man is in a dynamic metabolic equilibrium, i.e., it is being replaced in a period of days or a few weeks. These earlier experiments were largely concerned with the proteins of the liver, kidney, and other internal organs, and I wondered whether this concept of rapid turnover also applied to structural proteins such as collagen, which forms about a quarter of the total protein of mammals. It was found that the collagen obtained from different parts of the body was in general relatively inert compared with the mixed proteins of liver and kidney, or even muscles. However, there were differences between the turnover rates of collagen obtained from the liver, bone, skin, and tendon; the collagen obtained from the tail of the rat being particularly inert. In most of these experiments the turnover was measured not by the incorporation of radioactive glycine into collagen but by the disappearance of radioactivity from the protein with time. On this basis the complete replacement of the collagen of the rat tail was found to be equal to the whole lifetime of the animal [24].

These experiments also led my colleagues and me into the general field of collagen chemistry. We found, apart from the mixture of insoluble collagen, a more soluble fraction which could be shown by isotope experiments to be an intermediate in the biosynthesis of this protein, and this soluble material was later converted into insoluble fibres.

I had been concerned for some years about measuring protein metabolism in the whole animal by isotopic means. There are two main difficulties. One consists in the existence of different compartments in the whole organism and even in cells with different permeability characteristics. The other difficulty arises from the fact that in protein synthesis the cell uses both amino acids derived from the diet and imported into the cell, as well as amino acids liberated by the continuous turnover of proteins. In work carried out together with O.B. and S.B. Henriques [25] an attempt was made to obviate some of these difficulties by measuring the isotope content of orally administered glycine in liver, muscle and plasma. The isotope content of amino acids obtained from protein hydrolysates was related to the isotope content of free amino acids in different tissues. This work showed that blood flow and the rate of penetration into the cell are important factors in the interpretation of isotope experiments in the whole animal. However, it was not fully realized at that time that the pool of free glycine in the cell is by no means homogeneous.

Whilst further work in this field has taken place since our experiments were undertaken, not all the problems in this difficult field have been solved.

Together with C.W. Crane, I carried out investigations into protein absorption and metabolism using ^{15}N-labelled yeast protein [26]. To our surprise the urea and ammonia were labelled in the urine sample witin 30 min after giving the labelled protein, and the isotope content reached its maximum after approximately an hour. If the protein was first hydrolysed by a mixture of enzymes, the labelling in the urea and ammonia was somewhat higher, but the later results were almost identical with those obtained with whole protein. This indicated that, at least under the conditions used, enzymic hydrolysis in the gut was extremely fast and a similar very fast rate of hydrolysis also could be obtained by using concentrations of crystalline trypsin and chymotrypsin almost as high as the protein. We concluded therefore that the normal situation in vivo must be somewhat similar, i.e., that the concentration of proteolytic enzymes in the gut was very high. Another interesting finding was that, after giving this labelled yeast protein, the ^{15}N content of hippuric acid was indeed higher than that of the urinary ammonia, suggesting that glycine which is produced in response to benzoic acid ingestion, to a large extent must have been derived from the ingested protein. This technique was later used in various clinical conditions, particularly coeliac disease, and the results indicated that it could be used in studying deficient absorption in a variety of gastrointestinal diseases.

Porphyrin biosynthesis

In 1946 or 1947 stable and radioactive isotopes of carbon, nitrogen, and hydrogen became available in England, and this opened considerable possibilities in metabolic research. A mass spectrometer was installed at Hampstead and Dr. Ronald Bentley, who had been working with Rittenberg in Columbia, was put in charge of this instrument. I decided to investigate porphyrin biosynthesis, having been greatly impressed by the paper of Shemin and Rittenberg [27] which had just appeared showing that the nitrogen of glycine seemed to be a specific precursor of at least some of the nitrogen atoms of protoporphyrin. At this time, Dr. Helen Muir joined me, and we showed in a paper published in 1949 that all four nitrogen atoms are specifically derived from glycine [28]. About the same time, Shemin's group reported similar findings and for a few years our work and that of Shemin's group were running along parallel lines. We also showed that the carboxyl carbon of glycine was lost in the conversion of this amino acid into protoporphyrin, but that each pyrrolic ring of the porphyrin contained one carbon atom derived from the α-carbon atom of glycine. The latter also provided the four methyne bridges linking the four pyrrolic rings [29].

In 1950 I suggested [30] that the first step was a condensation between a molecule of glycine and two molecules of α-oxogluterate to give a pyrrole closely resembling porphobilinogen, but lacking an aminomethyl group. This already assumes a Knorr-type condensation. I suggested, which turned out to be wrong, that the meso-carbon atom may be added afterwards to either side of the ring, thus accounting for the two types of porphyrin isomers occurring in nature, viz. types I and III. However, even in 1949 I considered the possibility of aminolaevulinic acid (ALA) being involved, and in 1952 I started, together with J.J. Scott, on an extensive series of experiments hoping to devise the synthesis of ALA which would allow labelling of specific carbon atoms. My co-worker was something of a perfectionist and was anxious to raise yields at various steps of the synthesis, and thus the work proceeded perhaps more slowly than seemed reasonable. However, we described several methods of synthesis and this work was submitted for publication at the end of 1953 and published in 1954 [31]. I remember that in 1949, before the NIMR moved to Mill Hill, I discussed with Dr. J.W. Cornforth the possibility that a substance like porphobilinogen might be the precursor of protoporphyrin. He pointed out the difficulty of explaining how type-III porphyrin could be formed, and I suggested to him the possibility of an inversion, or even of a migration, of the one carbon substituent in the α-position of a pyrrole.

In 1953 Cookson and Rimington published their important paper on the structure of porphobilinogen [32]. This was a significant development because it now seemed that this substance would be the right precursor for the four pyrrolic rings. Condensation of two molecules of ALA was always in our minds as the most likely reaction, but the only reason why we had not accepted this earlier was that it did not in itself explain the formation of both types I and III of porphyrin. This indeed was the reason why we assumed that the addition of one carbon atom would occur after the pyrrolic ring was formed. However, with the publication of this paper by Cookson and Rimington it seemed likely that the rearrangement affecting one pyrrolic ring would occur later, i.e., during formation of the macrocyclic structure. Indeed, this problem has only been solved comparatively recently.

D. Shemin made an important suggestion in 1952 that an asymmetrically substituted succinate, such as succinyl-CoA, might be involved. Moreover, in a preliminary paper, which was published in 1953 [33], Shemin could assign the origin of each carbon atom of protoporphyrin to the various precursors in an unambiguous manner. Most of the work of Shemin and that of ourselves had suggested the mechanism which is now accepted by experiments either in whole animals or in whole cells. The work of Shemin was more complete than ours, as he definitely proved the metabolic origin of each carbon atom in the protoporphyrin molecule.

Our studies in porphyrin metabolism had also been extended in several directions. Together with C.H. Gray, we studied the incorporation of isotopic nitrogen into haemoglobin and into the porphyrins excreted from a subject suffering from congenital porphyria. We found that the uroporphyrins were maximally labelled at about the sixth or seventh day after the administration of labelled glycine, the coproporphyrin having a consistently higher isotope content than the uroporphyrin. We also studied the bile pigment excretion in faeces and discovered that, apart from the expected peak at about 120 days after administration of the labelled amino acid, there was an early peak within a few days of administration which suggested that some bile pigment was formed from sources other than red cells, which were degraded at the end of their normal life span of approx. 120 days [34]. Similar findings were reported by our American colleagues about the same time. This so-called 'early peak' of bile pigment excretion has given rise to much further work in the last 25 years. Haematological investigations, together with isotopic studies of haemoglobin and bile pigment, showed that congenital porphyria is associated with sporadic attacks of haemolytic anaemia, the occurrence of which is somewhat unpredictable. Our patient later underwent a splenectomy, which apparently produced some benefit, as shown by later isotopic studies.

In 1951 work of a purely physicochemical character on a variety of porphyrins was carried out together with J.J. Scott [35]. Ionizaton constants were measured by spectroscopic methods and interpreted in terms of electrostatic effects and resonance. Evidence for the existence of mono-cations was obtained, and it is of particular interest that the mono-*N*-methyl porphyrins, the behaviour of which we had studied, appear to be of particular importance in relation to the disturbance of porphyrin metabolism produced by certain drugs.

Porphyrin work continued at St. Mary's and Charing Cross

After my transfer to St. Mary's Hospital Medical School I continued the work on porphyrin biogenesis, which was later expanded into an investigation of the formation of chlorophyll, especially bacteriochlorophyll. During the period 1956–60 we had some visitors to the department, such as Sidney Udenfriend, who later became head of the Roche Institute of Molecular Biology; Nat Berlin, who is now director of the Cancer Centre at Northwestern University; J.B. Neilands on leave from the University of California; and W.G. Laver, now at the John Curtin School for Medical Research, Australian National University.

Our first effort was the purification and isolation of an enzyme responsible for the formation of porphobilinogen from δ-ALA. This was, I believe, the first characterization of this enzyme, and we showed that after the early stages of purification its activity became entirely dependent on the addition of compounds such as cysteine and glutathione, and it was inhibited by the presence of Tris buffer [36]. We suspected that it contained a metal, but we did not identify zinc, which was done much later by others. We did some work on ALA synthetase, showing that it required for its activity pyridoxal phosphate, which became firmly attached to the enzyme. We synthesized α-amino-β-oxoadipate and its esters: the latter were quite stable but the free acid had a half-life of less than a minute at pH 7.0 and at 0°C. We also prepared α-amino-β-oxobutyric acid which showed similar characteristics. The free acid here was somewhat more stable than the corresponding adipic acid compound.

Porphyrin biogenesis in bacterial systems

Dr. June Lascelles had studied porphyrin formation in *Rhodopseudomonas sphaeroides.* We felt that this organism provided better opportunities to study in detail the formation of porphyrins and ultimately of chlorophyll than was possible with other systems, such as those obtained from plants or animals. We concentrated our activities on this organism, which can grow under low oxygen pressure and produces (apart from porphyrins and haem proteins) chlorophyll. It can also grow in air, when it is nonpigmented, and produces haem compounds, apart from very small amounts of vitamin B_{12}.

Dr. June Lascelles had already shown that haem suppressed the formation of ALA-synthetase in this organism at a very low concentration, but we found that at a somewhat higher concentration it inhibited the already formed enzyme. It is now generally agreed that suppression and inhibition by haem are important controls in porphyrin biogenesis, both in bacteria and in some animal tissues such as the liver. We also observed that in *Rps. sphaeroides* another important control is associated with the formation of bacteriochlorophyll; thus, if the organism is grown in the presence of ethionine, an antagonist of methionine, there is an immediate cessation of chlorophyll synthesis and a large accumulation of porphyrin in the medium [37]. We assumed that this was caused by the inhibition of the methylation step. We could demonstrate, and this was largely due to the efforts of Kenneth Gibson, the presence of an enzyme which catalyses the conversion of magnesium protoporphyrin into its methyl ester [38]. Thus, the inhibition of the methylation step of chlorophyll biosynthesis led to accumulation of intermediates, of which coproporphyrin is one. This puts methionine into a central position in the control of chlorophyll synthesis.

Another strong inhibitor of chlorophyll synthesis is the amino acid, threonine. This was explained by the fact that threonine and methionine share a common pathway which involves homoserine as an intermediate. Threonine was shown to be a strong inhibitor of the homoserine dehydrogenase and thus inhibiting the synthesis of homocysteine, a methionine precursor. Furthermore, none of the other enzymes concerned in this pathway were either inhibited or repressed.

We also investigated the insertion of magnesium into protoporphyrin, but in spite of considerable efforts we could not demonstrate the occurrence of such an insertion in cell-free systems. I believe a similar lack of success has been experienced by many other workers in the field. It appeared that the magnesium incorporation is closely linked to the methylation step.

Aminomalonate as an inhibitor

In 1961-62 Margaret Matthew and I showed that aminomalonate was a specific inhibitor of ALA synthetase. This property was also shared by the monoethyl compound. In addition, we found that aminomalonate reacts with formaldehyde to give serine in the absence of an enzyme, but this reaction only occurs in the presence of pyridoxal phosphate. It was also of interest to find that one carboxyl group of aminomalonate is specifically removed by a decarboxylase which is active in the presence of pyridoxal phosphate.

γ,δ-Dioxovalerate

This compound is a transamination product of ALA and we wondered whether this transamination step might be of importance in the control of porphyrin synthesis. We indeed found an enzyme which was partially purified, which promoted the reaction of this acid with L-alanine, leading to ALA [39]. However, the equilibrium was entirely in favour of ALA formation, and this indicated the possibility that it might be involved in an alternative pathway of ALA synthesis which has been demonstrated in plants.

Uroporphyrinogen synthetase

In investigations carried out jointly with R.C. Davies an enzyme was purified from *Rps. sphaeroides* which produced uroporphyrinogen I from porphobilinogen [40]. If this reaction was done in the presence of ammonia or methoxyamine, no porphyrin or porphyrinogen was formed, but we obtained a linear tetrapyrrole which contained the amine. It was assumed that the amine affected the reaction by catalysing the condensation of the monopyrroles but the linear tetrapyrrole formed could not cyclize owing to the fact that it contained an amine substituent. This work was of special relevance in relation to the later work of Scott, Jordan and Battersby, who elucidated the pathway leading to uroporphyrinogen III.

Control of porphyrin and chlorophyll formation in Rhodopseudomonas sphaeroides

It has been known for some time that porphyrin and chlorophyll formation are controlled to a large extent by the activity of ALA synthetase. We could show that this enzyme existed in high-activity and low-activity forms, and the conversion from one type of enzyme to the other appeared to depend on sulphur compounds, in particular cysteine and glutathione and their derivatives. We showed that if the organism is growing semi-anaerobically in the presence of light it contains cysteine trisulphide which is a powerful activator of ALA synthetase [41]. Oxygenation of the organism in the medium in which it has grown produces a marked disturbance of sulphur metabolism, resulting in the gross depletion of glutathione, cysteine, and cystine, and the virtual disappearance of cysteine trisulphide.

However, this attractive interpretation, which was put forward at a meeting organized by The Royal Society [42], did not completely stand up to further experimentation, and work on these lines had to come to a stop when I left this field in 1976, about three years after my retirement from St. Mary's Hospital Medical School. The activation of ALA synthetase and the control of haem and chlorophyll biosynthesis are greatly affected by the state of the electron transport chain. They are likely to be influenced by oxidation and reduction of sulphur compounds, but the exact mechanisms have still to be elucidated.

Lectins

Our efforts in this field were stimulated by Prof. Nathan Sharon of The Weizmann Institute, who spent six months with me when I was still at St. Mary's. We then published a paper describing the purification of wheat-germ agglutinin, which gave a homogeneous product, and showed it was not a glycoprotein [43]. We also proposed that the binding site consists of three or four subsites with differing specificities, which is in a cleft in the molecule resembling that proposed for hen's egg-white lysozyme. Over the next 12 years or so the lectin field was explored in close collaboration with Dr. A.K. Allen. We were also helped particularly by Dr. D. Ashford and Dr. N.N. Desai, and in some aspects of the work we co-operated with Dr. R.R. Selvendran of the Institute of Food Research, Norwich. Apart from the wheat-germ lectin, we purified and described the properties of lectins from a variety of plants, such as the broad bean (*Vicia faba*) and the thorn-apple (*Datura stramonium*). One of the more interesting findings was that the lectins from pea, lentil, and broad bean are strongly inhibited by 3-*O*-methylhexoses, indicating that there is a strong hydrophobic interaction in at least some lectins.

Of particular interest was a lectin obtained from potato (*Solanum tuberosum*). This is a most unusual glycoprotein: it contains a large amount of L-hydroxyproline which is linked to furanosides of L-arabinose, forming about half of the total mass of the protein [44].

There are between one and four arabinofuranose groups, all of the β-type, with the exception of the fourth nonreducing residue which is of the α-arabinofuranose type [45]. For every four hydroxyproline residues, there is one serine which is associated with one galactopyranose group in the α-linkage. This glycoprotein resembles in many ways the insoluble hydroxyproline-rich glycoproteins of the cell wall of many plants. However, the cell-wall material and the potato lectin appear to be coded by different genes, and the exact relationship, if any, between the two types of protein is still uncertain. Potato lectin contains two different types of domains: one, which has been mentioned, resembling the cell-wall glycoproteins, and the other nonglycopeptide domain which contains a lot of cystine, glycine, and relatively large amounts of tryptophan. Antibodies can be prepared against the specific glycopeptide portion, and others directed against the other domain [46]. It is also possible to remove practically all the sugar residues from the potato lectin by trifluoromethane sulphonic acid under anhydrous conditions. This changes the conformation of the protein to a significant extent but does not abolish the lectin activity [47]. Measurement by circular dichroism indicates that the potate lectin has a collagen-like polyproline II structure which largely disappears on removal of the sugars.

General activities concerned with biomedical science

This account of my scientific activities would be incomplete if I did not mention my various endeavours which were not directly concerned with my own research. I have always felt that an academic scientist is in a privileged position in that he is paid for doing exactly what he wants to do, and this I believe imposes a duty and a responsibility to give some of his time to work which might be beneficial to society as a whole or to other scientists and academic colleagues. I felt this particularly when I was engaged in full-time research with little or no teaching.

Principalship of the Wright Fleming Institute

The Institute was started in 1909 by Sir Almroth Wright, who was a somewhat eccentric genius who enjoyed the friendship of Koch, Ehrlich and Metchnikoff, and had also been quite intimate with George Bernard Shaw. In due course, Fleming succeeded Sir Almroth Wright as Principal and the outstanding events of this period were the discovery of lysozyme by Fleming in 1922 and of penicillin in 1928. Reading the two papers by Fleming now on both penicillin and lysozyme, one feels that they are classical papers, in spite of the fact that the structures of these substances are very incompletely defined. It is still uncertain whether Fleming realized the clinical importance of penicillin, and it is quite clear that he did very little to explore the systemic use of this new compound, nor did he take any effective steps to establish its structure. However, it is clear that the two papers are important milestones in the development of microbiology, enzymology, and chemotherapy.

Fleming was a good clinical microbiologist but a bad administrator, who was happy to work by himself and did not try to exert any influence or guidance on members of his institute. I knew him during the last few years of his life: he appeared to have lost interest in recent developments in bacteriology and enjoyed in a somewhat naive but attractive manner the honours which came to him from all over the world. Fleming was succeeded by Robert Cruikshank who, again, was a competent clinical microbiologist without full appreciation of modern developments in this field of science.

For some years there had been a growing feeling amongst many members of St. Mary's Hospital Medical School that the Wright Fleming Institute was living on its reputation and not making an adequate contribution to modern microbiology. This dissatisfaction with the general management of the institute led to Cruikshank's resignation, when he accepted the post of professor of bacteriology in Edinburgh. I was asked to take on the Principalship of the Institute whilst retaining my Chair in chemical pathology at St. Mary's. I was anxious to continue with my own research and I accepted the post on a part-time basis, on the understanding that my tenure of the principalship would not go on for more than four or five years. I had the task of doing what amounted to an exchange transfusion: there were some members of staff who were doing useful work, but others were past their prime and had to be persuaded to accept suitable posts elsewhere. In this way I was supported by Mr. Harry Sporborg, who was at that time Chairman of the Council and who had a distinguished war record. I was also greatly helped by Robert Carr as Treasurer, who later became Home Secretary and is now Lord Carr. I managed to persuade an old colleague of mine, Rodney Porter, to accept the newly created post of Professor of Immunology. Porter did most of his chemical work on the structure of immunoglobulins, for which he received a Nobel Prize some years later, at the Wright Fleming Institute. It was also possible to attract Pat Mollison to head the new department of experimental haematology. We also created a new chair of virology under Keith Dumbell. An allergy department under Dr. Franklin continued the old tradition set up by Noon and Freeman. The existing chair of bacteriology was filled by R.E.O. Williams, who later became Dean of the Medical School.

To finance the creation of new Chairs and departments the Council of the Wright Fleming Institute decided to launch a public appeal under the name of The Fleming Memorial Fund for Medical Research. When this proposal was first mooted, Howard Florey, in my view rightly, expressed the opinion that the naming of this fund implied a lack of recognition of the very important work done at Oxford, both by Florey and Chain, in obtaining pure penicillin and using it effectively in clinical medicine. Indeed, without the Oxford contribution the original discovery by Fleming might never have been used in the treatment of patients. I believe that Florey underestimated Fleming's contribution, but he was understandably annoyed by the one-sided adulation given to Fleming by the general public. In somewhat difficult negotiations it was finally agreed to set up a board of independent trustees who, with the help of a scientific advisory committee, produced funds for a variety of research programmes. However, it was agreed that special recognition in the form of financial support should be given to the three laboratories involved in the discovery of penicillin. One of these was a new department of biochemistry set up for Ernst Chain at Imperial

College. The appeal was only moderately successful but it helped, together with a liberal allocation of funds from the University of London, to create the new departments mentioned above.

Association with The Lister Institute of Preventive Medicine

In 1968 I was invited to become a member of the Governing Body of the Lister Institute, which is the oldest research institute in the United Kingdom devoted to biomedical research. I was elected its chairman in 1971. The institute was loosely connected with the University of London, but its financial support came entirely from its investment income and from commercial activities consisting of the manufacture of a variety of vaccines. The institute had a distinguished record of scientific achievement in the biomedical field: amongst its staff might be mentioned Harden, who was one of the discoverers of phosphorylation of sugars which play an essential part in glycolysis; or in more recent times Walter Morgan and Winifred Watkins who made a fundamental contribution to our knowledge of the structure of blood-group substances.

In the mid-1970s it became clear that the financial situation of the Institute was unsatisfactory. The Institute could of course apply for research grants to the various research councils, but was not otherwise supported by public funds. The increased sophistication of science and continuing inflation were such that the income from investments and commercial activities was not sufficient to meet the needs of an active research organization. It was obvious that the Lister Institute could not continue to operate on its traditional basis. After a lot of heart-searching the decision was taken: to dissolve the existing institute, to help staff (which had declined in numbers) to get appointments in the public sector, to realize all our assets, and to become a grant-giving body. We were able to make generous redundancy payments to staff who could not be further employed. We disposed of our vaccine laboratories at Elstree in 1979 and our buildings in Chelsea in 1980 on relatively favourable terms, and were left with a capital of around £5 million. Owing to intelligent financial management our liquid resources now amount to £16 million.

It was decided in 1980 that we would use our funds to finance biomedical research of a high calibre, and we have been able to attract some of the best younger people in the field. We were greatly helped by our scientific advisory committee to make good appointments, and at present we have over 20 Research Fellows working on grants extending over five years or more. One of our Research Fellows has recently been elected to The Royal Society, and others have been appointed to senior university posts. The Institute has also been conscious of encouraging the commercial application of suitable scientific findings, and it has recently been particularly successful in the DNA fingerprinting method which is being exploited under a patent granted to the Institute and used by Imperial Chemical Industries plc. This complete re-organization was greatly helped by the extraordinary devotion and competence of Mr. Gordon Roderick, who has been the Institute's chief executive officer during this period of transition.

Work in Research Councils and other public bodies

In 1962 I was appointed a member of the Medical Research Council: this was a time when medical research obtained increasing support from public funds. Membership of the Council involved visits to various laboratories, such as the molecular biology laboratory at Cambridge, as well as overseas units operating in the Caribbean islands and Africa. From 1968–1969 I served on the Council for Scientific Policy, and in 1969 I was appointed a member of the Agricultural Research Council (ARC), now the Agricultural and Food Research Council. This work brought me into contact with a whole new field of biological research and I greatly enjoyed the mixture of fundamental and applied research which the Council promoted. I was particularly closely involved with some of the research institutes, in whose work I took a special interest. I retired from the ARC in 1979.

In 1971 the Medical Research Council and the ARC set up a joint committee to advise on research in the field of food and nutrition and on the promotion of research in that field, and I was asked to chair this body. We established five separate sub-committees to investigate all aspects of nutrition and produced a Report in 1974, which to an extent guided future policy in this field.

Whilst nutrition was not a major field of active research on my part, I was sufficiently interested in both its scientific aspects and its practical application to accept an invitation of The Royal Society to represent that body as a scientific governor of The British Nutrition Foundation. I served in that capacity for some years, later to become Honorary President of the Foundation from 1982–1986.

Activities in India and Israel

In 1944 I was invited to go to India and serve as consultant in nutrition to the medical directorate, GHQ India Command. I did not actually travel to India until the early part of 1945, when I spent four months travelling all over the country and became familiar with a variety of medical problems with special emphasis on nutrition. This was one of the most interesting periods of my life. The Indian civilization impressed me greatly and I soon realized that the problems of nutrition are linked with the whole culture as well as social factors. I became intellectually and emotionally involved with India, and that persists to the present day. I re-visited India for two lengthy visits, in 1977/78 and again in 1986, when I was asked to advise the Indian Medical Research Council on a variety of research activities, and later the Nutrition Foundation of India on their various research efforts.

In 1950 I was invited by Ephraim Katzir to visit his department at the Weizmann Institute as we had at that time similar research interests. This visit started my involvement with the academic life of Israel. A few years afterwards I was asked to become a Governor of The Hebrew University of Jerusalem, and I became closely involved with its various activities. I enjoyed my contacts with Faculties other than those concerned with the experimental sciences, thus I valued my involvement with the Faculties of Law and of the Humanities, and for some years I served as chairman of the Academic Committee of the Board of Governors of the Hebrew University.

Editorial work

In 1947 I was invited to join the editorial board of *The Biochemical Journal,* and in 1952 I was elected chairman of the board. At that time the board was relatively small and editorial work took quite a lot of one's leisure time: for many years I read the proofs of almost every paper which appeared in the Journal. Later on (1965) I served on the committee of The Biochemical Society, and was for two years chairman of the committee. I was elected an Honorary Member of The Biochemical Society in 1973.

In 1968 I was asked to join the editorial board of *Biochimica Biophysica Acta* and served until 1981, the last 13 years as an associate managing editor. I worked closely and enjoyably with my colleagues, who represented a group of European biochemists with diverse interests. I was also involved with the publication of North Holland Research Monographs entitled *Frontiers of Biology,* in which I shared responsibility with the late Prof. E.K. Tatum of New York. I am still concerned in the general planning and editing of *Comprehensive Biochemistry,* which is published by Elsevier.

In 1964 I was elected to Honorary Membership of the American Society of Biological Chemists, and in 1972 I was also elected a Foreign Honorary Member of the American Academy of Arts and Sciences.

Looking back on my career, I can appreciate the comment made by one of my sons, who felt that I had spread myself too widely, and this probably expresses the diversity of my interests and the multiplicity of my loyalties. It may be an expression of smugness, of which I am sometimes accused, to say that if I had my life over again I would not wish to plan it very differently.

REFERENCES

1 H. Blaschko, A biochemist's approach to autopharmacology, in G. Semenza (Ed.), Comprehensive Biochemistry, Vol. 35, Elsevier, Amsterdam, 1983, p. 189.
2 H. Krebs (in collaboration with Anne Martin), Reminiscences and Reflections. Clarendon Press, Oxford, 1981.
3 N.W. Pirie, Sir Frederick Gowland Hopkins (1861–1947), in G. Semenza (Ed.), Comprehensive Biochemistry, Vol. 35, Elsevier, Amsterdam, 1983, p. 103.
4 E.F. Hartree, Keilin, cytochrome, and the concept of a respiratory chain, in G. Semenza (Ed.), Oxygen, Fuels, and Living Matter, Part 1, Wiley, Chichester, 1981, p. 161.
5 A. Neuberger, Proc. Roy. Soc. Lond. A 893 (1937) 158, 68.
6 C.R. Harington and A. Neuberger, Biochem. J. 30/5 (1936) 809.
7 J.L. Crammer and A. Neuberger, Biochem. J. 37/2 (1943) 302.
8 A. Neuberger, Biochem. J. 32/9 (1938) 1435.
9 R.C.G. Moggridge and A. Neuberger, J. Chem. Soc. 139 (1938) 745.
10 A. Neuberger, J. Chem. Soc. 5 (1940) 29.
11 P.G. Johansen, R.D. Marshall and A. Neuberger, Biochem. J. 78 (1961) 518.
12 R.D. Marshall and A. Neuberger, Biochemistry 3/10 (1964) 1596.
13 G.S. Marks, R.D. Marshall and A. Neuberger, Biochem. J. 87 (1963) 274.
13a A. Neuberger, in H.W. Schultz (Ed.), Carbohydrates and Their Role. Avi Publishing, Westport, CT, 1969, pp. 115–132.
14 A. Neuberger, Past and present concepts of glycoproteins, in Glycoproteins of Blood Cells and Plasma, Ch. 1, 4th Ann. Sci. Symp., American National Red Cross, Lippincott, Philadelphia, 1971.
15 A. Neuberger and A.P. Fletcher, J. Chem. Soc. (B) (1969) 178.
16 A. Neuberger and W.A. Ratcliffe, Biochem. J. 129 (1972) 683.
17 R.C. Davies, A. Neuberger and B.M. Wilson, Biochim. Biophys. Acta 178 (1969) 294.
18 R.C. Davies and A. Neuberger, Biochim. Biophys. Acta 178 (1969) 306.
19 A. Neuberger and T.A. Webster, Biochem. J. 39/2 (1945) 200.
20 D.F. Elliott and A. Neuberger, Biochem. J. 46/2 (1950) 207.
21 H.R.V. Arnstein and A. Neuberger, Biochem. J. 55/2 (1953) 259.
22 A. Neuberger, J. Chem. Soc. 107 (1945) 429.
23 C.S. Hudson and A. Neuberger, J. Org. Chem. 15/1 (1950) 24.
24 A. Neuberger, J.C. Perrone and H.G.B. Slack, Biochem. J. 49/2 (1951) 199.
25 O.B. Henriques, S.B. Henriques and A. Neuberger, Biochem. J. 60/3 (1955) 409.
26 C.W. Crane and A. Neuberger, Br. Med. J. ii (1960) 815 and 888.
27 D. Shemin and D. Rittenberg, J. Biol. Chem. 166 (1946) 621 and 627.
28 H.M. Muir and A. Neuberger, Biochem. J. 45/2 (1949) 163.
29 H.M. Muir and A. Neuberger, Biochem. J. 47/1 (1950) 97.
30 A. Neuberger, H.M. Muir and C.H. Gray, Nature 165 (1950) 948.
31 A. Neuberger and J.J. Scott, J. Chem. Soc. 4918 (1954) 1820.

32 G.H. Cookson and C. Rimington, Biochem. J. 57 (1954) 476.
33 D. Shemin and C.S. Russell, J. Am. Chem. Soc. 75 (1953) 4873.
34 C.H. Gray, A. Neuberger and P.H.A. Sneath, Biochem. J. 47/1 (1950) 87.
35 A. Neuberger and J.J. Scott, Proc. Roy. Soc. Lond. A 213 (1952) 307.
36 K.D. Gibson, A. Neuberger and J.J. Scott, Biochem. J. 61/4 (1955) 618.
37 K.D. Gibson, A. Neuberger and G.H. Tait, Biochem. J. 83 (1962) 550.
38 K.D. Gibson, A. Neuberger and G.H. Tait, Biochem. J. 88 (1963) 325.
39 A. Neuberger and J.M. Turner, Biochim. Biophys. Acta 67 (1963) 342.
40 R.C. Davies and A. Neuberger, Biochem. J. 133 (1973) 471.
41 A. Neuberger, J.D. Sandy and G.H. Tait, Biochem. J. 136 (1973) 477 and 491; J.D. Sandy, R.C. Davies and A. Neuberger, Biochem. J. 150 (1975) 245.
42 E.A. Wider de Xifra, J.D. Sandy, R.C. Davies and A. Neuberger, Phil. Trans. R. Soc. Lond. B 273 (1976) 79.
43 A.K. Allen, A. Neuberger and N. Sharon, Biochem. J. 131 (1973) 155.
44 A.K. Allen and A. Neuberger, Biochem. J. 135 (1973) 307.
45 D. Ashford, N.N. Desai, A.K. Allen and A. Neuberger with M.A. O'Neill and R.R. Selvendran, Biochem. J. 201 (1982) 199.
46 D. Ashford, A.K. Allen and A. Neuberger, Biochem. J. 201 (1982) 641.
47 N.N. Desai, A.K. Allen and A. Neuberger, Biochem. J. 211 (1983) 273.

G. Semenza and R. Jaenicke (Eds.) Selected Topics in the History of Biochemistry: Personal Recollections, III. (Comprehensive Biochemistry Vol. 37) © 1990 Elsevier Science Publishers BV (Biomedical Division)

Chapter 3

Wladimir Engelhardt: the Man and the Scientist

LEV L. KISSELEV

Engelhardt Institute of Molecular Biology, The U.S.S.R. Academy of Sciences, 32 Vavilov Street, Moscow 117984 (U.S.S.R.)

Translated from the Russian by Dmitri Agrachev

I first saw Wladimir Aleksandrovich (W.A.). Engelhardt in the now faraway year of 1946, soon after the war, when the whole Engelhardt family came to dinner in our house. My father* and W.A. had been undergraduates together, and they had managed to keep their friendship in spite of the striking difference in temperament and the disparate roads they took through life. I was ten years old, and my impressions were obviously of the most superficial kind: I marvelled at the beauty of W.A.'s wife, Militsa Nikolayevna Lyubimova, and was intrigued by the extraordinary looks of the man himself: tall, thin, with slow languorous movements, blue eyes, grey hair and a most enchanting and easy smile. At dinner, W.A. would try the various delicacies prepared by my hospitable mother and pay her exquisite complements, he would even ask for detailed recipes of the dishes he liked best.

---

* Lev Aleksandrovich Zilber (1894–1966), Member of the USSR Academy of Medical Sciences, Honorary Life Member of the New York Academy of Sciences, microbiologist, immunologist, virologist, oncologist. He discovered the agent and carrier of tick-borne encephalitis (1937) and developed the virogenetic concept of the origin of cancer (1945–1961).

Plate 2. Wladimir Alexandrovich Engelhardt at his country house near Moscow (Nikolina Gora). (Photograph of 1979, not published before.)

In 1957-1958, a student of the Department of Biochemistry at Moscow University's Biological Faculty, I attended W.A.'s lectures on enzymology. His manner was totally devoid of 'stage effects', the dramatic gestures or exclamations that lecturers and other public speakers like to use for emphasis.

W.A. spoke slowly, often glancing at his notes, and would occasionally rephrase something he had said earlier. An attentive listener could not only absorb every bit of the explanation, but savour the elegant logic of enzymology. Of course, we were very young and could hardly appreciate Engelhardt's remarkable lecturing skills;

only much later, when I had heard a variety of learned speakers, did I realize how good he had been. He was always trim and well-dressed, always on time, it was plain that he prepared for every one of his lectures, and there was a curious personal touch to everything he said: one of the architects of enzymology could hardly be detached about it, and though his personal attitude was always played down, never emphasized, one could feel it in every word. W.A. had certain biases of which we were aware and some of which seemed almost totally irrational. In biochemistry, for instance, he treated nitrogen metabolism, lipids and membranes with cool indifference, but was very fond of the metabolism of phosphorus compounds and carbohydrates.

Now I think I understand that W.A. used his vast knowledge and amazing powers of intuition to peel off everything secondary, unessential or unreliable in order to expose the crystal-clear core of enzymology, its beauty and its stark logic.

Having completed the course of lectures, W.A. presided over our examinations and was very generous with good marks; he did not allow himself as much as to flinch when a female student admitted she did not know proteins consisted of amino acids. After the exams I was offered a job at the Institute of Radiation and Physicochemical Biology, then just forming in Moscow under the auspices of the USSR Academy of Sciences, where I enrolled in August 1959. For the next 25 years my scientific career developed under W.A.'s supervision, seldom close or continuous, mostly loose and sporadic.

At the time I joined the Institute something rather unusual happened which, I believe, was typical of W.A. It came to my notice that one of the Institute's 'socially active' members had expressed concern about my character reference obtained from the University. It was indeed an extraordinary document: one paragraph written by the Department of Biochemistry referred to me as a good student who had shown a genuine interest in science, while the next paragraph indicated that I had opposed the so-called teaching of Lysenko, had advocated Mendelism–Morganism and, moreover, had failed to repent. This second paragraph had been written at the instance of the administration and 'social organizations' of the University's Biological Faculty because a number of second-year students including myself had attended classes in genetics and statistics at the home of Prof. Alexei Lyapunov, a prominent mathematician who had professional knowledge of genetics. Later, Lyapunov moved to Academy City, newly set up at Novosibirsk, was elected Corresponding Member of the Academy and went on to make a substantial contribution to cybernetics as well as to the development of the school system and college education in Siberia.

A man of great will and dedication, Lyapunov could not reconcile himself with the fact that his two daughters, fellow students of mine, were getting no knowledge of genetics, while the so-called teachings of Lysenko, of which I see no need to write here, were rammed down their throats. In an attempt to reduce the damage, he arranged lectures by many geneticists who had lost their teaching jobs and in many cases were denied all access to scientific work after the memorable session of the All-Union Agricultural Academy that took place in August 1948. These classes initiated us into the fundamentals of genetics and proved a good school of professional decency at the outset of our scientific careers. When the Lysenkoists of the Biological Department found out what was going on, they were enraged. An attempt was made to throw the most active renegades, including the Lyapunov sisters and myself, out of the University, but the then Head of the University, I.G. Petrovsky, did not allow it.

Anyway, my character reference found its way to Engelhardt's desk, and he was asked to consider whether he really wanted to contaminate the new collective with so shady a character. When I walked into the Director's office I knew nothing of all this and thought that he simply wanted to talk to me about the best way to start my work at the Institute. W.A. slowly rose from behind his desk, stepped around it and advanced on me with a outstretched hand. This was so unexpected that I think I started to back away, but he seized my hand and began shaking it. He shook my hand for a long time, saying things like

> 'Well, well. Who could have thought that someone so young . . . It's amazing what you did, my dear fellow. Tell me all about it, I want to know every detail . . .'

I was asked to sit on the sofa, always a sign of particular informality, and submitted to a veritable interrogation about the genetics classes in Lyapunov's flat. W.A. demanded to be told every bit of the story, from a summary of the lectures to the nature of the reprisals (or rather, attempted reprisals).

W.A. was largely responsible for not letting the Lysenkoists into the Academy of Sciences, he did a lot to restore the normal development of biology in our country in spite of Lysenko and his myrmidons. The very existence of Engelhardt's Institute at that time was a great blow to Lysenkoism. We witnessed a number of attempts to discredit the Institute and to do serious harm to its Director. Both the Director and the Institute survived. Molecular biology's right to existence was finally recognized in 1965 when M. Keldysh, President of the Academy of Sciences, agreed, on Engelhardt's insistence, to rename the Institute of Radiation and Physicochemical Biology into the Institute of Molecular Biology, which now (since 1988) bears Engelhardt's name.

I remember showing my first manuscript to W.A. in 1961. It was about the macromolecular properties of transfer RNA. By that time I had already co-authored a number of papers on oxidative phosphorylation (it had been the subject of my 'diploma work', supervised by Vladimir Skulachev), but of course those papers had not been written by me. W.A. took my typewritten sheets with obvious interest and said he would certainly read them. A few days later he gave the manuscript back to me saying that it was fine but could be made even better. I saw that the text had been heavily amended, from the replacement of certain words to the deletion or substitution of whole paragraphs. In addition, there was a page typed by W.A. himself (we always recognized W.A.'s personal typewriter by its peculiar type) with three or four versions of one sentence from my paper, the difference being almost too subtle for me to grasp. Having rewritten the manuscript, I returned it to W.A. and the whole thing happened again. This time there was less criticism and more praise, but I was advised to work on the text a little longer. To cut a long story short, after four of five revisions W.A. generously conceded that it would probably do, considering it was the author's first attempt. It was quite a mediocre paper, too, and I was amazed at how much time and effort W.A. was willing to give it. It remains to be said that, as almost invariably happened in such cases, he absolutely refused to put his name on it, even though his contribution had to do with the paper's substance as well as its form.

He has always been that way. Says S. Burnasheva, an old co-worker of Engelhardt's:

> 'Wladimir Aleksandrovich was civilized and modest to a fault in all things, including the matter of co-authorship. It was very difficult to persuade him to put his name on a joint paper. Even though I was his post-graduate student and then a member of his laboratory, I have, regrettably, only one joint publication with him. He used to say jokingly "You did it, so accept the full responsibility".'

I often had the occasion to see how in the late afternoon, when the day's hustle and bustle had died away, he would sit down to his beloved typewriter and plunge into work. It is not just that he was amazingly persistent and industrious, he was a man driven by the quest for excellence, for perfection. He would welcome criticism, always thank his critics, and then carefully consider what to take notice of and what to reject. He appreciated criticism as an expression of interest in his work. He would look disappointed if a paper was returned to him with nothing but praise. When that happened he would say something like, 'I don't think you've read it' or, ironically, 'Is it perfect, then?'. He was patient and tolerant even when his work was criticized by people who were by far inferior to him in terms of intellect and experience. This is extremely rare, and I think very few people can claim to be as tolerant as Engelhardt.

This is not to say that W.A. always accepted criticism. I remember something that happened in 1960. An important decision had to be made as to whether or not the new Institute needed a vivarium. W.A. asked my opinion and was at first surprised and then annoyed at my firm belief that a vivarium was necessary.

> 'My dear Lev! Molecular biology is about molecules, not mice and rats! What do you need them for? You must work with bacteria, bacteriophages and macromolecules, I never imagined you could be so blinkered and conservative.'

I stood my ground and our conversation ended on a rather chilly note. As a result of all this we were forced to use a vivarium that belonged to another Institute. Though W.A.'s decision proved wrong, he never admitted it, nor did he like his staff to remind him of the difficulties it had caused. I think here we came up against a particular trait of W.A.'s character which in most cases proved a blessing: his persistent, even dogged, determination to move ahead, his absolute refusal to be fettered by routine. Occasionally, as in this episode, it would inconvenience those around him.

Everyone who worked with Engelhardt over the years appreciated his ability to listen to a cleaning woman as attentively as to an academician. That was no ostentatious 'democratism', but the natural good manners of a profoundly civilized person. Like many deserving people of his generation, W.A. could sometimes yell at his superiors, even the most exalted of them, but never at a technician. Deplorably, this excellent trait of the old Russian intelligentsia is rapidly disappearing: now it seems to be the thing to sweet-talk to one's bosses and rage at a lab assistant.

W.A had a very peculiar view of scientific schools. Though his pupils are legion and many of them are successful biochemists and molecular biologists, Engelhardt never thought of himself as the head of a scientific school. To him, the notion of a scientific school had nothing to do with so many followers copying their master's methods as applied to the same problems. Rather the opposite. Above all he valued an original approach, a sense of novelty, the ability to stop a certain avenue of research when it no longer produced any major results. He always knew when a point had been reached beyond which there could be no breakthrough, just a comfortable sort of drifting along, and that was the time to drop everything and take up a new problem. I think W.A. passed his passion for the unknown and the untapped to many of his coworkers, and that is why Engelhardt's Institute has so often launched major research projects in totally new areas, without much scientific groundwork, and often proved successful. It is hardly surprising to me that this pillar of biochemistry suddenly abondoned the field that had brought him glory and, at a fairly advanced age, hurled himself into a just emerging sphere of molecular biology. Now that I have been a scientist for 30 years, I marvel more than ever at Engelhardt's ability to stop in his tracks, safe and well-trodden as they were after years of spectacular work, and set out on a labyrinthine journey with not so much as a decent reputation, let alone fame, guaranteed at the end.

Now I know how rare, almost unique, this ability really is.

What W.A. valued most in his students and coworkers was an unconventional mentality, a sense of the paradoxical, an openness to new perspectives, a readiness to start from scratch. In this respect he was very much like my father who did not hesitate to switch avenues of research and even scientific disciplines in the 50 years of his career: having started out in bacteriology and microbiology, he then turned to immunology and virology and ended up as an experimental oncologist. Since my father's example was always in my mind, I found it easy to understand this quality of W.A.'s which many people found strange and 'difficult'.

In 1984, a few months before his death, W.A. arranged a seminar on oncogenes for several of the Institute's laboratories. He addressed the seminar with a 40-min talk, which amounted to a lively and highly personal exposé of the available data, presented in a vivid idiosyncratic way, totally devoid of 'standardized' concepts. He was 89, and the only fitting term to be used here is the Engelhardt Phenomenon.

Maybe none of Engelhardt's deeds over the past 25 years has so much reflected his character, his way of dealing with a scientific challenge, his flair, his attitude towards his fellow workers and his ideas as to how scientific work should be organized as the project that came to be known as 'reverse transcriptase (revertase)'.

Molecular biologists surely remember the spring of 1970 when Howard Temin and David Baltimore discovered an enzyme that could catalyse DNA synthesis from an RNA template. There are two basic reasons why this process of RNA-directed DNA synthesis (reverse transcription) immediately caught the attention of the scientific world: for one thing, the revertase (Engelhardt's name for the enzyme) could synthesize genes in a test tube, thus laying the foundations of genetic engineering; for another, the existence of reverse transcription explained the replication of RNA-containing tumour viruses (retroviruses) and the mechanism whereby they integrated their genetic material into cell DNA. W.A., with his perfect flair for new perspectives, quickly realized the significance and the far-reaching ramifications of that discovery. I am sure many people thought along the same lines, but what distinguishes Engelhardt from all those people is that he turned his thoughts into deeds by planning and bringing off a remarkable campaign that led to the emergence of a new field of molecular biology in the Soviet Union, Czechoslovakia and the German Democratic Republic.

What did he actually do? To begin with, he used his wide scientific and personal contacts for a kind of preliminary networking: the problem was discussed with a variety of experts representing different areas and trends of science. Gradually he came to believe that a number of scientific centres in several countries could be effectively brought together within a single coordinated research programme. Assisted by nobody (and he was 75 years old!), W.A. toured the Research Institutes of the Soviet Union and made trips to Berlin and Prague, talking about his project to key people everywhere. The response was overwhelmingly enthusiastic; W.A. himself had not expected to find everybody so willing to join the programme. Now we understand that the enthusiasm was as much due to Engelhardt's personality as to the scale and importance of the scientific challenge. People were irresistibly attracted by Engelhardt's supreme intellect, his genuinely civilized manners and downright charm. His human qualities cemented this remarkable enterprise whose participants were a motley group in terms of scientific background, age and past experience.

The project proved so stimulating that after its successful completion the participants decided not to part company and set up a new association, as informal as the first one, under the name of 'Revertase-Oncogene'. This programme was launched in 1980 with W.A.'s approval and support and continues to function after his death, on the basis of the same organizational and moral principles that Engelhardt had laid down and elaborated over the years.

W.A. believed in the utmost importance of direct communication and interaction among scientists. Therefore the Molecular Biology Research Council, set up by Engelhardt, sponsored countless national and regional 'schools' in molecular biology. When W.A. attended these schools, his presence invariably created a festive atmosphere and his interest in talking to as many people as possible always lent a special zest and excitement to these events.

W.A. had a rare sense of humour and the ability to laugh at

himself. He tolerated fairly barbed jokes about his person and liked to give vent to his own wit in appraising people and events. In the first years of the Institute's existence, when I used to edit our 'wall newspaper', I once asked Engelhardt's opinion of the current issue just posted in the hall outside his office. I had seen him peruse our 'local press', so I inquired what he thought of it. 'I think it's too bad', came the totally unexpected reply. Shaken, I let my eyes roam over the paper trying to guess what it was that had displeased the Director. Obviously relishing my confusion, W.A. repeated with some emphasis, 'It really is too bad', making me almost frantic with distress. After a very long pause, when he had made sure I was totally speechless and helpless, he continued unperturbed:

> 'Do you know, Lev, what Macmillan (then Prime Minister of Great Britain) said when he found no cartoon of himself in the morning paper? He said his popularity had obviously waned and it was time for him to retire. Your paper hasn't a single cartoon of the Director, so the Director has probably reached the point of retirement as well.'

Having said that, he could contain himself no longer and burst into that radiant smile that we loved so much. I could only laugh in return and promise that we would atone for our guilt.

Engelhardt was a man of amazing erudition, his knowledge was not only profound but embraced an extraordinarily wide range of subjects. I remember how a multilingual delegation came to see him in his office in 1980 or 1981. What followed is a performance worth recounting. Engelhardt addressed his guests in four languages, asking them which language he should use to tell them about the Institute. It turned out that they did not have any one language in common, though each understood at least one of the four languages. Next it transpired that none of those people were biologists, so they probably did not know the difference between a protein and a DNA molecule. To me the situation seemed hopeless, but Engelhardt began talking about the Institute in Russian with complete equanimity and, having spoken for quite a long time, repeated the whole thing, almost word for word, in English, French and German. I understand these languages (though my knowledge of them is far from fluent), so I could appreciate the accuracy of W.A.'s self-translation and his ability to remember big chunks of his own presentation. He spoke for about an hour, and I do not think anyone of those present had ever seen or will ever see again anything like it.

It was one of Engelhardt's many hobbies to translate Russian poetry into other languages. A lot of his friends have seen his English translations of Fyodor Tyutchev, a poet he admired.

Engelhardt's longevity was an enigma to many people, for it was not a mere physical thing, but an active, creative kind of longevity. That again is part of the Engelhardt Phenomenon, a riddle I cannot solve, but I think that one of the reasons, and maybe the mainspring of his amazing energy, was W.A.'s unflagging interest in everything that happened in science and in the world around him. He would listen with the same fascination to someone's impressions of an article in the latest issue of *Nature* or *Cell* as he would to the story of a summer's canoeing adventure or to a new recipe. A brilliant experimenter, he extended this passion of his to cooking, an art that is curiously akin to biochemistry. He loved to invent new recipes and would sometimes treat his close friends to the material results of these experiments; then he would closely watch the taster's face, trying to guess from the expression in his eyes what he really thought of the new creation. He loved it when people praised his culinary skills. It also pleased him very much when someone told him he had a nice new tie that went extremely well with his suit. Occasionally I would resort to these little tricks to put W.A. in a good mood: share an exotic recipe or admire a new garment.

I have been W.A.'s travelling companion on several trips abroad: to the United States, to Spain and to France. Some of his qualities that had been blurred and fragmented in our workday existence at home became especially visible then. His profound knowledge of the history, culture and languages of the countries we visited was never ostentatious or overpowering. It seemed as natural as his respect for our hosts. New cities and new acquaintances fascinated him, and his genuine interest enriched his conversation. One almost forgot his age: it was as though he left all those years at home and looked at the world with boyish eagerness and curiosity, shaming those of us who were only half as old but often 'lazy and uncurious', to use Pushkin's phrase. In spite of his hearing problem, which would sometimes make him miss half a lecture if the speaker's diction was not very good or if the room had poor acoustics, W.A. could always tell a good paper from a dull or mediocre one and, miraculously, would comment on the thing that was indeed the most interesting of all. I think his tremendous intuition, on which he relied a lot, and his great scientific experience made it possible for him to get the whole picture together from a few pieces, and then to pass judgement.

I believe Engelhardt has had not one but two lives: the life of a brilliant practicing scientist laced with major discoveries that have long since found their place in biochemistry textbooks, and the life of a model organizer who created an institute, a journal, a research council and the 'Revertase' project, and set the example of honourable behaviour in science and in society for a host of molecular biologists and biochemists.

I have been trying to describe Engelhardt as I saw him over the last 25 years of his nearly 90-year-long life. I witnessed the upshot, but to understand the scientist and the nature of his discoveries it is much more important to know the beginning, the source. W.A. himself pointed this out in his brief autobiography [1], which the reader can find without difficulty, so I will only recall some essential facts that seem to have influenced his further development.

W.A. was not a 'born' biologist. His early interests embraced such disparate fields as electrical engineering, mathematics, chemistry and medicine. What pushed him in all these directions was, I think, the desire to experiment, the urge to 'test' nature. The Russian word yestestvoispytatel (naturalist) literally means 'nature-tester' and I think it is an excellent description of W.A.'s motivation as a scientist. Moscow University's Medical Faculty, which he attended in the years of the Russian revolution, had little effect on his professional development. What seems to have played an important part were the lectures of N. Koltzov, an outstanding forerunner of molecular biology. Here is what Engelhardt said, addressing the Second All-Union Congress of the Vavilov Society of Geneticists and Breeders in 1972 [2]:

> 'I don't think I can refer myself to any particular "school", as we are wont to call them, but I sincerely regard Nikolai Konstantinovich as my first teacher, whom I have remembered all my life."

Koltzov was a Professor at Popular Shanyavsky University, and it was there that W.A. attended his lectures and accomplished his first experimental project under Koltzov's supervision: a study of the effects of pH on the uptake of India-ink particles by microorganisms. It was still a far cry from biochemistry, but Koltzov's lessons remained with him for the rest of his life.

In the 1930s N. Koltzov was one of those who opposed Lysenko's devastating crusade against Russian biology; Koltzov's Institute was routed and Koltzov himself lost his job, and only his sudden death in 1940 kept him from the tragic fate of Nikolai Vavilov.

Koltzov's lessons were not only a school of experimental biology but a school of civic courage and commitment to genuine scientific values. It is hardly surprising therefore that in the 1950s, when Engelhardt found himself at the head of the Biological Division of the USSR Academy of Sciences, he took every possible action against Lysenko, who, of course, saw to it that W.A. was removed from his post. The decision to set up an Institute of Molecular Biology was passed by the Academy in 1957, but Lysenko succeeded in thwarting it, and it was only two years later that this research centre appeared camouflaged as the Institute of Radiation and Physicochemical Biology, although no serious work in radiational biology was ever done in it. In the early 1960s, when Lysenko still had enormous power, Engelhardt publicly opposed the nomination of two lysenkoists for an Academy election, a move that might have cost him his newly created Institute. If today the Institute of Developmental Biology of the USSR Academy of Sciences bears the good name of Koltzov, it is among others, W.A.'s doing, too.

Engelhardt's scientific career began in Moscow in 1921, at the Biochemical Institute of Public Health organized by Prof. A.N. Bach. Curiously enough, W.A. started out in a field that was somewhat distant from Bach's own research interest and had to do with the nature of immune reactions: W.A. was to obtain antibodies against certain enzymes, including phenolase, an enzyme that interested A.N. Bach [2a].

W.A. himself later admitted that those experiments had not yielded any major results in terms of a new understanding of the nature of enzymes or the mechanism of immune reaction, but there

In 1927 he spent two months at Peter Rona's laboratory in Berlin, two very important months that allowed him access to some of the most spectacular biochemical projects. At Rona's laboratory he literally worked side by side with David Nachmansohn, who became a lifelong friend. On the floor above them H. Weber worked with muscular proteins. Through Rona Engelhardt met O. Warburg, C. Neuberg, C. Lohmann and O. Meyerhof. Berlin was then the world's centre of biochemistry, at least as far as anaerobic metabolism and phosphoric acid metabolism were concerned, which could not but have a profound effect on the 33-year-old scientist from Moscow.

In1929 Engelhardt accepted the invitation to occupy the Chair of Biochemistry at Kazan University and left Moscow for a few years. was one observation that neither the scientific community nor, I think, Engelhardt himself properly appreciated at the time. W.A. discovered [3,4] that antibodies could bind to the antigen not only in solution (as was already known) but after they were transferred from the solution into a heterogeneous state, adsorbed or immobilized by some kind of carrier. In those early experiments it was kaolin and aluminium hydroxide. Later W.A. called it 'the fixed-partner principle' [5]. At that time this totally new approach enabled W.A. to use 'immobilized' antibodies to fish out haemoglobin and invertase, proteins that did not produce insoluble antigen-antibody complexes in solution.

It is certainly right to regard those experiments as early precursors of the innumerable versions of immobilized systems that we now refer to as immunoaffinity chromatography, ELISA, immunosorption, immobilized enzymes, bioaffinity chromatography, etc. It is interesting that W.A. 'revisited' those early experiments several decades later and, together with his coworkers, applied the same approach to new problems [5].

Engelhardt's first publications proclaimed his special love of enzymes, which stayed with him for life, and his affection for immunology, which was to remain largely platonic. Still, it is hardly coincidental that in the last year of his life he set up a new laboratory at his Institute for the express purpose of studying immunoglobulin genes.

He did not yet know that Kazan would inscribe a brilliant page in the book of his life.

The first year was taken up by the task of organizing a laboratory, virtually from scratch, and putting together a course of lectures, but the next year W.A. started intensive experimental research into the phosphorus metabolism, a domain he had been thinking about for some time.

By that time ATP had been discovered and, moreover, it had been proved to form as a result of a non-oxidizing transformation of glucose through fermentation and glycolysis. Though respiration was regarded as an energy-producing process, nothing was known about the possibility of a relation between respiration and phosphorus metabolism. Trying to understand in retrospect why respiration remained almost forgotten alongside the highly successful studies of glycolysis and fermentation, Engelhardt arrived at the simple explanation [6] that, while fermentation and glycolysis were mostly studied in yeast and in the liver, neither was a very good object for the study of respiration. He thought it was pure luck that he had stumbled upon avian erythrocytes which, unlike mammalian red blood cells, have a large nucleus and are characterized by intensive respiration. All this became clear later, but what mattered at the time, as W.A. himself told me, was that the town of Kazan was full of pigeons and pigeon blood was easy to come by without upsetting the laboratory's extremely modest budget. W.A. found that the erythrocytes kept a virtually constant high ATP level, while there was access to oxygen and respiration was maintained [7,8]. However, when respiration was poisoned by potassium cyanide or when nitrogen instead of oxygen was pumped through the erythrocyte suspension, one observed the degradation of ATP and accumulation of inorganic (mineral, as it was called then) phosphate.

After an hour's incubation at 37° C all the acid-labile phosphorus of ATP was degraded down to inorganic phosphate. Though extremely simple and graphic, these results could be interpreted in two different ways. Indeed, the fact that ATP got degraded in the absence of oxygen could mean either that respiration prevented the degradation of ATP and stabilized it or that respiration did not prevent the degradation of ATP but provided the conditions necessary for its resynthesis. The situation would be static in the first case and dynamic in the second. Now this problem could easily be solved with the help of ^{32}P, but the year was 1930 and radioactive isotopes were still a thing of the future.

Engelhardt found an amazingly simple experimental solution [7,8]. He made oxygen accessible to erythrocytes whose ATP content had been depleted after anaerobiosis. Restored respiration led to the reesterification of the inorganic phosphate and brought back the original ATP level. Engelhardt very accurately termed this process 'respiratory resynthesis' of ATP.

By alternating aerobiosis and anaerobiosis it was possible to change the ATP concentration in pigeon erythrocytes many times over [8]. The 'respiratory resynthesis' of ATP is now known as oxidative phosphorylation, an essential part of bioenergetics. The discovery of oxidative phosphorylation assigned a very special place in bioenergetics to ATP, as both respiration and glycolysis turned out to be pathways of ATP synthesis, ATP being the cell's chief energy-donating molecule.

The discovery that respiration was coupled with phosphorylation opened the floodgates: I cannot even begin to summarize all the numerous, or rather innumerable, studies, that followed.

Ironically, in my University years, working under the guidance of V. Skulachev, a talented disciple of Engelhardt's who went on to make a major contribution to bioenergetics, we staged special experiments with the purpose of dissociating respiration and phosphorylation instead of coupling them. As a matter of fact we succeeded [9,10] having applied certain tricks, and, incidentally, we also used pigeons, but those experiments only confirmed the universal nature of Engelhardt's discovery.

Engelhardt was instrumental in solving one of biochemistry's great enigmas: the Pasteur effect. What it basically comes down to is that 'respiration inhibits fermentation'. In other words, the anaerobic glycolytic pathway of carbohydrate degradation in the presence of oxygen is inhibited for the benefit of the aerobic metabolic pathway, i.e., oxidative phosphorylation. The biological expediency of the Pasteur effect is obvious, for the aerobic breakdown of the carbohydrate (hexose) molecule can release many times more energy (stored in the macroenergetic bonds of ATP) than anaerobic glycolysis. The Pasteur effect was the most important regulatory link in the energy metabolism, and it was only natural that it intrigued and worried Engelhardt, the discoverer of respiratory phosphorylation.

To explain the Pasteur effect W.A. advanced what I think was the simplest of all possible concepts. He suggested that the inhibition of fermentation by respiration was attained by the direct action of oxygen on some component of the anaerobic pathway. According to Engelhardt's hypothesis, fermentation was inhibited in the aerobic conditions through oxidative inactivation of some mechanism involved in the enzymic chain of fermentation and glycolysis. The simplicity of this explanation - the direct action of oxygen turning off the glycolytic pathway - was probably the most unexpected thing about it. One of Engelhardt's favourite pupils, Nikolai Sakov, was charged with the task of putting the hypothesis to the test.

The idea was to investigate the sensitivity toward oxidation of different enzymes (hexokinase, isomerase, aldolase, etc.) involved in the initial steps of the anaerobic breakdown of glucose. Compounds with different redox potentials were used as oxidizing agents. All of the tested enzymes (with one exception) were insensitive. The exception was phosphofructokinase, which is known to catalyse the synthesis of fructose-1,6-diphosphate from fructose-6-phosphate and ATP: this enzyme turned out to be extremely sensitive to compounds with a redox potential of 0.05 to 0.25 V. The same inhibition was caused by a number of other oxidizers: hydrogen peroxidase, iodine, dihydroascorbic acid. Reduced forms of those compounds had no effect on phosphofructokinase. Thus, a number of disparate agents, totally alien to the cell and its enzymes, had something in common, viz. the ability to inactivate one of the enzymes of the anaerobic pathway of carbohydrate metabolism.

This would seem to fully corroborate Engelhardt's initial hypothesis. However, there was the potential objection that this effect might not be directly linked to the Pasteur effect as none of the tested compounds was a natural component of the anaerobic metabolism of carbohydrates. Therefore, decisive experiments were carried out in which cytochrome and cytochrome oxidase served as oxidizing components. Oxidized cytochrome proved to be able to inactivate phosphofructokinase in the presence of an intermediate carrier. Furthermore, even catalytic amounts of oxidized cytochrome were active when cytochrome oxidase was present. When exposed to air, an almost complete inhibition of enzymic activity occurred in this system. Thus, there is no doubt that those experiments essentially reproduced the Pasteur effect in a simplified model system. The experiments were completed in the Spring of 1941. In June the war between Germany and the U.S.S.R. broke out. N. Sakov was drafted; Engelhardt and his family left Moscow and found themselves in Frunze, the capital of the Central Asian republic of Kirgizia. W.A. took the laboratory protocols with him and waited for news from the front. The news, when it came, was tragic: Sakov had perished at Stalingrad, and the work was published posthumously [11]. The war made it impossible to get it to an international journal and, of course, the Russian-language publication, in wartime, went unnoticed. Speaking at a conference in Paris in 1946, on the occasion of the 50th anniversary of Pasteur's death, Engelhardt expounded his interpretation of the Pasteur effect. But Europe was busily recovering from wartime devastation; the proceedings of the conference were never published, at least not to my knowledge, and somehow the facts did not register with the biochemical community. As sometimes happens, they were rediscovered 20 (!) years later by J. Passoneau and O. Lowry [12] in a paper entitled 'Phosphofructokinase and the Pasteur effect'. The authors made no reference to Sakov and Engelhardt, probably because it never occurred to them that a study like that could have appeared in Russian 20 years earlier, in the heat of a terrible war. Besides, Engelhardt had abandoned that field of research a few years after the discovery was made.

Still, W.A. finally got his chance to tell the world about that pioneer work. In 1973 he attended a conference on bioenergetics held at the American Academy of Arts and Sciences in Boston and presented a brilliant exposé [13] of the experiments accomplished 30 years earlier.

While Engelhardt's previous and subsequent discoveries (oxidative phosphorylation and ATPase activity of myosin) are known to every educated biochemist, his contribution to the study of the Pasteur effect is virtually unknown in the West. Maybe this article will at least partly correct that injustice.

In my presentation of Sakov and Engelhardt's experiments on the mechanism of the Pasteur effect I have somewhat violated the chronological order of events, for the most important thing that happened in the late 1930s and early 1940s in the professional lives of W.A. and Militsa Nikolayevna Lyubimova-Engelhardt (Plate 3), a brilliant experimenter and W.A.'s coworker as well as his wife, was undoubtedly, the discovery of the ATPase activity of myosin.

W.A. has spoken to me on many occasions about different episodes of that story, and though the facts were always unchanged, each time he would alter the emphasis somewhat, for as he grew older and switched scientific interests, his memory would highlight this or that particular page of that exciting chapter in the development of biochemistry. Some experts in the field may have a different version of the story, but I am going to rely on W.A.'s own writings and on what he told me.

It was only natural that in the mid-1930s Engelhardt began to show an interest in muscular contraction. For one thing, it was obvious even then that the muscle, which performs a lot of mechanical work, cannot but fall within the scope of bioenergetics. Besides, after the ATP molecule, a universal accumulator and donor of energy, was shown to have a key role both in glycolysis and in oxidative phosphorylation, there was no avoiding the question of the relation between function of muscle and ATP. Finally, a very important idea had been advanced and a very important experiment had been accomplished by that time. O. Meyerhof postulated that in any machine that is not a heat engine the chemical process supplying the energy must act upon the machine's structure, causing changes which result in its performance. At the time this postulate was not experimentally elaborated either by Meyerhof himself or by others. Lundsgaard found ATP to be the immediate source of energy for muscle work. It would seem that Lundsgaard and Meyerhof had all the groundwork that was necessary to discover the ATPase activity of myosin. H. Weber, E. Lundsgaard and O. Meyerhof must have

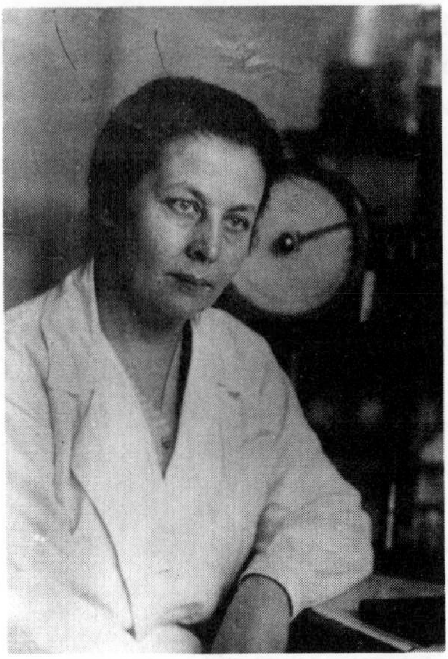

Plate 3. Militsa Nikolayevna Lyubimova-Engelhardt in the laboratory of the Bach Institute of Biochemistry (Moscow) when and where the ATPase activity of myosin was discovered. (Photograph of 1939.)

been within a hair's breadth from that discovery, having obtained the first purified preparations of myosin. Weber pointed out that a functioning muscle has myosin in a solid gel-like form and not in solution. However, the researchers' fascination with glycolysis, oxidative phosphorylation and other metabolic reactions in cell extracts made myosin appear as something extraneous to biochemistry, a part of muscle physiology, so that it eluded the attention of enzymologists. Meanwhile, enzymologists working with water-salt muscle extracts did not detect any ATPase activity, or found that the activity was too low to warrant its association with muscular contractions. In other words, the groundwork had certainly been prepared for the discovery of the ATPase activity of myosin, but no one actually discovered it before Engelhardt and Lyubimova, even though a number of scientists were ready for it both theoretically and experimentally. I would even claim that Engelhardt was not quite as close to that discovery as some of the others, for, as he himself admitted, his knowledge of muscle contraction was not very deep at the time, nor was his experience in enzymology on a par with that of the luminaries working in Germany. And in spite of all that, myosin ATPase was discovered by somebody who did not seem quite ready for it. I see two explanations for this. First, Engelhardt had been thinking relentlessly about ATP ever since his Kazan interlude, and his knowledge of the works of Meyerhof, Lohman, Lundsgaard, Weber and others led him inevitably to the search for the ATPase. Had not oxidative phosphorylation been discovered a few years earlier precisely because ATPase destroyed ATP while respiration restored it in pigeon erythrocytes? Second - and I feel this is the more important reason of the two - Engelhardt was a heretic, a man with a paradoxical mind who loved to stun those around him with ideas that were totally unexpected and unconventional. He hated all manner of rules and canons in science and enjoyed demolishing them. He would not follow the beaten track and look for muscle ATPase in extracts, where others had looked before him, but tried the insoluble sediment, i.e., that part of the muscle extract that had been carefully removed and discarded by his predecessors. Says Engelhardt:

> 'Ours was a very modest exploit at that stage. We simply dipped into the garbage, figuratively speaking, and got out the material that others wouldn't even look at.' [14]

That was not yet *the* discovery, for one could easily claim that the ATPase in the insoluble part of the extract, mostly made up by myosin, was firmly associated with it. The logic of the experiment demanded that one should try and remove this activity from the sediment by extraction. Engelhardt and Lyubimova applied a concentrated salt solution which was known to convert myosin into the soluble state. I resume Engelhardt's own story:

> 'Great was our astonishment when after treating the residue with solutions of higher ionic strength, as used for the isolation of myosin, we found the full

enzymatic activity present in the myosin containing extract. This result was exactly opposite to our expectations.' [15]

These words are important. They show that Engelhardt had not foreseen specifically the ATPase activity of myosin, he had not set out to reveal it as belonging to that particular protein, but the experiment logically led him to this result, which was, for him, unexpected. After this, numerous attempts were made to separate myosin and ATPase activity, all to no avail: myosin and ATPase were inseparable. Moreover, the temperature sensitivities of myosin and ATPase were found to coincide.

Here I should like to digress a little. Engelhardt loved to coin new terms. So in a paper submitted to *Nature* [16] he referred to myosin's activity as 'ATPase', though only the complete term 'adenosine triphosphatase' was accepted at the time. W.A. told me that *Nature* had rejected this 'poetic licence'. Of course later ATPase gained currency and today's readers of *Nature* have long since grown accustomed to the term. Engelhardt also coined the term 'codase' for amino acyl-tRNA-synthetase [17], and though it is not yet generally accepted, one cannot but recognize that it is as accurate as it is concise. Finally, after H. Temin and D. Baltimore had discovered the reverse transcriptase, he condensed it into 'revertase', in which form it obtained currency in Russian scientific literature.

Why did W.A. refer to the ATPase activity of myosin as 'unlikely'? Myosin accounts for a considerable part of the muscle's dry weight, it is a structural protein specially designed for muscle work. Enzymes, for their part, are catalysts that are present in infinitesimal amounts in the cell's soluble part. Clearly, therefore, bringing together a structural contractile protein and an enzyme must have been heresy to the biochemists of the time. Engelhardt and Lyubimova's discovery was so patently anti-dogmatic that traditional mentality simply rejected it. A certain psychological barrier had to be overcome. No wonder many scientists, including major ones, were doubtful or even openly sceptical. Some objected and attempted to refute the revolutionary results. As time went by, however, the work of Engelhardt and his co-workers, as well as a host of other researchers, proved beyond the shadow of a doubt that myosin itself, and not an admixture associated with it, possessed ATPase activity. Since the late 1980s, we know the structure of the catalytic centre within the myosin molecule, which is responsible for ATP hydrolysis.

The discovery made by Engelhardt and Lyubimova was clearly fundamental. Never before had a primarily structural protein, a protein that constituted a considerable proportion of the muscle fibre, proved to have catalytic capability. All the previously discovered enzymes accounted for a small, often negligible, part of the total protein mass and had no function other than their catalytic activity. Furthermore, ATP hydrolysis activated by myosin meant a direct conversion of chemical into mechanical energy in which the motor itself (muscle) provided its own energy source through the splitting of ATP. These experiments by Engelhardt and Lyubimova started a new field: mechanochemistry, in which thousands of projects have since been accomplished. The year 1989 marks 50 years since the discovery of myosin ATPase.

Mechanochemistry is a borderline field where biophysics and the physiology of muscle contraction, enzymology and bioenergetics come together. It would be wrong to assume that the discovery of myosin's enzymic activity was the most important in the emergence of mechanochemistry. One should not forget the other side of the coin: if myosin causes the hydrolysis of ATP, what is the effect of ATP on myosin? Here decisive results were attained by two parallel research teams: J. Needham's lab at Cambridge (U.K.) [18] and Engelhardt's group at the Institute of Biochemistry, the USSR Academy of Sciences, in Moscow [19].

Needham and co-workers found that ATP affected the properties of myosin molecules in solution, specifically viscosity and flow birefringence. These changes were associated with ATPase activity. ADP produced no such effect. There was no effect after ATP hydrolysis.

These experiments were extremely important, for they demonstrated the ability of ATP to alter essential physical properties of the enzyme associated with its contractile function. J. Needham called myosin 'the contractile enzyme', and W.A. liked to use this apposite term. Engelhardt always gave credit to the Cambridge group and even suggested that the ATP-caused change of myosin viscosity be called the Needham effect.

Engelhardt and his co-workers pioneered another branch of mechanochemistry, namely the study of myosin threads that change their properties under the influence of ATP. To this day mechanochemists conduct parallel investigations of solutions and model threads or monomolecular layers of 'contractile enzyme'.

It turned out at a somewhat later date that myosin had not been an individual compound in any of the above experiments: it had been part of a very strong complex with actin, another muscle protein, discovered by F. Straub. It appeared to be actomyosin, rather than pure myosin, that had the ability to change its physical properties under the influence of ATP, though it was still myosin that possessed ATPase activity. The 'impurity' of the myosin preparations used by Needham, Engelhardt and others proved to be a blessing: it led to the discovery of a new functional phenomenon, viz., the ability of a low-molecular metabolite (ATP) to cause major changes of the macromolecule (actomyosin).

The interrelationship of conformation, function and catalytic activity is a fundamental and commonly accepted concept of today's molecular biology, but it was a truly revolutionary approach for the biochemistry of the 1940s.

Next Engelhardt's laboratory concentrated on what W.A. called 'the chemical basis of the locomotory function of cells and tissues' [20]. They used sperms, which are known to possess a motility of their own, ciliate (flagellar) microorganisms, mobile plant (mimosa) cells. In every one of these cases they identified proteins which, like the actomyosin complex, could perform the enzymic hydrolysis of ATP. Thus the locomotory activity of virtually any cells, not just muscle, obeys the fundamental principle of 'myosin ATPase', viz., a combination of the structural basis and physical changes upon the enzymatic release of ATP energy in the course of its hydrolysis.

Obviously, it is not my intention here to describe the further development of mechanochemistry and bioenergetics after the Engelhardt–Lyubimova breakthrough, but any serious researcher in that field must be aware that the whole thing started with the publication in *Nature* [16].

Now, five decades later, we have much more evidence to the effect that mechanical rearrangements in cells play a key role in mitosis for chromosome segregation, in the compartmentalization of a number of processes, and in other cases. The principles of mechanochemistry have a significance that goes far beyond the muscle, where it all began.

It is probably true that any fundamental discovery (if it is genuine) takes on a long and multifaceted life of its own, as it were independent of its author. It is certainly true of oxidative phosphorylation and the myosin ATPase activity. W.A. used to say jokingly:

> 'I must really be a classic, for they no longer make any reference to my work on respiratory phosphorylation and myosin ATPase.'

The discoveries found their way into textbooks and now belong to science with its tendency to make all knowledge impersonal.

Engelhardt's 'love affair' with ATP was to grow into a life-long passion. In the last 25 years, when he did no experimental work himself, burdened as he was with strategic authority, he still thought about it and read all the relevant publications, following every major project that directly or indirectly involved ATP.

In 1960, when I first began working with W.A., I volunteered to take up the study of aminoacyl-tRNA synthetases. If my suggestion was enthusiastically accepted, I think it was because ATP is one of the three substrates for those enzymes. In almost all our work at the time Engelhardt's interest was the strongest incentive, his praise the most cherished reward, not at all easy to come by. In our study of aminoacyl-tRNA synthetases W.A. got interested in a hypothesis advanced by L. Frolova and myself [21] on the role the tRNA anticodon plays when codase recognizes its substrate tRNA among all others. This concept, formulated back in 1964, was the first concrete hypothesis about protein-nucleic acid recognition, even though it had no direct proof at the time, mainly because the primary structure of tRNA was as yet unknown and the very existence of the anticodon had not been proved.

W.A. joined in this project, took part in the discussions and suggested new experimental options, as reflected in one of the papers [22]. At that time a special volume was in the works to mark the 70th birthday of F. Lipmann, and W.A. was invited to contribute to it. Engelhardt knew Lipmann and admired him, they had corresponded for years, so the offer was gratefully accepted. After giving it some thought, he suggested that he and I write an article together about protein-nucleic acid recognition in the case of aminoacyl-tRNA synthetases and tRNA. After some intense discussions, involving several versions of the text, the paper was written and published [17]. Its ideas were proved to be long-lived by the further development of molecular biology and biochemistry [23]. To the best of my knowledge, we were the first to clearly delineate the two types of binding between a protein and a nucleic acid: nonspecific (binding) and specific (recognition). Though the involvement of the tRNA anticodon in enzyme recognition did not find any direct experimental evidence for a long time and was disbelieved by many researchers, the advent of new techniques, particularly RNA engineering, has furnished strict proof of the old hypothesis, at least for the series of synthetase-tRNA pairs [23].

The other time W.A. became directly involved in our work was in the late 1970s, when his interest was caught by the role of metal ions in enzymic reactions, and he asked us to find out whether aminoacyl-tRNA synthetases were metal enzymes. We did indeed demonstrate that bovine tryptophanyl-tRNA synthetase was a Zn^{2+}-containing protein, and the removal of the Zn^{2+} ion not only deprived the enzyme of aminoacylating and activating properties [24] but, as we later showed [25], stimulated a new GTP/ATPase activity, which was completely inhibited by Zn^{2+}. It was the last experimental project to be co-signed by Engelhardt [24].

In the last decade of his life, after his 80th birthday, which was sparkingly celebrated at the Institute, he continued to perform the duties of the Institute's head and Chairman of the Molecular Biology Research Council and gave what little free time he had to writing down thoughts about the place of modern biology within the gamut of natural sciences, the interrelationship of science, technology and humanities, the synthetic and analytical approaches in science. However, these writings, in which amazing intellectual depth is

Plate 4. W.A. Engelhardt, H. Theorell and F. Lipmann (from left to right) at a FEBS Meeting (Varna, Bulgaria, 1971).

matched by stylistic brilliance, are not yet available in English. To give the reader a taste of these essays, I will permit myself several longish quotations which, I feel, need no comment.

'A scientist is naturally vested with what I can only call universal responsibility. He is responsible for the quality of his scientific "product", he is expected to be infinitely demanding as to the authenticity of his material, irreproachably proper and aboveboard in his use of other people's work, rigorous in his analysis, totally sure of his conclusions. These are the very basic things, a scientist's personal ethic. His responsibility is much wider than this when it comes to the possible uses and consequences of a scientist's work in the economic and technological spheres. It would be naive to think that the behaviour of an individual scientist can prevent or alter a crisis. What matters is the voice of the scientific community, the professional posture of all men and women of science.' [26]

'Success, a major discovery or a relatively modest achievement which uplifts and inspires a scientist, may come in different ways. Very disparate ways. I think there are two basic scenarios. Everything may happen as expected. Then one feels gratified for having chosen the right objective, used appropriate methods and effectively overcome the difficulties. In the other scenario everything that happens is unforeseen, sometimes the opposite of what one expected. I believe that totally unexpected results, unpredictable discoveries, can give one an especially strong sense of satisfaction. The joy of consummation mingles with the sweet poignancy of luck, of marvellous good fortune. I suppose it does make some sense to talk of a "fluke" or a "happy accident", but one thing is absolutely certain: flukes and happy accidents happen to those who are prepared for them, who have deserved them by dedicated work and who can read the correct meaning into the unexpected.' [27]

'I think creativity is the most important of all the gifts that nature has bestowed on man. It is rooted in a continuous probing of the unknown. I pray to be forgiven for a crude simile, but the compulsion to solve a crossword puzzle is akin to the scientist's compulsion to rack his brains over nature's enigmas. They are a part of the same endless quest for answers. I believe the creative instinct is closely related to the natural need to quench one's thirst. Remember Pushkin's famous line, "Tormented by spiritual thirst"? . . . The poet's spiritual thirst is not unlike the thirst for knowledge that plagues every scientist. (Valery Bryusov once called it a "sleepless craving for knowledge".)' [28]

'Mankind does not know what problems tomorrow will bring. But it must be prepared to deal with them. The number of discoveries is growing, the flow of information is increasing in snowball-fashion. Who will process all this information, who will confirm and elaborate these discoveries? Man the Creator, of course. I'd like to believe that the instinct of never-ending quest, the need to constantly gain ground in the confrontation with the Unknown is inherent in every human being. Everybody likes a beautiful landscape. But to look for it, one person will come down to a river bank, while another will climb a snowy peak.' [29]

'Getting back to creativity, I have to emphasize yet again that it cannot flourish or naturally develop without freedom. Hence the management of human creativity is a very fine and challenging social problem.' [30]

'One often hears that so-and-so has served science all his life. Is that really so? Wouldn't it be more correct to say, in the case of a truly creative scientist, that science has served him, giving him the greatest happiness and satisfaction? I can certainly claim this, looking back upon the long road behind, and I would like to plant this faith in all our young scientists who are just setting out on the great journey.' [31]

The mention of a 'snowy peak' is not accidental. In his youth Engelhardt had done a lot of mountain-climbing: he and Militsa Nikolayevna had been to the Central Caucasus, the Pamir glaciers and the Tyan-Shan range on the Chinese border. W.A.'s love of the mountains led him to admire the art of Roerikh, on which he wrote a beautiful essay. Here is an excerpt:

'I have no doubt at all that anyone who understands beauty and has spent some time in the mountains cannot but be swept off his feet by the amazingly rich and heady gamut of colours and shapes in Roerikh's paintings. Roerikh would not be an artist if he had not been irresistibly and relentlessly drawn towards beauty. Having experienced the spell of the mountains, he gave himself up to them in the latter part of his creatively exuberant life, so most of his work is imbued with their enchantment. It is all the more remarkable therefore that, however important the richness and diversity of colour in his mountain landscapes, we never perceive, and the artist himself never presents, this orgy of colour as beauty per se, only as a means of leading us towards the understanding of what lies hidden beyond the beauty. What we experience here is the unity of beauty and knowledge, for beauty breeds thoughts. Quite often thoughts will have been put there by the artist himself, but almost equally often they will be elements of the spectator's creativity, "co-creation" being an inalienable property of all genuine art.' [32]

Artistic sensitivity, a profound understanding of beauty in art and in life and the ability to accurately convey his perceptions were certainly a part of Engelhardt's talents, and I am not the only one who has commented on this.

Here is a short extract from the reminiscences of Natalia Yeltsina, W.A.'s co-worker and friend:

'He wanted very much to meet Natalia Krandievskaya-Tolstaya. It finally became possible when she came to Moscow, to the Uzkoye holiday centre. I told him about it. W.A. went to see her and presented her with a bottle of French perfume. A few days later she rang me up at home and, her voice trembling with excitement, read me a poem that she had dedicated to him.

Here is the poem.

To W.A. Engelhardt

Now that my heart is weary and stiff
And all desires are safely held at bay,
This fragrant magic hits me like a whiff
Of happiness from quite another day.

It is the fragrance of a life long past,
A youth that comes and goes oh so swift -
You brought it back, my friend, and made it last
A while longer, though the die is cast,
What little time I have is running fast,
I thank you, O Magician, for your gift.'

And here is a fragment of W.A.'s letter to Yeltsina, which he wrote in hospital in the early sixties.

> 'I enjoyed your letter very much. At one point I nearly leapt off my hospital bed – that was when I read your wonderful description of the monument to Vrubel. I saw it so clearly and at once I said to myself, "Why, this must be a Rodin". Then, half a page later, I read that the sculptor is indeed a pupil of Rodin's, and of course Rodin is, to me, the God of Sculpture. There are two kinds of envy, good and bad, and it's with good clean envy that I think of your meeting with Sherwood; too bad that I will never see his works, for there will hardly be a place for them other than this brief personal exhibition.
>
> Last night we had the first frog concert on the pond. The frogs are amazing. Why don't they perform on the radio? If the Philharmonic Society could hear them, they would surely say, "What delicate young voices!" I applauded them, but they were too shy and simply vanished.'

Engelhardt's involvement with art was never separate from his scientific endeavours. This is demonstrated by his works and is confirmed by his associates. Says S. Burnasheva, his long-time co-worker:

> 'W.A.'s formulation of a scientific problem was as elegant as it was profound, he always looked for beautiful and simple solutions.'

A brief but graphic description of Engelhardt, the man and the scientist, can be found in the foreword and afterword to the book *Understanding Life's Phenomena* [2], which was compiled from W.A.'s works when he was still living, but came out after his death. The foreword and afterword were written by A. Bayev, a man who had been very close to W.A., indeed had been a student of his at Kazan, then followed him to Moscow and made a substantial contribution towards the development of the Institute of Molecular Biology. It was Bayev and his co-workers who carried out the now well-known studies of the primary structure of transfer and ribosomal RNAs. Using tRNA, they developed the very elegant 'dissected molecule' technique. I will quote briefly from Bayev's afterword to Engelhardt's book.

> 'There are different kinds of experiment; sometimes the instrumental component dwarfs the underlying idea and the experimenter himself. W. Engelhardt was a great master who had at his fingertips the magic of a terse and irresistibly convinc-

ing experiment, as demonstrated by his studies of glycolysis, cell respiration and muscular contraction.

Engelhardt embodied the inseparable fusion of an experimenter and a thinker. No matter how one labels the peculiar intelligence or manner of a scientist – Pasteur's "bias", Aristotle's mentality or Galilei's method – for Engelhardt an idea and its experimental realization came together in one scientific act. His thought freely journeyed through the maze of biological phenomena commanding experimental facts to make up a logical thread that led up to the truth.

Engelhardt was a mentor for generations of biochemists, setting as he did an example of experimental mastery and scientific logic. He did not have a favourite method or a preferred experimental object, a plant or a microbe, that he would force on his pupils: everything was determined by the researcher's goal.'

Engelhardt loved Russia and was deeply attached to his home town, the old Russian town of Yaroslavl on the Volga river. (He was born in Moscow on 4 December, 1894, but was taken to the family house two months later and lived there throughout his childhood and adolescence.) In 1978 he visited Yaroslavl again and walked in its streets and on the waterfront overwhelmed by emotion. I know this because he told many people about it, and there were tears in his eyes as he spoke. But he was also a citizen of the world, for science is truly international and a good experiment is good in New York, in Moscow and in Tokyo, while a bad one is just as bad anywhere. Not only did he speak all major European languages, had travelled all over the world and corresponded with scores of leading scientists, but he was a profoundly humane person who hated war and totally rejected it as a way of enforcing ambitions of any kind. It is therefore quite natural that for years he was Vice-President of the International Council of Scientific Unions and an active participant in the Pugwash movement. To W.A., the international cooperation of scientists was not a political slogan or a public relations exercise but a necessary and practical thing that he pursued with his customary diligence and persistence, for which all of us should be grateful.

The uniqueness of Engelhardt the scientist is, to me, symbolized by the fact that while he was totally and completely a biochemist (indeed, I do not think it is an exaggeration to say that Engelhardt **is** biochemistry), he had the amazing mettle, having reached the official retirement age, to start a second life in science as a molecular biologist.

And he was never on the sidelines in this new science. He laid down the guidelines for molecular biological research at his Institute; in the 1960s he arranged for a series of monographs to be published under the title *The Foundations of Molecular Biology,* which was instrumental in attracting young talents to the new science; he founded the journal *Molekularnaya Biologiya (Moscow),* which brought together the molecular biologists of the vast country; he created the Molecular Biology Research Council, which organized countless conferences, symposia and workshops and thus set up the much needed network for exchanging scientific information, promoting cooperation and joint projects, raising the general level of research. Finally, as I have already mentioned, he took part in a number of experimental studies which aroused his interest.

I know I have not exhausted my subject: Engelhardt, the Man and the Scientist. He was simply too big to be summed up in any article, even a longish one like this. If, however, this article conveys some idea of his uniqueness, his charm and his amazing talent, to those who have never seen or heard the man but know him as a classic of biochemistry, I consider my task accomplished.

REFERENCES

1 W.A. Engelhardt, Annu. Rev. Biochem. 52 (1982) 1-19.
2 W.A. Engelhardt, Understanding Life's Phenomena (in Russian), Nauka, Moscow, 1984, p. 128.
2a W.L. Kretovich, in G. Semenza (Ed.) Comprehensive Biochemistry, Vol. 35, Chapter 11, Elsevier, Amsterdam, 1983, pp. 353–364.
3 W. Engelhardt, Biochem. Z. 148 (1924) 463–472.
4 A. Bach, W. Engelhardt and A. Samysslov, Biochem. Z. 160 (1925) 117–126.
5 W.A. Engelhardt, L.L. Kisselev and R.S. Nezlin, Monatsh. Chemie 101 (1970) 1510–1517.
6 See [1], 7.
7 W.A. Engelhardt, Biochem. Z. 227 (1930) 16–38.
8 W.A. Engelhardt, Biochem. Z. 251 (1932) 243–368.
9 V.P. Skulachev and L.L. Kisselev, Biokhimiya 25 (1960) 90–95.

10 V.P. Skulachev and L.L. Kisselev, Biokhimiya 25 (1960) 452–458.
11 W.A. Engelhardt and N. Sakov, Biokhimiya 8 (1943) 9–36.
12 J.V. Passoneau and O.H. Lowry, Biochim. Biophys. Res. Commun. 7 (1962) 10–15.
13 W.A. Engelhardt, Mol. Cell. Biochem. 5 (1974) 25–33.
14 See [2], p. 270.
15 See [1], 11–12.
16 W.A. Engelhardt and M.N. Lyubimova, Nature 144 (1939) 668–669.
17 W.A. Engelhardt and L.L. Kisselev, in N.O. Kaplan and E.P. Kennedy (Eds.), Current Aspects of Biochemical Energetics, Academic Press, New York, 1966, p. 214.
18 J. Needham, A. Kleinzeller, M. Miall, M. Dainty and D.M. Needham, Nature 150 (1942), 46–49.
19 W.A. Engelhardt, Adv. Enzymol. 6 (1946) 147–191.
20 See [2], p. 34.
21 L.L. Kisselev and L.Yu. Frolova, Biokhimiya 29 (1964) 1177–1189.
22 L.Yu. Frolova, L.L. Kisselev and W.A. Engelhardt, Dokl. Acad. Nauk SSSR 164 (1965) 212–215.
23 L.L. Kisselev, Progr. Nucleic Acids Res. Mol. Biol. 32 (1985) 237–266.
24 L.L. Kisselev, O.O. Favorova, M.K. Nurbekov, S.G. Dmitrienko and W.A. Engelhardt, Eur. J. Biochem. 120 (1981) 511–517.
25 G.K. Kovaleva, N.B. Tarussova and L.L. Kisselev, Mol. Biol. (Moscow) 22 (1988) 1307–1314.
26 See [2], p. 253.
27 See [2], p. 268–269.
28 See [2], p. 271–272.
29 See [2], p. 274.
30 See [2], p. 279.
31 See [2], p. 299–300.
32 See [2], p. 288.

G. Semenza and R. Jaenicke (Eds.) Selected Topics in the History of Biochemistry: Personal Recollections, III.
(Comprehensive Biochemistry Vol. 37) © 1990 Elsevier Science Publishers BV (Biomedical Division)

Chapter 4

Autobiographical Notes from a Nomadic Biochemist

HERMAN M. KALCKAR

Department of Chemistry, Metcalf Center for Science and Engineering, 590 Commonwealth Ave., Boston, MA 02215 (U.S.A.)

Introduction

Around 1934–1935 when my younger brother Fritz Kalckar (1910–1938) had already established very fruitful contacts with the great scholars in theoretical physics in Copenhagen (Niels Bohr, Christian Moller, George Hevesy and others), I was just beginning to find an approach to cell physiology which seemed really creative. I was lucky, too. My mentors who inspired me through the next four years were three big L's: Lundsgaard, Lipmann and Linderstrøm-Lang (which actually makes four L's).

Ejnar Lundsgaard, our young and brilliant professor of physiology at the Faculty of Medicine, University of Copenhagen, had already won international recognition by his description and masterful experimental interpretation of the energetics of muscular contraction without lactic acid formation. He was able to demonstrate this aberrant state of affairs in a reproducible way by exposing the skeletal muscle to surprisingly low concentrations of iodoacetic acid.

It remained uncertain whether Otto Meyerhof, who had developed the lactic acid hypothesis of muscle contraction, was willing

Plate 5. Herman M. Kalckar.

to accept this revolution in muscle physiology; in any case, he did appreciate that Lundsgaard informed him personally about these remarkable findings and an invitation from Meyerhof to spend a few months at the new Kaiser Wilhelm Institute in Heidelberg was accepted with pleasure by Lundsgaard. At the new Institute, Lundsgaard found one young scholar who fully grasped the enormous perspectives offered by the discovery of the 'contraction without lactic acid formation': the young biochemist Fritz Lipmann.

Since at that time it seemed clear that Germany was to become infested with the dictatorial Hitler regime and its sinister programs, Lundsgaard and his teacher and mentor, Valdemar Henriques, also a member of the Carlsberg Foundation, acted swiftly to secure Fritz Lipmann for the development of Danish science. In one of the laboratories for basic research supported by the Carlsberg Foundation, Lipmann had the opportunity to develop his new visions of future metabolic biology.

At that time, after I had received my degree in Medicine in 1933, Lundsgaard accepted me as a part-time researcher in his Department of Physiology some time in 1934. This was a special opportunity, so much the more since his trusted friend, Lipmann, worked in the new Laboratory only a couple of blocks away. The constellation seemed most promising for my scientific career.

My first experiment, of which I still possess the records, was a repetition of the alactacid iodoacetate muscle contraction experiments of the Lundsgaard design of 1930. Using the Fiske and Subbarow phosphate method, I tried to account for the levels of 'phosphagen' which Cyrus Fiske at Harvard Medical School in 1927 had already identified as phosphocreatine. Since Lundsgaard's time was to become more and more occupied by administrative matters in the Faculty of Medicine, Lipmann became my main scientific mentor. As a mentor, Lipmann had a strong realistic and illuminating evaluation of new, loudly acclaimed discoveries. Whether it was the popular dogma of the role of lactic acid in muscle contraction, or the prevailing postulate about aerobic glycolysis as a specific characteristic of cancer tissue, Lipmann himself preferred more 'open-ended' interpretations. When phosphocreatine and ATP were discovered, he was never in doubt of the increasingly important role of biochemistry for the development of biology.

In my own first self-selected series of experiments, I tried to focus on the possible role of phosphocreatine in the retina before and after exposure to light. The exposure to light was in the yellow region: the dark control was red (near infrared) light. This was of course performed in the dark room, where it was difficult to maneuver fine adjustments, and on one occasion I managed to crack one of the thin glass plungers of the 1930 colorimeter. Lundsgaard helped me back on the track with spare parts, in his usual discreet way; he was fuming and probably not only from pipe smoke. Further efforts on retina preps brought me only fuzzy data and the state of affairs did not improve. It was time for a change.

The phosphatase-phosphatese approach to active glucose transport

In 1933 W. Wilbrandt and L. Laszt, two physiologists, tried to tackle the problem of the mechanism of the highly efficient absorption of glucose from the intestinal mucosa in mammals, such as rabbits. In this tissue they noted the high activity of phosphatases, enzymes which catalysed the release of phosphate from hexose phosphates. They assumed that these enzymes could also catalyse the formation of hexose phosphates from inorganic phosphate and glucose. This activity they called the 'phosphatese activity'.

Lundsgaard had also been interested in the active transport of glucose but he focused on the function of the reabsorption of glucose in the primary tubules of the kidney. Both the intestinal and the renal transport systems were inhibited by phlorhizine.

The discovery of transphosphorylations and oxidative phosphorylation

Around 1935 some new observations changed the general perspectives in biochemistry. E. Adler and H. von Euler, in their attempts to purify one of the yeast dehydrogenases described by Otto Warburg, found, as a side product, an enzyme which was able to catalyse a transphosphorylation from ATP to glucose or fructose. Curiously enough, they listed this transphosphorylation under the name 'heterophosphatese'. Very quickly Meyerhof became aware of the possible identity of 'heterophosphatese' with the so-called 'hexokinase'. Keeping in mind the Lundsgaard review of Fiske and Subbarow's work and correspondence with Lohmann, I turned my mind more and more towards phosphorylation systems.

The Lipmann dialogue

Fortunately Lipmann encouraged me as a starting point to study the newer literature on carbohydrate and phosphate metabolism in isolated mammalian tissue extracts or tissue particle preparations. Hans Krebs seemed to have had some luck in disclosing different types of respiratory metabolism in tissue particle preps from a variety of organs, especially pigeon-breast muscle. This was done about the time when he formulated the tricarboxylic cycle, also called 'the Krebs cycle'.

For a variety of reasons, I selected to study the metabolism of phosphate and carbohydrate in extracts of kidney cortex. First of all the kidney cortex, as mentioned, contains the primary tubules which perform the highly effective reabsorption of glucose and show high oxygen consumption. One might argue that pigeon-breast muscle, bristling with enzymes catalysing the Krebs cycle would be superior. However, to obtain this tissue, I would have to train myself in the 'art' of cutting the heads of pigeons, and I greatly preferred not to kill animals, probably due to my lack of courage and skill. In any case, I was lucky to obtain fresh mammalian kidney from the liver perfusion experiments on cats and rabbits, which were conducted at least once a week in the physiology department.

As mentioned, my studies on phosphorylation and respiration in kidney cortex extracts turned out to be rewarding. In the presence of oxygen (aeration in Warburg manometer vessels), inorganic phosphate was esterified to various substrates such as glucose or glycerol. Addition of 5′ AMP (a precious gift from the late Dr. Pawel Ostern in Lwow, Poland, shortly before the Nazi invasion of his city) became of special importance, since it enabled me to show ATP formation in the preps. If phosphatase activity was arrested by addition of sodium fluoride the P/O ratio approached 1. Under anaerobic conditions no phosphorylation was detectable, even in preps well 'spiked' with fluoride. I also noted that addition of dicarboxylic acids, like fumarate or succinate, stimulated oxygen consumption as well as phosphorylation. However, I never managed to get P/O ratios higher than 1.5 and then only if succinate was added to the cortex prep [1].

It even turned out that oxidative phosphorylation remained at about the same order of magnitude, if I merely added fumarate or succinate to the extract, i.e., in the absence of glucose. This observation seemed very puzzling and it induced me to pursue this aspect.

I tried to dialyse the cortex preps and whenever I obtained an active prep (at that time cellophane bags contained impurities which often inactivated the preps, especially with respect to phosphorylation capacity) I found that fumarate (or malate) in the absence of glucose not only stimulated respiration, but also gave rise to a phosphoric ester with different properties from hexose phosphate or glycerophosphate. What was the nature of this phosphoric ester? In pursuing this question, I once more received encouragement and advice from Lipmann.

In 1934–1935, work by Gustav Embden opened up new perspectives regarding phosphoric esters and their role in glycolysis; this new panorama was further expanded with the discovery of *phosphoenol-phosphoric acid (PEP)* by Lohmann and Meyerhof. It was customary at that time to use the stability of phosphoric esters in acid or alkali as a first criterion of the nature of the different esters. The properties of the phosphoric ester formed in my fumarate experiments did not correspond to those of ATP. Since it was known that PEP was able to serve as a phosphate donor and could convert ADP to ATP, it attracted my special interest. PEP was known to be dephosphorylated rapidly at room temperature in alkaline iodine and even more rapidly by mercuric chloride. I used these highly specific criteria and found that the ester formed from fumarate or malate fulfilled all of these criteria. PEP seemed indeed to be formed in the cortex preps from oxidation of malate in the presence of inorganic phosphate.

This type of generation of PEP could scarcely have originated from hexose phosphate, since the presence of sodium fluoride barred not only phosphatase activity, but, as shown by Lohmann and Meyerhof, enolase activity as well; in this way the conversion of 2-phosphoglycerate to PEP was arrested. I therefore believed that, with fumarate or malate in the experiments, PEP must have originated from one of the dicarboxylic acids, including oxalacetate, all members of the Szent-Györgyi–Krebs di-tricarboxylic acid pathway. My new observations were first published in *Nature* [2].

Lipmann's response to my discovery of PEP generation from dicarboxylic acids in respiring kidney cortex dispersions was unreserved praise, and made me feel good indeed. My observations and those presented a year later by Belitser (see next section) gave us international recognition as pioneers. In a comprehensive review by F. E. Hunter, he declared 1939 'the banner year' for development of the concept of 'Oxidative Phosphorylation' [3,4].

Biochemical observations made in the U.S.S.R.

As mentioned, in spite of my repeated efforts, I never succeeded in obtaining P/O ratios significantly above 1.5. The further development in this direction we owe to V.A. Belitser in the U.S.S.R. In an extensive paper on oxidative phosphorylation he showed that pigeon-breast muscle or rabbit-heart muscle, in the presence of oxygen, is able to bring about phosphorylation of creatine to phosphocreatine. Belitser and Tsibakowa were able to show that in the presence of arsenous acid and fluoride, the oxidation of succinate to fumarate did in fact generate P/O ratios as high as 3–4. The paper appeared shortly before the outbreak of World War II in 1939 [5]. Much later, in June 1960, I had the privilege to meet Belitser in Kiev, when the National Academy of Sciences at the beginning of the scholarly exchange sent me to the USSR Academy (Akademi NAYK).

International Congress of Physiology in Zurich, 1938

Returning to my report on events in 1938, I was an active participant at the International Congress of Physiology in Zurich. I presented some of my recent findings on the oxidative phosphorylation of glycerol as well as the generation of PEP from dicarboxylic acids.

My text and communication were presented in German, since I did not trust, as yet, my capacity to lecture in English. Prof. R.A. Peters from the University of Oxford was chairman of the session and he as well as Dorothy and Joseph Needham from the University of Cambridge were actively interested in some of the metabolic aspects of my topic, and they asked questions. I therefore had to switch to English, enjoying a very reassuring dialogue. I can still see and hear the very tall Joseph Needham rising up and introducing himself to the audience: 'Joseph Needham, Caius ('Kee-ees') College, Cambridge; can I ask a few questions?' My English apparently sufficed.

I had the great pleasure to meet Severo Ochoa for the first time. He was now working in R.A. Peters' department and, independent of the work of Belitser in Moscow, was able to obtain P/O ratios, in dialysed brain dispersions from pigeons, as high as 2.8.

Severo was not only deeply committed to science but loved the arts and especially music. Our friendship started 50 years ago and it still persists.

Preparing for a Ph.D. degree in Physiology

In the autumn of 1938, I collected my experimental data on phosphorylations in kidney cortex extracts with the purpose of obtaining a Ph.D. degree from the Faculty of Medicine, University of Copenhagen; this title went under the name 'Dr. med.'. In the Institute of Zoophysiology, thanks to the initiative of the world-famous Prof. August Krogh, it was possible to earn a 'Dr. phil.' degree by collecting two or three articles written in English or German and accepted by the inter-Scandinavian journal *Skandinavisches Archiv für Physiologie*. In contrast, the Medical Faculty at that time required the thesis text in Danish.

Be that as it may, I was able to add a 14-page 'Overview' in English with one large table and several references to the 123 pages of Danish text, references and tables. The entire task was not only time-consuming but also an expensive way of striving to get a Ph.D. degree. Fortunately, my mother rendered generous financial help.

An unanticipated special Dedication

While trying to finish my Ph.D. thesis, we lost my younger brother Fritz, just six months after his exciting journey to California in 1937. He died suddenly in 'status epilepticus' in January 1938. According to his California friends, Fritz had suffered a 'grand mal' in 1937. His fatal attack happened barely a year before effective treatment of epilepsy became available in the U.S.A. and in Denmark as well. My mother never got over this loss. The Bohr family, who also loved Fritz dearly, rendered as active comfort as possible. Niels Bohr spoke at the modest cremation ceremony and, besides his speech, we were also touched by his request to bring with him young Aage Bohr who had enjoyed a mutual warm friendship with Fritz. This all happened in January 1938. My Ph.D. thesis became dedicated to the memory of Fritz Kalckar.

My attempts to achieve transphosphorylations by the use of radioactive phosphate

Fritz' wisdom continued to shape much of my planning. During the late 1930s George Hevesy at the Bohr Institute was able to offer some of the first preps of radioactive phosphate, ^{32}P. These early ^{32}P preps were made from carbon disulfide bombarded with radium, obtained from the Radiology clinic at the well-known Finsen Hospital in Copenhagen. After several exposures to radium bombardment, the carbon disulfide was distilled off and the residue, after enrichment with sodium phosphate, was co-precipitated repeatedly as crystalline ammonium magnesium phosphate.

These preps were made available to me as well as to other biologists. Unfortunately, I had no luck with these preps. The cortex extracts showed only very low respiration and no phosphorylation in the presence of the ^{32}P preparations. Years later I was able to present some intriguing data on the ^{32}P turnover of the three phosphates in ATP in intact muscle; they eventually were reported, in a Festschrift for Martin Kamen.

Although I was admittedly depressed during 1938, I had much encouragement and inspiration as well. Lundsgaard and Lipmann were formidable mentors and generous personalities. Funding for my research came from three sources: the University of Copenhagen, financed by the Ministry of Education; The Carlsberg Foundation and, finally, US dollars from the Rockefeller Foundation.

My personal salary for teaching during the last years as 'assistant professor' came from the University of Copenhagen. The medical students who came to my afternoon lectures in physiology numbered 30 to 40, only about 20–30% of those who showed up at the Professor's morning lectures. I felt that I had the cream of the crop of medical students, and that we had generated a mutual dialogue. Our texts were Lundsgaard's excellent textbook interwoven occasionally with the British classical 'Starling's'.

I was also in charge of some of the morning lab demonstrations; preparations started at 8 a.m. For this it was a great help that I had the privilege of having an apartment at the institute.

Getting through an old ornate way of defending a degree of 'Doctor medicinae'

In the late 1930s the defense for a Dr. med. degree was a highly ceremonial event, whether at Danish, Swedish, or Norwegian universities. In my case the ceremony took place in the old original section of 'København' under the auspices of the almost 500-year-old university. The candidate (and defender, 'Praeses') had to appear in 'white tie and tails', much like at the ball in 'Fledermaus'. The opponents appear either in black 'Diplomatfrock' or in 'Jacquette' with striped trousers, in either case with black tie. The entire affair started at a quarter past 2 pm. It was the middle of January 1939.

My formal defense of my thesis was not formidable as compared with Prof. Lundsgaard's concentrated and stylish opposition. But once he was over the critique of the language (mainly my casual Danish text and some of the tables) the scientific dialogue proceeded very well indeed. Lundsgaard agreed with Lipmann, that the thesis revealed several new and important features regarding a relatively new field, metabolic biology, barely 20 years old.

The Linderstrøm-Lang touch

During 1938 I came to know a most interesting scholarly chemist and biochemist, Kaj Linderstrøm-Lang at the Carlsberg Laboratory. Lang attracted scores of bright young biochemists, especially from Great Britain and the U.S.A. Among the latter I came to know one of Lang's favourite American scholars, Rollin Hotchkiss from the Rockefeller Institute. Already in 1937 he and Lang developed protein chemistry along new fundamental lines.

Through Lipmann, Lang had developed a strong interest in the bioenergetics of protein synthesis and wanted to learn more about phosphorylation systems. Lang fully agreed with Lipmann that the postulate about the role of peptidases catalysing hydrolysis as well as synthesis of peptides in proteins was not the answer to the problem of cellular protein synthesis. I was mainly a listener to these fascinating discussions and cherished the ideas that transphosphorylations and ATP might well be involved in cellular protein synthesis. Lipmann's visionary ideas began to take shape in 1938 by his bold experiment in which he generated ATP from his 'homemade' acetylphosphate, catalysed by the bacterial enzyme. All this was performed in early 1939, when I had left Denmark, and only six months before the outbreak of World War II. It was Lipmann's solo performance taking place during his last months in Denmark.

In 1938 Dr. Warren Weaver, Director of The Rockefeller Foundation was visiting Copenhagen and my candicacy for a research fellowship was being discussed. I assume that the Big L's were instrumental in the cheerful outcome and I was appointed a Rockefeller Research Fellow in January 1939, shortly after my defense of my Ph.D. degree at The University of Copenhagen. Now came the problem how to select a laboratory in the U.S.A. which was stimulating and which would widen my horizon in general biology.

My choices were once more influenced significantly by my late brother's exciting letters from Berkeley and Pasadena, California. He spent five to six months at the University of California, Berkeley and later at CalTech, Pasadena's Physics Department. His close friends were Robert Oppenheimer and C.C. Lauritsen. They in turn were close friends, since Oppenheimer taught theoretical physics at both places. Niels Bohr's visit to Ernest Lawrence was carefully prepared by Fritz. This visit is described by Martin Kamen in his refreshing, honest and humorous book *Radiant Science, Dark Politics*.

Linderstrøm-Lang knew well the Department of Chemistry with Linus Pauling and Charles Coryell and the Biology Department with the great geneticists, T.H. Morgan, A. Sturtevant and Calvin Bridges. The latter had shown my brother the *Drosophila* mutants and their classification.

Lang was also very familiar with Henry Borsook who was interested in thermodynamics and the bioenergetics of peptide formation, although he still operated with the old-fashioned notion of reversal of peptidase activity to reclaim polypeptides and knew nothing about phosphorylations and their potentials.

England and the United States

In January 1939 I visited London for the first time; my purpose was to visit Sir Robert Robison, an outstanding biochemist at The Lister Institute for Preventive Medicine. Robison, as an editor of *The Biochemical Journal*, had been very helpful to me in getting my fumarate-PEP paper [3] published in this journal. I believe that I also met Dorothy Needham at that time. The London fog was very dense and full of soot and while visiting some favourite family near Reading, they drove me to Southampton, bound for New York. I was lucky enough to get to New York on the S/S 'Queen Mary' which fared well even in a heavy storm. It was the first time I crossed the Atlantic.

In New York City, I was lucky to be the guest of the pre-eminent immunochemist, Michael Heidelberger, whom I knew through my family in Denmark. Michael was also an excellent clarinetist, whether he played the quintettes by Mozart or Brahms; he found the latter more of a challenge.

Although I had decided to spend most of 1939–1940 at CalTech's Biology Department, I wanted to make a brief stop at Washington University, St. Louis, MO, where Carl and Gerty Cori worked and where in 1936 they had discovered the first phosphorolytic fission with the isolation of glucose-1-phosphate generated from glycogen and catalysed by a muscle enzyme which they now called phosphorylase. I arrived in the bleak and sooty city of St. Louis in February 1939, and was well received by the Coris. Unexpectedly it turned out that they were very interested in my articles on oxidative phosphorylation but were unable to repeat them. Gerty wanted me to make a few demonstrations of the experiments but Carl cut it short and simply asked a few questions. I saw that they employed the old-fashioned Meyerhof extract techique, using test tubes. I explained to them that kidney cortex extracts show very high oxygen consumption and that it is mandatory to subject the preps to plenty of aeration, optimally in Warburg vessels which I used throughout my work to record oxygen consumption for P/O ratios and which therefore were shaking except during the brief readings.

Sidney Colowick who had witnessed this encounter summarized it 35 years later in a brief dinner speech as:

> 'Dr. K. said, "just shake them" and everything was o.k.!'

The Coris also wrote to express their thanks. This may well have prepared the way for Carl Cori's generous invitation to me to come to their lab after my first year's fellowship ran out.

I went the cheap way to California by Greyhound bus, via New Mexico. I stopped at Grand Canyon, Arizona, and went down and up on a mule seeing winter at the top turning to spring and summer at the bottom with cactus flowers.

California, 1939

On the first of March 1939 I arrived in Pasadena and was received by two young students who soon became my good friends: David Bonner, originally from Salt Lake City and his Danish friend Erik Hegaard. The other friend whom I had met briefly in Copenhagen at Bohr's institute was Max Delbrück. They all loved the desert and, before I got my own car, we went camping in the desert using Max's car; even when he did not join us, Max was always generously lending us his car which was a 1937 Ford. This was springtime and I enjoyed especially the small oases with beautiful cactus flowers and other blooming plants, not to forget the hummingbirds.

Thanks to the Bonner brothers, James and the younger David (Dave), I managed to find a bungalow shortly after the arrival of my wife Vibeke. We agreed that the CalTech faculty club, Athenaeum, was a nice but too expensive domicile for a Research Fellow. Dave and James were also most helpful in introducing me to a superb auto mechanic, Jose Plannes (of Basque origin); he had helped James and Max, not only to find some good rugged used cars, but also had personally inspected these cars; Jose was THE auto mechanic all the young CalTech scholars turned to at that time. Jose helped me superbly too. A Ford coupe, vintage 1934, was for sale and I bought it. It held up, not only on desert roads but in the mountains as well. After I received my driver's license, I had the nerve in early April to drive Theo (Hugo Theorell from Stockholm) the whole way up to Mt. Wilson observatory, which he desired to see. The snow was just beginning to melt. Through Theo, I came to know Linus Pauling who later played a very positive role in my career.

In the Biochemistry Unit of the Biology Division, I was encouraged by Henry Borsook, head of the unit, to give some seminars. I chose as a subject transphosphorylation which was a fresh topic to them. At that time I received an electrifying letter from Fritz Lipmann in which he suggested that acetylphosphate might not only be a phosphoryl donor but also an acetyl donor. The observations appeared in *Nature* a few weeks later. Lipmann's thoughts added to my own topic on PEP formation from malate. All this evoked much interest in our discussions, among others, with Max Delbrück and a brilliant graduate student, Norman Horowitz.

In over four to five years Borsook had built up a programme around the thermodynamics of peptide formation and I felt that I should not ignore it, notwithstanding my preference for what Lipmann later called 'phosphate bond energy'. Borsook's expert in obtaining thermal data was Hugh Hoffman, formerly of the G.N. Lewis School of Thermodynamics, University of California, Berkeley. To determine heat as well as entropy data on peptides, one had to first burn tablets of benzoic acid in a calorimeter. As a beginner I had to weigh tablets of something as repulsive as naphthaline and determine the weight with a precision of eight decimals and burn them in a calorimeter recording the galvanometer deflection with the same order of accuracy. It all took place in the basement, popularly called the 'crematorium'. So, while at the time of sunset and twilight, the mountain with the Mt. Wilson observatory was bathing in a serene last orange glow, here I was sitting in the 'crematorium', merely managing to obtain thermal data on benzoic and hippuric acid, albeit only with six-decimal accuracy. Hugh H. was a hard taskmaster requiring eight-decimal accuracy and he soon felt that he should not 'turn me loose in the entropy forest'. Upstairs they thought I was kind of daft or masochistic to volunteer for the 'crematorium'. They enjoyed Hugh H., especially in the evening when the strict taskmaster got a good portion of Bourbon and showed the World War I navy tattoos on his arms.

Fortunately, I returned to the seminars and began to write a large review, much inspired by one of Linus Pauling's former coworkers, Charles Coryell, who was very interested in the energetics of phosphoryl donors, such as phosphocreatine and ATP (adenyl pyrophosphate). Pauling's ideas about resonance had greatly influenced Coryell's thinking and I incorporated some of this in my review. At that time I did not know about the intriguing review which Lipmann had started through his good friend Dean Burk, collecting thermal data. Lipmann's review had a great impact in biochemical circles.

The second microbiologal course by Van Niel, 1939

For the summer of 1939, I was lucky to be enrolled in a combined lab and lecture course provided by an unusual microbiologist, Cornelius B. van Niel who, perhaps more than anybody else I had encountered at that time, sensed what he called 'the unity of Life'. Van Niel was Dutch and he graduated from Delft University, Biotechnological School, a highly distinguished school. Here he wrote his Ph.D. thesis about a new type of sugar fermentation catalysed by certain bacteria which generated propionic acid and carbon dioxide. The fermentation which generated propionic acid was therefore called propionic acid fermentation. Van Niel reminded me that a Danish cheese expert had first sniffed the existence of such a type of fermentation when he found out that the gas in the eyes of Swiss cheese was pure carbon dioxide. The Dane's name was Orla Jensen. I had not paid attention to these matters while in Denmark, perhaps worse, I was ignorant of some other important work at the Carlsberg Lab. O. Winge had made great basic contributions to yeast genetics, by describing the existence of diploid as well as monoploid yeast. Clearly, my transfixation on the four big L's had made me overlook the great tradition of Danish genetics at that time, founded by Winge's great teacher and researcher, Wilhelm Johannsen.

Although we were very near the Pacific coast, we discussed rather sparingly the role of the microorganisms in the ocean except algae and most important, the chemosynthetic microorganisms. Photosynthesis was of course broadly covered and various postulates were critically discussed. This part inspired me to expand the review which I had started at CalTech.

From Pasadena to Saint Louis, early 1940

The autumn of 1939 was unusually hot, even for Pasadena, in stark contrast with the climates of Pacific Grove and Berkeley. Fortunately, the hospitable Lauritsens invited us over to their swimming pool. November was sunny and pleasant as was December 1939. My wife and I were invited to a real traditional American

Christmas dinner at the Morgans (T.H. and Lilian) in the stylish old house built and designed by the founder of The Kerckhof Institute for The Biological Sciences. The founder was of course Thomas Hunt Morgan himself and the multidisciplined institute had also served as my homebase. As far as I remember, my wife and I, as newcomers, were the only ones invited. T.H. asked about Denmark and more specifically about Niels Bohr.

*On my way to the Cori Lab, February 1940**

Even before the outbreak of World War II, in September 1939, I received a formal invitation from Carl Cori, encouraging me to join his lab as an independent investigator. This was great news to me. German scholars who did not want to return to Hitler's Germany were in a difficult position, since they were declared 'stateless'. Fortunately, the American granting agencies and universities disregarded such types of decrees. The resourceful Thomas Hunt Morgan found a teaching and research position for Max Delbrück at Vanderbilt University.

In early 1940 I said farewell to my friends at CalTech. Linus Pauling generously offered to get my extensive review on bioenergetics published in *Chemical Reviews*, provided I could manage to send him a finished copy by the end of 1940.

We sold our vintage 1934 Ford in January 1940 and travelled more or less comfortably by train towards the Midwest. It was more trying for my poor wife than for me, since I sat up most of the night writing page after page of the review [6]. After a couple of days in colourful New Orleans we continued North arriving late February 1940 in the damp and smoky Saint Louis. It was a depressing city, but I had the benefit of working in a lab which had developed a new type of enzymology which now became my main interest.

I had partly lost my enthusiasm for pursuing P/O ratios in various

---

* See H.M. Kalckar, in G. Semenza (Ed.), Comprehensive Biochemistry, Vol. 35, Chapter 1, Elsevier, Amsterdam, 1983.

tissue particle preps. Carl Cori and his able coworkers had managed to obtain good P/O ratios in heart tissue dispersions, keeping in mind my warning against using unshaken test tubes. By now I would rather learn enzymology 'à la Cori'. Nevertheless, Carl did persuade me to participate in collaborative work with him and a brilliant student, Sidney Colowick and also encouraged us to delve into enzymology on our own. Thus began some of my happiest months in research.

Ventures in enzymology with Sidney Colowick

With his 25 years Sidney Colowick possessed wisdom, a warm sense of humor and a brilliant analytical insight. Sidney and I were interested in studying the enzymes of trans-phosphorylations and our first attempt centered around further fractionation and purification of yeast hexokinase. We soon found that the semi-purified hexokinase preps catalysed the transfer of only 1 mol of P from ATP to hexose with the release of ADP; in crude hexokinase preps ATP was converted to 5′AMP(5′adenylic acid), and 2 mol of hexose phosphate were formed per ATP consumed. Sidney and I were therefore groping for a factor in yeast, or in skeletal muscle, which might give us a more efficient balance in terms of hexose phosphate generation.

Although we did not find the suspected factor in yeast preps, we had luck with testing extracts from skeletal muscle. Moreover, a simple and effective purification of muscle extracts, merely by acidification with HCl, heating to 90°C for 1 to 2 min, neutralizing to pH 6.5 and spinning, we obtained a supernatant containing only 1% of the original protein content. This fraction turned out to contain the complementary activity. In other words, the mixing of the fractionated yeast prep and the heat-resistant muscle fraction brought about the formation of 2 mol hexose phosphate per ATP consumed and the appearance of 5′AMP. We called the active complementary fraction from muscle 'myokinase' [7]. We tried of course insulin as a substitute for myokinase, but insulin had no effect on this transphosphorylation.

Carl felt that this research project should be communicated to the annual meeting of The American Society of Biological Chemistry in Chicago, April 1941. Since Sidney was a graduate student, I was supposed to present our observations at the big meeting. This was a great opportunity for me, but it also scared me. At CalTech I had only faced a workshop club of one dozen scholars and participated in informal discussions. As mentioned above, at the International Congress in Zurich in 1938, I deemed my English too poor and rattled off my communication in German. This time I had no choice, but I had the help of Sidney Colowick, who had often reminded me of the comfortable simplicity of the English language. I did not need to elaborate on changes like 'more warm' or 'more steep', just say 'warmer' or 'steeper', etc. His patience with me was heartwarming.

My active participation in the biochemical meetings in 1941

Otto Meyerhof, who was recently brought to the U.S. from Nazi Germany, was one of the major speakers at the big annual meeting in Chicago, held at the Stevens Hotel (later renamed Conrad Hilton) on the stately Michigan Avenue near the waterfront. After Meyerhof's address, I was scheduled for a 10–15-min communication and then to participate in the discussion. My report on the work with Sidney on myokinase actually went well. However, in trying to answer questions about my previous work on oxidative phosphorylation, it became clear that I was indeed a freshman in English. When oxidation-reductions became the topic in further discussion, I used Sidney's language lessons in an awkward way, describing the redox potentials as becoming 'negativer' or 'positiver'. The audience received my phrasing smilingly and almost everybody woke up. But poor Sidney felt kind of guilty, not to have covered these territories in our language lessons.

At the beginning of the summer 1941 both Sidney and I had to prepare ourselves for major changes in our scientific programmes. Sidney had to collect sufficient scientific material to qualify for a Ph.D. degree. Carl Cori was at that time still Professor of Pharmacology and with the arrival of a promising young M.D., by the name of Earl Sutherland (later to become a Nobel Prize winner) it was decided that Earl, Sidney and Carl join forces on a project combining pharmacology, physiology and biochemistry. This venture started with a seemingly simple question: the mechanism of glycogen synthesis in the liver from glucose. In my own case, my Rockefeller grant was not further renewable, but Carl had managed to get a new research grant for my work from The Commonwealth Fund. It had one important extra stipulation, the use of radioisotopes as a tool in tracing various metabolic pathways. The fund provided travel money, giving me a chance to visit two or three biochemistry laboratories engaged in isotope techniques.

Visiting Columbia University, Biochemistry Laboratory

In the summer of 1941, I visited three biochemistry laboratories engaged in biochemical isotope research projects. I used most of my summer vacation to visit them, mainly in the Eastern U.S. At the University of Rochester, NY, the Department of Biochemistry and Pharmacology had at that time a programme on phospholipid turnover in mammalian organisms. As far as I was concerned, I must confess that I did not pick up more than I had learned in George Hevesy and Hilde Levi's course on the handling of radioactive phosphate in 1937 in Copenhagen.

The most inspiring visit was at Columbia University, College of Physicians and Surgeons, on the west side of New York City. In the Biochemistry Department under Rudolph Schoenheimer's leadership they used ^{15}N-labelled amino acids to monitor the rate of turnover of proteins from various mammalian organisms. I learned a lot from Rudi and his splendid co-worker Sarah Ratner, about the refined methods to follow the trails of protein catabolism. Rudi expressed great interest in the new Lipmann concepts and he had also read with interest my review which appeared at about the same time as Lipmann's. Several hours were spent discussing bioenergetics.

My first encounter with Lipmann's Squiggle

Some time in May 1941 I found in the library of the Department of Biochemistry at P and S, volume 1 of *Advances in Enzymology*, the monumental article by Fritz Lipmann on bioenergetics [8]. His title was: 'Metabolic generation and utilization of phosphate bond energy'. This exciting review was well over 60 pages and had just been published. My own review was, as mentioned, inspired by Lipmann's demonstration that acetyl phosphate can act as an acetyl donor as well as a phosphoryl donor. At CalTech I was schooled by the physical chemists to invoke the expression of free energy changes with great caution. For instance, the formation of chemical bonds usually generates energy. 'Phosphate bond energy' would not convey any meaning to them since in biological reactions the medium is water, which is present in the order of 58 M. It was clear that Lipmann, a veteran in basic thermodynamics had used the expression 'phosphate bond energy' as slang, going back to Meyerhof's slang in 1927. This seems clear from Lipmann's introduction of the 1941 review; he starts as follows:

> 'The more or less pronounced breakdown of creatine phosphate during muscular contraction, early suggested its connection with energy supply. Interest in the compound became stronger after O. Meyerhof and J. Suranyi found that unexpected large amounts of heat were released by enzymatic decomposition. The biochemistry of the energy-rich phosphate bond was, in fact, herewith opened.'

Even before a reunion with Lipmann, a month later in Boston, I enjoyed his forceful formulation of the types of phosphate bonds which upon hydrolysis release high amounts of heat; they were marked by a squiggle ($\sim$ P). Phosphoryl bonds of this type, such as pyrophosphates or ATP or guanidine phosphates, he called energy-rich 'bonds' [8]. Phosphate ester bonds such as hexose-6-phosphates or glycerophosphates which, upon hydrolysis, release only one third of the heat per mol phosphate released were classified without the squiggle sign.

One of Lipmann's most intriguing strokes in this review was to invoke entropy changes, such as in the conversion of phosphoglycerate to PEP. Lipmann moved with swiftness through Dean Burk and others to collect thermal data and calculate possible changes in entropy. Unlike me, they did not spend time to obtain calorimetric data. Anyhow, thanks to my former contacts with Coryell, I was able to enjoy this sophisticated argument about entropy changes as well.

It was therefore a special pleasure to have a personal reunion with Lipmann in his research unit at Massachusetts General Hospital (MGH) in Boston in June or July 1941 and listen to his thoughts. Obviously he agreed with me that the expression 'phosphate bond energy' was old slang, not to be used in formal teaching without pointing out the background.

In July I visited Harvard Medical School, Department of Biological Chemistry with Baird Hastings as the hospitable chairman. The young able research group was actively engaged in tracing metabolic cycles in various mammalian organisms, using the shortlived ^{11}C (^{14}C was discovered later in 1941 by Martin Kamen in Berkeley).

In September, both Sidney Colowick and I were participants in a Symposium on Respiratory Enzymes at the University of Wisconsin, Madison. Carl Cori, Meyerhof and Lipmann were among the main speakers. I presented a short communication on myokinase. Later Marvin Johnson, a very able and seasoned biochemist from the University, asked Sidney and me, whether we had any ideas about the mechanism of action of myokinase. After a long discussion, Marvin Johnson conceived an intriguing idea which might be called a 'dismutation of ADP'. I shall try to spell out this idea, in conjunction with my experimental game plan to show its correctness.

Experimental proof of myokinase catalyzing ADP dismutation

My main occupation during the months of autumn 1941 centered around the strategy of proving (or disproving) the concept of 'ADP dismutation'. At that time ATP and 5′AMP were available in small amounts in the Cori lab; only a few other labs possessed these nucleotides. I had first to make ADP from ATP and then, after addition of myokinase (in the order of 1–2 μg protein per ml) and

5-min incubation (with magnesium chloride), fractionation with barium acetate at pH 8 showed that about 50% appeared as ATP and the other 50% as 5′AMP. Moreover, the use of a highly specific deaminase from muscle (which Gerhard Schmidt described already in 1928), combined with myokinase, converted all the ADP to 5′inosinic acid [9]. The enzyme myokinase was later called adenylate kinase, indicating that it (also present in other tissues) catalysed just as well the phosphorylation of 5′AMP to ADP, with ATP as the phosphoryl donor.

The unattractive radioisotope project on muscle

In the autumn of 1941, I had to put my mind to a project that would be consonant with the requests stipulated by the funding from the new grant. I took up the examination of phosphorus turnover in intact skeletal muscle as traced by ^{32}P-labelled phosphate, especially the turnover of phosphorus of ATP and of phosphocreatine (PC). The Physics Department of Washington University provided me with carrier-free preps of ^{32}P and the Radiology Department of the Medical School gave me ample space to conduct the experiments. I was lucky to have two able young coworkers, Jean Dehlinger, a medical student, and Alan Mehler, a graduate student (at present Professor of Biochemistry).

The challenge to examine the turnover of phosphorus of ATP and PC in intact muscle came from some claims in the literature, even a recent one, that neither ATP nor PC could act as phosphoryl donors, nor could ADP and creatine act as phosphoryl acceptors since injection of ^{32}P preps into the blood stream of rabbits or rats did not significantly label ATP or PC. The critics seemed to overlook the fact that resting skeletal muscle has a very low pool of inorganic phosphate. Using their simple technique one could not issue any statements about the trail of phosphorus in the intact muscle, since most of the phosphorus stemmed from the blood stream. They also seemed to have forgotten the hard fact, stated by Hevesy in 1940 that ^{32}P injected into the blood stream enters the skeletal muscle very sparingly.

We tried to design a more careful plan in which we perfused the leg artery of an anaesthesized animal with saline to remove the blood and replaced it with phosphorus-free saline into which we now injected carrier-free radiactive phosphate. We found that the resting muscle only permitted transport of a small amount of ^{32}P and this was mainly found in ATP and PC. To make a long story short, the rate of entry was found to be very slow and was the rate-limiting factor [10]. This largely confirmed Hevesy's early warnings. In a later review on muscle, Meyerhof gave us high marks for our burdensome analyses.

In that paper [10] we used myokinase as an analytical tool to study the distribution of a modest amount of ^{32}P in ATP of the perfused muscle preps. The outcome of these analyses gave us some unexpected results in that the middle phosphate frequently showed higher ^{32}P enrichment than did the terminal phosphoryl. Perhaps the terminal P was diluted for ^{32}P from the large pool of PC, being less enriched with the isotope. This odd distribution of ^{32}P in ATP was encountered in resting muscles from Missourian bullfrogs [10].

In spite of all our efforts, I could not warm up to these projects. My warm feelings still remained with the imaginative and artistic approach to biochemistry pursued in the Cori department. The Coris, together with Arda Green (a protein chemist from Harvard), were able to crystallize the enzyme glycogen phosphorylase. This pure enzyme preparation enabled them to study the topography of the biosynthesis of glycogen and starch. I do not hesitate to rank these events among the striking advances of molecular biology. My friends Sidney Colowick and Severo Ochoa felt much the same as I did.

An attractive offer for a research associate position in New York City

Oliver Lowry, whom I first met during my visit to Baird Hastings' department at Harvard Medical School in the summer of 1941, wrote me a letter in the spring of 1942, with an offer from him and his chief Otto Bessey, chairman of a research unit at the Public

Health Institute of The City of New York. The offer was for a Research Associate position at the Department of Biochemistry and Nutrition at the Institute, an institution funded by The City of New York and from private sources. One of the overseers close to Dr. Bessey was Dr. Charles Glen King, co-discoverer of ascorbic acid and chairman of the Nutrition Foundation. One of the other distinguished overseers was Michael Heidelberger, the master detective of bacterial polysaccharides and my old friend from the 1930s.

The lab unit was very well equipped, including a brand-new ultraviolet (UV) spectophotometer, one of the early models of the useful Beckman UV. This enabled me to perform enzymatic spectrophotometry in the region of 240–260 mμ (now called nm). Enzymatic spectrophotometry was first introduced by Otto Warburg; his analyses operated in the wavelengths around 340 nm, monitoring formation of NADH and NADPH.

Using the same principles I was now able to monitor adenylic acid deaminase activity by UV spectrophotometry, operating mainly at 265 and 250 nm, the latter wavelength showing maximum absorption for inosinic acid. Comparing the spectrophotometric method with the distillation test of deaminase which I used earlier, one might estimate an increase in sensitivity of 10 to 20 times in favour of the UV method, without losing the high specificity of the enzyme test.

A report [11] from 1935 stating that liver nucleosidases are activated by phosphate or arsenate caught my interest during early 1943. Although nucleosidase catalysed fission of the ribosyl nitrogenous linkage, I was obviously influenced by the Coris' discovery of glycogen phosphorylase and its catalytic fission of one to four maltosidic linkages in glycogen; I soon found that inorganic phosphate was consumed and that inosine nucleosidase is a phosphorylase which catalyses the reversible reaction: inosine (hypoxanthine riboside)+phosphate $\rightleftharpoons$ hypoxanthine+ribose-1-phosphate. This equilibrium greatly favours the synthesis of the nucleoside. To arrive at this conclusion, one had to subject a prep from rat liver to several steps of fractionation (first ammonium sulfate fractionation, then further removal of other proteins by isoelectric precipitation at pH 6). When trying to demonstrate enzymatic synthesis of guanosine, we found that the undesirable impurity, guanase, could be removed from the nucleoside phosphorylase prep by adsorption to a suspension of barium sulfate [12]. By this primitive device, we were able to demonstrate the enzymatic synthesis of guanosine from guanine and ribose-1-phosphate, presumably catalysed by the same enzyme. Thanks to my next-door neighbour and friend, Oliver H. Lowry, the highly acid-labile ribose-1-phosphate could be characterized by one of his early ingenious analytical methods, the Lowry-Lopez method [13]. This method, which operates around pH 4.5–5, could also be used for analysis of the acid-labile PC.

The phosphorolytic release of hypoxanthine was monitored by a variation of our newly designed enzymatic spectrophotometry by myself and my able assistants (Manya Shafran and Alice Newman Bessmann). In this case the enzyme xanthine oxidase was used; this enzyme catalyses the oxidation of hypoxanthine via xanthine to uric acid. It was known from chemical data collected by organic chemists between 1930 and 1943 that the UV spectra of hypoxanthine and uric acid are strikingly different, with uric acid showing the absorption maximum at 290 nm, whereas hypoxanthine has practically none. In playing with these methods, I found that nucleoside phosphorylase incubated with ribose-1-phosphate and hypoxanthine catalysed the biosynthesis of the nucleoside inosine and that the equilibrium was in favour of nucleoside synthesis. Likewise, ribose-1-phosphate incubated with guanine in the presence of nucleoside phosphorylase yielded predominantly guanosine [12]. This project turned out to be rewarding in several directions. Later I shall elaborate on these aspects, but now I want to tell something about my life in New York City.

The years 1943–1946 were stimulating for my wife and myself. Vibeke had rewarding contacts with The New York Public Library of Music. In general the level of musical performance was at its highest and most exciting. In addition, the Ochoas and the Buedings, some of our closest friends, shared our interest in music.

During my stay in New York, a reunion after six years with Rollin Hotchkiss in 1944 became particularly rewarding. Rollin introduced me to the renowned old microbiologist Oswald T. Avery of the Rockefeller Medical Research Unit. At that time Avery and his able coworkers were formulating a paper which was to overturn the common belief in protein as the only agent of conveying cellular information. What they found, which is now classical knowledge, is that the transforming factors of pneumococci are protein-free large molecules of DNA.

What became so memorable that day in 1944 was the full hour which the generous Dr. Avery spent with Rollin and myself, once more formulating the novel arguments. The grand old microbiologist, while carefully rolling his own skimpy cigarettes, said to us that he was not worrying anymore about protein impurities, but rather about traces of pneumococcal capsular polysaccharides, substances which he, together with Michael Heidelberger and others, had analysed to near perfection over the last 25 years. Yet, the carefully delivered verdict about the nature of the transforming factor was: DNA!

It is especially pleasant to recall that my friend and companion of that day, Rollin Hotchkiss, was one of the rare biochemists who, in the next few years, was to push the Avery discovery forward. Not only did he provide the ultimate elimination of the idea that protein might be present in the DNA-transforming factor, but he achieved quantitation of several new pneumococal genetic markers, linked and unlinked (R.D. Hotchkiss Harvey Lectures 1953–1954, Vol. 49, pp. 124–144).

My renewed friendship with Rollin Hotchkiss encouraged me to join him in 1945 to register for the first Delbrückian bacteriophage course at Cold Spring Harbor. To get enrolled, one had to solve three logarithmic equations to base 2. This seemed very reasonable since growth of *Escherichia coli*, the host of the phages, normally follows a log curve to base 2. An elderly chairman of a big lab of medical bacteriology who had also applied, did not manage to solve the log equations, but Rollin and I managed to persuade Max Delbrück to give the busy chairman permission to participate.

As I recall, in 1945 there were only four or five participants, but over the years Max's course attracted more and more people approaching a 'logarithmic curve'. Needless to say, it was fun to be a participant in the first phage course at Cold Spring Harbor. Plating was not new to me, but the quantitative aspects, watching through the use of many identical plates, the huge fluctuations in phage-resistant mutants was a novel experience. This was just a year after the masterful paper by Salvador Luria and Max Delbrück invoking the principle of the Poisson distribution to prove that occurrence of phage resistance is a random event.

Contemplating a return to Denmark

Until 1943 contact with Denmark was possible provided the letters reported only 'neutral' events, since all the letters were opened and censored by the German occupation army. Following the attempt to kill Hitler, Denmark was apparently, as so many other European countries, in the hands of the SS army and the Gestapo (joined by a small but vicious band of Danish Nazis). I received a laconic message that my mother had died near the end of August 1943. It was very disturbing news, the more so since I did not know what happened to my older brother or to my brave friends who had actually resisted the Nazi infestation of Denmark.

In late 1945 one of these great friends, Kaj Linderstrøm-Lang (Kaj LL), came to New York, invited by Warren Weaver, director of the Rockefeller Foundation; Weaver, a distinguished mathematician, had been greatly involved with 'operation research' teams during the battle of Britain. After their meeting I had a wonderful opportunity to spend time with Kaj LL, in spite of his immense popularity with American scientists. He told me in his low-keyed way about his encounters with the Gestapo and their brutalities which led to the loss of many lives, including members of his own family. He knew about my mother's death and he expressed warm empathy, so characteristic of Kaj.

Then we came to talk about the future. He had followed my scientific projects from my bioenergetics paper in early 1941 in *Chemical Reviews* to my most recent paper on nucleoside phosphorolysis which appeared as a full paper in *Federation Proceedings*, 1945. Kaj expressed the hope that I would be a frequent guest lecturer in Copenhagen, but that I should pursue my highly promising career in the U.S.

Curiously enough, it happened at that time, without Warren Weaver's knowledge, that a very zealous assistant manager for Fellowships of The Rockefeller Foundation in a letter addressed to me, expressed his sincere feelings that a former Rockefeller Foundation Fellow (whose fellowship had terminated in early 1941) still ought to return to his homeland, i.e., Denmark. Fortunately, Ollie Lowry, Kaj and Dr. Weaver found this absurd, nor did I want to be 'bamboozled' by such a message. My future decisions were to be influenced by other circumstances.

Louis Rapkine, a cherished friend of Warren Weaver's and of Kaj LL's, came to New York and I had the good fortune to meet him. Rapkine had also lent his energy and knowledge to Weaver's programme of 'operation research', also covering France. Rapkine was a French Canadian from Montreal and decided to join the Free French forces under De Gaulle. Before the war, Rapkine had done excellent biochemical work at the University of Cambridge, England, had won many friends, among them the Nobelist Sir Frederic Gowland Hopkins. In 1945 Louis Rapkine and his wife Sarah felt that, by returning to France, they would be able to help initiate new creative projects, important for the young generation of French scientists, especially in Paris.

In another 14 months after this memorable meeting in New York in 1945 Kaj LL and I were participants in Louis Rapkine's efficiently and graciously organized International Conference in Paris, commemorating the 50th anniversary of Pasteur's death, just one year after the liberation of Paris ('Commémoration du Cinquantenaire de la Mort de Pasteur').

Louis Rapkine's decision to return to Paris, together with LL's warm interest in my approach to biochemical research, made a deep impression on me. At the same time a letter from my inspiring mentor at the University of Copenhagen, Ejnar Lundsgaard, arrived in which he told me that he had decided to keep my professorship open for me and would try to expand it, in case I decided to return to Denmark. Moved by these events, and with Lang at Carlsberg and Rapkine in Paris, I made a bold decision to return to Denmark in early 1946, after some luck in obtaining research support from various American sources. In this way Charles B. Huggins, the famous Chicago surgeon (later a Nobelist for his skillful methods to combat prostate cancer and its metastases), became a staunch supporter of my research. On his way to Stockholm, Charles B.H. wrote to me: 'I made a brief stop at "Kalckar Ville", meaning Copenhagen. I was often invited to University of Chicago, mainly to "Huggins Ville".'

Return to Copenhagen, 1946

My wife and I returned to Denmark on a nice Norwegian ship, first anchoring in Bergen, an ancient city, which probably then was not quite in the mood to call itself an old 'Hansa City' after its bitter fight against five years of Nazi tyranny. After Bergen, we finally reached Copenhagen in April 1946. Among the survivors of the dark years was my older brother Emil (rescued by friends in Sweden), his wife and two children, Jorgen (10 years) and Birgit (7 years).

The years in Copenhagen up to 1950 and the succeeding years became memorable also in a personal sense. In spite of many cheerful events such as reuniting with my brother Emil and his family, the loss of my mother in late August 1943 during the Nazi infestation of Denmark, still burdened my view on life. I had to fight depression constantly. Fortunately my wife, Vibeke ('Vips') had a happy reunion with her entire family, some of whom had survived the last 22 months of the war in Sweden. Vips was able to resume her many activities, including that of an effective and charming hostess in a large apartment which we acquired through the help of Carlsberg Foundation in 1948. We lived near the harbour yet also close to the museums and The Royal Danish Academy of Sciences of which I became a member in April 1948. In the season, the members of the Academy met every fortnight; besides brief business, communications of scientific material were presented, both from the humanistic and the mathematical and nature-oriented sides. Linderstrøm-Lang was a frequent visitor of these meetings and the president, Niels Bohr, was also there.

My reunion with Lang after the Academy meetings, walking with him, talking freely to him about the darkest two years, relieved me in a way which I do not think any person would have been able to do. Much like Niels Bohr, Kaj LL had a bright belief in the struggle against 'evil empires' such as Nazi Germany. As a resistance fighter he was imprisoned during 1943 by the all-dominating Gestapo. Kaj could make jokes about these prison months and considered them nothing compared with the heartbreaking loss of young brave resistance fighters, three of whom in his own family, killed by Gestapo just a day before that evil empire capitulated.

My scientific life was rapidly restored, thanks to help from many quarters. First of all my mentor of some 12 years, Prof. Ejnar Lundsgaard saw to it that a sizable unit of his institute was set aside for my research, promoted my rank at the University, and gave starting funds for the continuation of my research. Already I had managed to obtain some research support from America and, together with the support from the Carlsberg Foundation, I was able to purchase equipment for my lab which I called The 'Cytofysiologisk Institute'.

Already in 1947, I was lucky enough to be approached by a young student, Hans Klenow ('studiosis magistrarum' as he was assigned, since he came from the graduate school, University of Copenhagen). Klenow was interested to learn and participate in research projects. As a starting point in the new lab, I had been intrigued by the enzymological puzzles of the activities observed in two oxidases: xanthine oxidase and xanthopterin oxidase. Another bright 'stud. mag.', Niels Ole Kjeldgaard joined us and this project soon disclosed quite a number of interesting biochemical problems. These were monitored not only by spectrophotometry but also by fluorometry (at that time with a Farrand fluorometer). We isolated xanthine and pterine oxidase from cows' milk, using a milk centrifuge of Danish manufacture. Our intention was to collect the enzyme from whey fraction and continue the fractionation with ammonium sulfate.

In short, the most noteworthy observations derived from this project were as follows. Most folic acid preparations were found to exert a marked inhibitory effect on xanthine or xanthopterine oxidase. This seems to be due to an inhibitory factor in most folic acid preps (pteroyl glutamate preps). Much of this was resolved when Dr. Thomas H. Jukes of Lederle Lab sent us a PGA prep of particularly high purity; this prep seemed free of most of the inhibitory effects on the two oxidases. In addition, Lowry, Bessey and Crawford from the dynamic nutrition lab which I had just left, identified the oxidase inhibitor as 2-amino-4-hydroxy-6-formylpteridine, i.e., the 6-aldehyde. This fully agreed with our observations that incubation with xanthine oxidase catalysed the conversion of the 6-aldehyde to a 6-carboxyl group, a derivative which does not inhibit the oxidases [14].

Aside from an exercise in biochemical methods, I cannot say that this project deserved further exploration and soon Klenow as well as Kjeldgaard showed their great abilities for independent research.

Longing to go back to my nucleoside phosphorylase project, I was very lucky to be approached by a brilliant young postdoctoral biochemist from National Institutes of Health, Morris E. Friedkin who wanted to spend a year in our lab. Previously, Friedkin had done outstanding work in Albert Lehninger's lab on oxidative phosphorylation - one of the first successful uses of radioactive phosphorus in mitochondrial and nuclear preparations of liver.

Morrie Friedkin, let me call him by his first name, Morrie, had also followed with interest my publications on nucleoside phosphorylase. Morrie was also aware of some recent reports by L.A. Manson and J.O. Lampen on phosphorolytic splitting of thymidine by enzyme preps from calf thymus.

In our studies on the effect of liver nucleoside phosphorylase on hypoxanthine- or guanine-deoxyriboside (-DR), we observed a pattern which seemed strikingly reminiscent of the phosphorolytic fission of purine ribosides. However, Morrie and I soon realized that the Lowry–Lopez method which operated so well for the acid-labile ribose-1-phosphate did not reveal any phosphorolytic fission of DRs. Yet, Morrie soon revealed the phosphorolytic fission by demonstrating the formation of inorganic phosphate by the old-fashioned precipitation of ammonium magnesium phosphate. Finally, Friedkin managed to isolate deoxyribose-1-phosphate as a crystalline cyclohexylimide salt [15,16].

We were fortunate to have in the neighbouring lab, Dr. Hoff-Jorgensen, an experienced chemical microbiologist. We now joined Hoff to see if the hypoxanthine-DR or guanine-DR, obtained from

D.R.-1-P and the purine bases by the purine nucleoside phosphorylase, would act as a growth factor for *Thermobacterium acidophilus* R26. This turned out to be the case and we were all gratified.

Let me add here: I learned after more than 25 years that lack of nucleoside phosphorylase in infants, a rare heritable disease in man, can seriously interfere with an early stage of T-cell division. Apparently, accumulation of DR-guanosine in the bone marrow gives rise to accumulation of DR-guanosine triphosphate. This in turn inhibits transcription which is critical for early T-cell division [E.R. Giblett et al., The Lancet, May 3 (1975) 1010–1013].

Another able young Research Fellow, Walter MacNutt who joined Hoff and myself in 1950, gave us an instant surprise by pursuing Hoff's microbiological work on growth factors. It turned out that enzyme preps from *L.helvetica* were able to catalyse a 'dry' exchange of purine and pyrimidine bases between the DRs without the intervention of phosphorolysis. These bacterial enzymes we called *trans-N*-glycosidases. To ascertain that the enzymatic conversion between DR-inosine and adenine was not merely a transamination of DR-inosine from adenine, we needed [^{14}C]adenine. Since [^{14}C]adenine might be needed for another project for which I also had some responsibility and interest (see later) I had to see whether this rare item could be made available to us. This problem was solved in an unusual way.

In early 1950 the great organic chemist of University of Cambridge, Alexander Todd, and his young coworker Malcolm Clark, had designed a new and more efficient method of obtaining [^{14}C]adenine with a high isotope yield. Clark arrived in Copenhagen a few months later. He soon became popular under his first name, Malcolm. I greatly enjoyed acting as his unskilled assistant at our fumehood of the lab. Malcolm and I first 'rehearsed' the method on about 10% of the ^{14}C-labelled material. At a certain stage of evaporation when crystals were supposed to appear, nothing happened. It escaped our attention that the evaporation step called for vigilance. Although formic acid has only a slightly higher boiling point than water, it is more 'expensive' to evaporate water, because of the hydrogen bonding, than to evaporate formic acid. In terms of dollars, it was the other way around – so 10% of our specified grant money had disappeared through the chimney as [^{14}C]formic acid.

After this flop, Malcolm and I had to strike a difficult balance: start with at least 25% of radioactive precursor and have faith that crystallization would ensue. This time crystallization of the labelled product did occur. Water presumably went through the chimney. Finally, the remaining 65% provided us with a nice yield of [8-^{14}C]-adenine.

The procedure was published in the British *Journal of the Chemical Society* on Alex Todd's recommendation [17]. He generously requested not to put his name on as co-author. The ^{14}C-labelled adenine was stored for use in future biological and microbiological projects.

Events 1949–1952

In 1949 I received an invitation from The New York Academy of Medicine to deliver a Harvey Lecture in late October entitled *Enzymatic Reactions in Purine Metabolism* [18]. I covered a broad field of purine biochemistry, including enzymatic biosynthesis of the purine ring, as described by John Buchanan and coworkers, as well as by David Shemin and his group. It was of course a special satisfaction for me to present the most recent work with Morrie Friedkin on the enzymatic phosphorolytic fission of deoxyguanosine and show his success in crystallizing the highly acid-labile deoxyribose-1-phosphate.

After New York, I stayed almost two months in Cori's lab. He was now Professor of Biological Chemistry, but his activities encompassed several disciplines, biochemistry as well as pharmacology.

In early January 1950, I was lucky to be invited as a member of a symposium on protein biosynthesis which took place at the Pebble Beach Lodge in California. At that time I had not contributed anything to the problem of peptide synthesis except the calorimetric burning of hippuric acid tablets in the basement of Borsook's department in 1939. I suspect that Henry Borsook had kindly suggested my invitation to this symposium, since I was scheduled for CalTech a few days later. One of the most interesting and delightful

Plate 6. From left to right: Irving Frantz Jr. (Huntington Laboratory, Boston), Henry Borsook (CalTech), Fritz Lipmann (Massachusetts General Hospital) and Herman Kalckar (University of Copenhagen). (Photograph taken at the Symposium on Protein Synthesis, Pebble Beach Lodge, California, January 1950.)

participants at the meeting was Paul Zamecnik from Harvard and MGH. Fritz Lipmann was also present and reported his new ventures in biological acetylations.

Back at CalTech in 1950, the Biology Department with George Beadle as chairman had further developed as a great center for advanced ideas. He brought Max Delbrück back from Vanderbilt University. Max had attracted several brilliant young investigators to join his lab. Gunther Stent was one of them, and perhaps the most stimulating. Through Max, he knew about Bohr and the Copenhagen spirit. Gunther also knew Ole Maaløe, the Danish microbiologist, who had joined the bacteriophage group during his visit with Max. Since I had been an active participant in the first 'Delbrückian' bacteriophage course at Cold Spring Harbor in 1945, perhaps they thought I could contribute some biochemical techniques to Ole and Gunther's phage genetics venture. In fact, I was able to tell Gunther that I had, with a young British chemist, synthesized ^{14}C-labelled adenine (cf. [17]) which could be used for various research projects.

While in Saint Louis, I received a letter from another bright young scholar, Barbara E. Wright, from Stanford University. She had received a masters degree from Van Niel's Microbiology Department, Hopkins Marine Station, Pacific Grove which was affiliated with Stanford University's Graduate School. Since George Beadle was Professor of Biology at Stanford University before coming to CalTech, I showed him Barbara's letter. Beadle had only heard favourable words about Barbara, who apparently would be getting her Ph.D. from Stanford. Beadle felt it worthwhile for me to plan a stop at Pacific Grove on my way back and interview Barbara Wright which I did. Her brightness and independence in scientific thinking impressed me, and after our meeting and conversation with her magnificent mentor C.B. Van Niel (also my great mentor in the summer of 1939) things looked very promising. In another six months I invited Dr. Barbara Wright to join our Copenhagen lab as a postdoctoral microbiologist.

Revisiting Cori's department and later that of Arthur Kornberg at NIH, I became aware of the many new and exciting concepts that had developed in biochemistry during my recent travels and these were close to my own research.

More American scholars arrive in Copenhagen

In September 1950 a young scholar, James (Jim) D. Watson from Bloomington arrived in my lab. He looked a bit lost and admitted that he was rather homesick for the Indiana countryside. Jim and his father were devoted bird watchers. During the first weeks I introduced Jim to two leading Danish scholars in Natural Sciences with whom I suspected he would find rapport: the geneticists Mogens Westergaard and Ole Maaløe. They appreciated Jim very much and he in turn seemed to warm up in their company. On took Jim to evening performances of the famous Danish Palestrina choir which was much admired in Europe and in England, but I am not sure if Jim was much taken by these Palestrina concerts. Gunther Stent and Ole would be more apt to enjoy this brand of music.

Later in the autumn of 1950 Jim, Gunther and Ole got together, and Ole's lab at the State Serum Institute became the working place for the bacteriophage geneticists. Like Hershey, their work centered around the transmission of ^{32}P-labelled phage to unlabelled *E.coli* hosts and to the phage progeny. This was much like Hershey's programme which also used ^{35}S for labelling the parental phage protein and monitoring the progeny. It is well known that Hershey found only phosphorus label in the progeny and no sulfur label, indicating that DNA rather than protein is transmitted to the progeny.

An additional contribution from our lab to the Watson–Maaløe phage project was the availability of [8-^{14}C]adenine. However, to use it, Jim and Ole had to isolate an adenine-requiring *E.coli* mutant. They soon isolated such a mutant and used it as a host in the proper growth medium. As expected, the phage progeny showed a high transmission of labelled adenine as well as guanine [19].

My contribution to the Watson–Maaløe phage project had been the synthesis of [8-^{14}C]adenine, à la Alex Todd, which Jim and Ole acknowledged. However, Jim felt that this somewhat distant collaboration was not what he had expected when he was urged to join me in Denmark. He expressed it later in his essay in the Cold Spring Harbor 'Festschrift' for Max Delbrück in 1966 [19a]:

> 'Though Kalckar was admittedly a biochemist, through his brother he knew Bohr and therefore should be receptive to the need of high-powered theoretical reasoning. Even better, Kalckar's interest in nucleotide chemistry should be immediately applicable to the collection of nucleotides in DNA.'

Frankly, in early 1951 I was not prepared, even to initiate a programme the type of which Jim was alluding to.

I have mentioned that Walt MacNutt had found a trans-*N*-glucosidase in *Lactobacillus helveticus* which seemed to catalyse the 'dry' exchange of free purines with purine deoxyribosides. In the interaction of hypoxanthine deoxyriboside (HDR) and adenine, ADR plus hypoxanthine were formed; one might imagine that this reaction is a transamination with adenine as the ammonia donor; yet the balance could also be due to an enzymatic exchange between the purines. The use of the [^{14}C]adenine showed that the enzyme catalysed the 'dry' nonphosphorolytic exchange between the purine-DR and the free purine [20]. The [^{14}C]adenine also became of interest to Gene Goldwasser who was able to show that the [^{14}C] adenine could be incorporated into RNA of pigeon liver homogenates under respiratory conditions.

In December 1950 Barbara Wright arrived, probably more receptive to 'high-powered theoretical reasoning' than I was. I introduced her to Kaj Linderstrøm-Lang as well as to the geneticist Mogens Westergaard and other Danish scholars, especially those who also had been in the resistance movement during the German occupation; they liked her bold 'Western' ways as well as her intellectual brightness. I myself responded much the same way to her. Barbara's presence made me forget my depression and I began to feel much younger and a spirit of mutuality developed between us.

Seeing my brother's two young children, I wanted to marry Barbara and raise a family. I went through the intense emotion of a divorce from my wife, Vibeke, married Barbara and raised a happy small family.

As always, Kaj LL was perhaps the only one who had followed my emotional trauma with real understanding and soon sensed my new situation. He provided Barbara with a place at The Carlsberg Laboratory, which she enjoyed very much both intellectually and, no less, because of his 'Jeffersonian' humour especially at lunches. Added to this circle were two spirited and able young biochemists, Paul Plesner and Mahlon Hoagland. Ten years later I had the pleasure of a reunion with Paul and Mahlon in the laboratory of Paul Zamecnik.

In 1952 I had the privilege of receiving two more young scholars. Paul Berg came from Western Reserve University and William Joklik from Australian National University. Together they solved in record time a metabolic puzzle, describing yet another type of transphosphorylation, nucleoside diphospate kinase, i.e., UTP+ADP $\rightleftharpoons$ UDP+ATP (nick-name, 'nudiki').

In 1952–1953 we were still fortunate to have Hans Klenow, one of our most independent and resourceful researchers, as an active member of our lab. Hans had established early contact with C.E.Cardini in the Leloir group. Using the enzyme phosphoglucomutase and glucose-1,6-diphosphate, Klenow showed that crystalline phosphoglucomutase catalysed the interaction of ribose-1-phosphate with glucose-1,6-diphosphate, generating ribose-1,5-diphosphate and glucose-6-phosphate [21]. Ribose-1,5-diphosphate was purified and characterized and its possible role in the biosynthesis of nucleotides was discussed [22]. Our [8-^{14}C-]adenine was incubated with ribosediphosphate and phosphoglucomutase to see whether 5′ AMP was formed [23]. It seemed encouraging, but the role of this pathway for nucleotide biosynthesis remained ambiguous.

I wished I had been more aware of the general importance of the discovery of pyrophosphorylation by Arthur Kornberg and his coworkers. Later they extended this new principle to other areas. In a monograph called *Reflections on Biochemistry*, dedicated to Severo Ochoa's seventieth birthday, Arthur Kornberg wrote a small chapter [24] titled 'For the love of enzymes', where he mentions:

> 'I have also had a favourite spot for the synthetase that makes phosphoribosylpyrophosphate (PRPP): ribose 5-P+ATP $\rightleftarrows$ PRPP+AMP.
> Most of us anticipated that ribosyl activation for nucleotide biosynthesis would use the same device of phosphorylation, so well known for glucose. But the novelty of pyrophosphorylation used by this enzyme (coupled with the elimination of inorganic pyrophosphate upon subsequent condensation) established my unalloyed awe for the ingenuity and fitness of an enzyme.'

Human galactosaemia

In 1951 Leloir and coworkers in Buenos Aires had already described the complex nature of the biochemical conversion of glucose phosphate to galactose phosphate [25]. This seemingly so simple positional 'swap' of a hydroxyl of the C-4 group of glucose with the hydrogen on the same carbon, thus forming galactose, was found to require at least one coenzyme, UDPGlc or UDPGal. It was suggested by us, that such nucleotides might be formed by an interaction

TABLE I
Metabolic scheme. The galactose pathway towards glucose metabolites

Reaction	Enzyme	(EC number)
(1) Gal + ATP → Gal-1-P + ADP	Galactokinase	(2.7.1.6)
(2) Gal-1-P + UDPGlc ⇌ UDPGal + Glc-1-P	P-Gal-uridylyl transferase	(2.7.7.12)
(3) UDPGal ⇌ UDPGlc	UDPGal 4-epimerase	(5.1.3.2)
Sum: Gal + ATP → Glc-1-P + ADP		
UDP Gal formation from glucose		
(4) Glc + ATP → Glc-6-P + ADP	Hexokinase	(2.7.1.1)
(5) Glc-6-P ⇌ Glc-1-P	Phosphoglucomutase	(5.4.2.2)
(6) Glc-1-P + UTP ⇌ UDPGlc + PPi	PP-uridylyl transferase	(2.7.7.9)
(7) UDPGlc ⇌ UDPGal[a]	UDPGal-4-epimerase	(5.1.3.2)
Sum: Glc + ATP + UTP → UDPGal + ADP + PPi		

[a] UDPGal is donor of galactosyl units to various low or high molecular weight acceptors: the transfer is catalyzed by various galactosyl transferases.

between galactose-1-phosphate (Gal-1-P) and UDPGlc, and by a transfer between glucose-1-phosphate (Glc-1-P) and UTP, as summarized in Table 1.

In 1952 we initiated our studies on the complex galactose metabolism in eukaryotic cells. These studies took place in our research unit of cell physiology at the University of Copenhagen. Along with Agnete Munch-Petersen and visiting research fellows, among them Evelyn Smith-Mills from Great Britain, studies were conducted on the enzymatic interactions of UTP, UDPGlc, Glc-1-P and Gal-1-P. The enzymes involved were prepared from the same type of yeast as the one used by the Leloir group, viz., *Saccharomyces fragilis* pre-induced by galactose. We were able to detect two types of 'uridylyl transferases' [26,27] (see equations):

Gal-1-P+UDPG ⇌ Glc-1-P+UDPGal	PGal-uridylyl transferase
Glc-1-P+UTP ⇌ UDPGlc+PP	PP-uridylyl transferase

In 1953 my contact with biochemical research in the United States was renewed, and let me stress, in a very creative way, 'The Visiting Scientists Program' at The National Institute of Health in Bethesda, MD, near Washington, DC. Thanks to Dr. Bernard L. Horecker, Chief of Research at the National Institute of Arthritis and Metabolic Diseases, as it was called in 1953, I received a generous invitation to conduct basic medical research in a field of biochemistry of my own choice. In 1954, I had not only Bernie's encouragement and support but also that of Dr. Hans DeWitt Stetten, incoming Institute director. Hans, as well as Bernie, soon became my trusted friends during these years.

At NIH I was able to pursue further research on the enzymology of the UDPglucose system, thanks to an enlightened and active group of biochemists and pharmacologists there. I joined forces with Jack L. Strominger and Elizabeth S. Maxwell, and for the first opus we had Julius Axelrod (from the National Institute of Mental Health) as an additional participant. The joint papers described a novel dehydrogenase pathway from liver by which UDPglucose, through a two-step dehydrogenation of the free 6-hydroxyl group of the glucosyl moiety, could be converted to UDPglucuronic acid ('active glucuronic acid'). In this essay I will merely emphasize a few aspects of this reaction in connection with our ongoing studies of galactose metabolism. UDPglucose dehydrogenase turned out to be highly specific for UDPglucose; it required NAD as a coenzyme, and was inhibited strongly by NADH. In my further studies on galactose metabolism, I used this specific and sensitive spectrophotometric enzyme test for UDPGlc, catalysing the reduction of 2 mol of NAD per UDPGlc oxidized (i.e., UDPGlc + 2 NAD → UDPglucuronic acid + 2 NADH).

In 1955 I tried to assay a haemolysate of my own erythrocytes as a possible source of the enzyme which catalyses the interaction between Gal-1-P and UDPGlc. The new UDPGlc dehydrogenase test for scoring the interaction, i.e., reaction 2, Table I, would register a consumption of UDPGlc if that transferase enzyme was present. I was lucky; UDPGlc was only consumed if Gal-1-P was present. And there was no indication of regeneration of UDPGlc through reaction 3 (see later).

At this point I will have to insert a few comments about the above-mentioned shift of the hydroxyl group of C-4 of glucose to galactose. This inversion went under the name of the 'glucose-galactose Walden inversion.' (The Walden inversion was supposed to operate through an anhydride intermediate.) In his 1951 review Leloir patiently presents the many attempts by organic chemists to explain the nature of the 'legendary glucose-galactose Walden inversion.' This expression was also used for the alleged conversion of galactose-1-phosphate to glucose-1-phosphate by medical investigators.

A research group in Manchester, U.K., demonstrated clearly the presence of active galactokinase (i.e., reaction 1) in erythrocytes from galactosaemic subjects; in vitro incubation of galactosaemic haemolysates with galactose and ATP brought about a marked accumulation of Gal-1-P. Against this background, the current view of galactosaemia centered mainly on defects in the so-called 'galactowaldenase', which supposedly is needed to convert Gal-1-P to Glc-1-P and glucose. The latter was sketched in an oversimplified way (not indicating how UDPGlc is involved); this version went as follows:

$$\text{Gal-1-P} \rightleftharpoons \text{Glc-1-P}$$

My own view on the question of the defective step in galactosaemia differed decisively from this formulation for several reasons. Some of my reservations stemmed from numerous reported observations on infants on galactose-free diets, some were based on biochemical and developmental observations.

Two of the more serious toxic, chronic symptoms which ensue if galactosaemic infants remain on an ordinary milk diet deserve special attention: cataract of the lens and development of mental retardation.

However, both the acute and chronic symptoms can be avoided by dietary therapy. If the galactosaemic infants or children are confined to the consumption of vegetarian lactose-free milk formula (such as soybean milk or related formulas), which contain only sugars, like glucose or sucrose, no toxic symptoms ensue. In short, galactosaemic infants kept on a galactose-free diet have an excellent chance of normal development.

Furthermore, it is known that the central nervous system, and especially the mammalian brain, contains large amounts of galactolipids in which almost half of the galactose is deposited after birth. In normal rats normal deposition takes place on a diet free of galactose, the galactosyl lipids originating from dietary glucose.

Both of these facts raise some basic questions. If the congenital galactosaemia which we were discussing was due to a defect in the hypothetical 'galactowaldenase' (Table I, reaction 3), a galactose-free diet might 'induce' the developing brain to synthesize glucolipids instead of galactolipids, since synthesis of galactosyl compounds requires reaction 3. This in turn may jeopardize normal brain development. Yet normal brain function is actually better preserved in infants on galactose-free diets.

A defect of the enzyme PGal uridylyl transferase (PGalUT) would not interfere with the synthesis of galactolipids, even on a galactose-free diet. In this case a normally functioning 4-epimerase would secure a normal conversion from glucose via UDPGlc to UDPGal and hence an unhindered formation of galactosyl compounds from glucose (reactions 4–7).

The disease symptoms observable in galactosaemic infants fed lactose are probably due to a 'traffic jam', i.e., accumulation of Gal-1-P which is unable to be converted to UDPGal due to the absence of the enzyme PGalUT. It was therefore my first priority to compare the levels of the PGalUT enzyme in the haemolysates of galactosaemic and nongalactosaemic subjects. The problem of the unspecific absence of reaction 3 in all the human haemolysates would have to be solved later.

Although galactosaemia is a rare disease, it was possible for us, thanks to the skill and efforts of Drs Kurt Isselbacher and Josephine Kety to acquire blood samples from eight subjects with galactosaemia and more than a dozen subjects who were either normal or afflicted with unrelated diseases. We found that the galactosaemic patients specifically lacked PGalUT (reaction 2) but did have normal levels of the other uridylyl transferase which reacts with pyrophosphate (PPUT), and galactokinase was also present [28]. (See metabolic scheme in Table I.)

We wanted to ascertain that the type of galactosaemia which we had been studying is an exclusive defect in the enzyme PGalUT and does not affect UDPGal-4-epimerase. The absence of the latter in haemolysates from galactosaemic as well as normal subjects was clarified through the following sequence of events in our laboratory, all within the same lucky year of 1956. Liz Maxwell was able to purify UDPGal-4-epimerase (abbreviated: epimerase) from calf liver. It turned out that the purified epimerase required NAD for its activity and was strongly inhibited by NADH [29]. The absence of epimerase activity in normal as well as in galactosaemic haemolysates in earlier experiments was due to an extra need for added NAD.

This disclosure not only put us on the right track for delivering a clear verdict or the specific enzyme defect in our galactosaemia cases; in addition, a more basic chemical concept also emerged: the nature of the epimerase reaction.

UDPGal-4-epimerase is apparently an oxidation-reduction enzyme. It was at this point that Maxwell and I decided to abandon the previous name ‘Waldenase’, which had been associated with various types of anhydride formation, and rename the enzyme ‘UDPGal-4-epimerase.’

In our renewed and improved type of enzyme assays of human haemolysates, we also fortified the haemolysates with NAD and found that epimerase was indeed present in both galactosaemic and normal samples. Hence the type of galactosaemia which we had encountered repeatedly constituted a total (or near-total) defect on one enzyme, PGalUT.

This was also found to be the case when liver biopsy samples or fibroblast cultures from such subjects were investigated. Other groups later reported that many of the PGalUT mutants are structural gene mutations.

The question about the levels of this enzyme in erythrocytes from heterozygotes (especially parents of galactosaemic subjects) was clarified by H.N. Kirkman, who joined our group in 1957. He showed that the heterozygotes had a partial defect of PGalUT, amounting to about 50% of the levels found in haemolysates from normal subjects. I reviewed all these findings in a CIBA Symposium [30].

Two different types of epimerase defect in infants have recently been observed: the mild type described by R. Gitzelmann and coworkers in 1972 and discussed in this essay, and a severe type of epimerase defect, so far observed only in two cases (Gitzelmann, 1985, pers. commun.). These seem to lack epimerase in most types of cells investigated. Exposure to lactose or galactose brings about accumulation of Gal-1-P as well as UDPGal in the erythrocytes. The disease symptoms are more severe, especially the cerebral disturbances. Limited intake of galactose may not mitigate the clinical symptoms, which in one case included not only cataracts but also impairment of brain development.

It is too early to draw any conclusions about the pathogenesis and developmental aspects of these types of epimerase defects in the realm of dietary therapy, as discussed above. At present we might make the circumstantial assumption that the so-called mild form, in which the absence of epimerase is merely confined to mature erythrocytes and leukocytes would tend to disappear since these cells are devoid of protein synthesis and hence unable to maintain a modified unstable epimerase. In contrast, one might suspect that in the severe cases in which epimerase activity is absent in dividing cells as well, no epimerase is synthesized.

In the latter case, effective therapy would become very difficult. A galactose-free diet would deny the developing brain of the newborn the formation of galactolipids and glycoproteins. Moreover, the addition of galactose to the diet would give rise to accumulation of the toxic Gal-1-P, which frequently leads to mental retardation. It can really be questioned whether a universal epimerase defect would permit adequate development by dietary therapy.

Out of the biochemistry 'stall' to a 'teach-in' against strontium-90 fallout

In 1957 I was hospitalized for a long time in the NIH hospital with mononucleosis volunteering to donate some of my lymph nodes. Being unable to be in touch with my lab work I thought it behooved me to break out of my biochemistry 'stall'. My thoughts became occupied with broad problems of general human concern which cropped up during this precarious new atomic age of huge atmospheric test explosions with ensuing radioactive fallout over most of the northern hemisphere. In this way strontium-90 contaminated especially the grain used for bread, and found its way into the teeth and bones of young children. Was it possible to make young families with children aware of this new precarious world? Perhaps one could enlist parents and children in an enlightening game. 'Do not give your milk teeth (baby teeth) to the tooth fairy, give them to science'.

I tried to formulate a sort of declaration which I entitled 'An International Milk Teeth Radiation Census' in which I suggested that the public health agencies, as a long-range project, organize a large-scale collection of dated milk teeth. I sent this article with the above title to *Science*, which politely turned it down, stating that the topic was a bit too touchy at the present time. This was also the feeling of Willard Libby of the U.S. Atomic Energy Commission, who wrote me in 1957 that, although he found the collection of milk teeth essentially a sound idea, he had a preference for avoiding publicity. In contrast, many of my scientific colleagues who felt very positive about my article urged me to send it to *Nature*, and thanks to my friend Hans Krebs, this journal accepted my article in the summer of 1958 [30a].

The article apparently generated more interest than I had anticipated. At that time I was working at McCollum Pratt Labs at Johns Hopkins University, having pretty well forgotten about my well-meant 'declaration' against radioactive fallout. I was then taken aback when I received letters from British girls who expessed interest in my 'Census' and of all things contained the prettiest milk teeth from their younger sisters (with dates, names and ages). The young ladies did not quote the *Nature* article, but *The London Sunday Pictorial*, which had popularized my Census declaration into an Andersen fairy-tale style story. All of a sudden I became a father-figure on the basis of their version: 'Do not hide your milk teeth anymore under your pillow for the tooth fairy, send them to Dr. Kalckar for Science!' So I became the lucky recipient of the cutest letters and milk teeth, without a practical framework of monitoring the teeth. Perhaps Willard Libby was right about too much publicity?

At that time Johns Hopkins University was so fortunate to have Milton S. Eisenhower (brother of President Dwight Eisenhower) as its President. Milton Eisenhower had always been keenly interested in general education in the Western hemisphere. Milton Eisenhower was popular amongst students and faculty. He always came to the faculty lunch room and made himself available for questions and conversation. In this way I had the opportunity for an occasional brief conversation, and it dawned upon me that I should try to make an appointment with him to get his opinion and advice about the general interest in the milk teeth census, which apparently was underway. I was delighted to receive his unreserved and enthusiastic support for initiating such a census for various American metropolitan and suburban areas. He was even prepared to contact a close friend who was heading a dental school in the South (I still have a copy of this letter).

My old university from the early 1940s, Washington University in St. Louis, MO, had already taken the initiative to start a 'Baby Tooth Census.' I received letters from Dr. A.Langsdorf from the Medical School, Mrs. Edna Gellhorn, Dr. Louise Zibolt Reiss from the Dental School and Dr. Barry Commoner from The Science School. I was pleased to receive these letters and for their generous citation of my paper in *Nature*. Dr. Z. Reiss was in charge of analysis and monitoring of the milk teeth collection which grew to a very great size, thanks to the great interest of children and family (further stimulated by issuing of decorative and attractive buttons). Dr. Reiss' article entitled 'Strontium-90 absorption by deciduous teeth', appeared in *Science* [30b]and it also started by citing my *Nature* article.

Thanks to the joint efforts and broad participation of the monthly St. Louis Magazine *Nuclear Information* and the *Code of Arms*, so attractive to kids, the participation in the census was so large that it permitted refined analyses. It became clear that milk teeth of 'vintage' 1956 showed ten times more strontium units than those of 'vintage' 1950. Teeth collected from bottle-fed children seemed to show a significantly higher strontium level than those from the breast-fed group. In the 1960s, during President John F. Kennedy's administration, a ban against atmospheric tests of large thermonuclear bombs was instituted with the cooperation of the Soviet Union (Nikita Krushchev?). A few years after the atmospheric test ban, I had the pleasure of learning of the report from St. Louis that the new vintages of milk teeth were 'past the culmination' of radioactivity and later collections showed less and less strontium-90 radioactivity.

I had the joy of getting my first D.Sci.h.c. from Washington University in 1964 with a citation of my *Nature* article of 1958. Gerty Cori had died some years before but Carl Cori was present with his second generous wife, Anne. It was a joyous occasion, indeed.

My personal encounters with developments in molecular biology

With one exception, my encounters with molecular biology and genetics have always been second-hand or third-hand, inspiring reading, to say the least.

Before I end up with a sort of personal encounter, although somewhat parochial, it would be weird, were I to avoid recalling that great chronicle *The Double Helix* and its poet, Jim Watson.

Thanks to Jim, I first came to read an early edition, in manuscript form, I believe some time in 1967. After merely a couple of days' reading I was invited by Jim to have dinner in Cambridge. He asked my opinion about the text, and whether I wanted some paragraphs removed about myself. I said that this script was as witty in style as young Charles Dickens' Pickwick papers and, in addition, I sensed a unique creative and educational touch for biology students; why should I censure the frankness and freshness of his chronicle? After just deleting a few irrelevant lines about my family, Jim and I felt in good spirit. As everybody knows, *The Double Helix* by Jim Watson is still being read and enjoyed in many countries, by 14-year-old youngsters to researchers and biology teachers well over 80 (my present age).

Coming back to *The Double Helix* 20 years later, I find it even more exciting reading. The dubious encounter, especially the one with that introvert and highly able crystallographer (Rosalind

Franklin) and Jim's blunt report of it, now feels to me more like a candid confession.

But didn't I have any confessions to make for distancing myself from Jim and Ole at the midcentury project? My late friend, Ole Maaløe, thought so at that time, and his thoughts were not so readily dismissed. But enough about all these matters. I now want to recall a direct personal encounter with molecular genetics around 1970.

In 1966, as Head of Biochemical Research at MGH and Professor of Biological Chemistry at Harvard Medical School, I had permission to accept graduate students. The graduate student who approached me in 1964, Henry C.P. Wu, soon turned out to develop into a first-rate young scholar and soon received his Ph.D. degree. In 1966 another brilliant young biochemist, Winfried Boos, from the University of Konstanz, Bodensee, joined our laboratory.

We were concentrating our efforts to explore a very puzzling type of regulation of galactose transport in *E. coli.* A bacterial mutant called *GalK,c* because it had a genetic defect in galactokinase, was 'notorious' to us because, as the 'c' was meant to indicate, the other genes of the Gal-operon operated constitutively. This meant that induction of the Gal-operon did not need addition of galactose to the growth medium [30]. This is in contrast to another GalK strain, *GalK,i*, which also was galactokinase-less, yet needed galactose (or an analogue, D-fucose) to induce the other members of the Gal-operon.

I described these mutants in collaboratorion with Elke Jordan and Michael Yarmolinsky when working at Johns Hopkins Biology Department [31].

To make a long story short, GalK,i releases galactose from its cellular UDPGal and seems unable to recapture this from the medium. In contrast, GalK,c seems to possess a galactose 'scavenger' mechanism, sufficiently high to maintain the induced state of GalK,c.

Winfried Boos now developed biochemical methods, trying to isolate the Gal scavenger from either of the GalK strains. By a skillful use of polyacrylamide-gel electrophoresis of cell extracts, obtained by osmotic shock devices, he found that the fractionated shock fluid from GalK,c showed a specific high-affinity Gal-binding protein, whereas shock fluid from GalK,i did not [32]. The Gal-binding protein was considered a periplasmic protein, residing in the space between the outer and inner membrane.

Shortly afterwards, listening to a fascinating lecture by Julius Adler from University of Wisconsin on galactose chemotaxis in various *E.coli* mutants, I noted and became excited by a certain parallelism. Adler had found a galactose chemotactic receptor located in the periplasmic space of *E.coli.* The characteristics of chemotaxis were only partly understood at that time, yet it was suspected of being related to one of the most primitive neurobiological responses known in nature.

I wanted Winfried and Julius to exchange the Gal mutants, which scored positive (+) or negative (–) in their respective tests, and for which purpose? I had exhilarated myself with the homespun idea that the GalK,c strain would score + in Julius' chemotactic test, and the GalK,i strain would score - in the chemotactic test.

Unfortunately, due to my habitual type of day-dreaming, as Winfried used to call it, my monomaniac mumbling (soliloquizing) was obviously not deciphered, neither by Julius nor by Winfried. Finally, after a good deal of cajolery, both friends began to understand me. They finally did exchange their bacterial mutants for executing their respective tests. The finding was in fact that only the strain with high-affinity Gal binders (GalK,c) scored positive in Gal chemotaxis, and this was also reflected in Winfried's tests of Gal binding vs. Gal chemotaxis. The receptors in Gal chemotaxis seemed indeed identical with the Gal binder.

As Winfried Boos mused in his generous and thoughtful chapter in the 'Festschrift' to my 75th birthday:

> 'I feel obliged to tell you on this occasion that it was really Herman Kalckar who first perceived the idea that a transport related binding protein is chemoreceptor in *E.coli*, even though the published record may suggest otherwise.' [33]

In an article in *Scientific American*, Julius Adler [34] also recalled our fruitful encounter of 1970.

Since one of my favourite pupils, Paul Plesner, recently challenged me to try to express some of my philosophical thoughts on my long trotting along the trail of biochemistry, why don't I try to make a philosophical parable to the story which I have just finished?

The grand old British biochemist, Sir Frederic Gowland Hopkins, ventured to answer a question posed to him through the philosophy of Alfred North Whitehead: 'Has the modern biochemist, in analysing the organism into parts, so departed from reality that his studies no longer have biological meaning?'

In a dialogue with himself, Hopkins wrote:

> 'So long as his analysis involves the isolation of events, and not merely of substances, he is not in danger of such departure.' [35]

Winfried, Henry, Julius and I actually managed in a 'parochial' way, somehow to make a transition from biochemical analysis to cellular events, like endogenous induction and sugar chemotaxis.

Our case, that a transport-related protein might also be used as a chemoreceptor, could only be postulated on account of the recognized specificity of mutational steps, selecting mutants by two widely different criteria. This makes it feasible to 'isolate' two different types of events and thus disclose one of the capricious ways of the economy of Nature.

My old friend, the late Max Delbrück, who had always teased me about my mumbling, seemed to have enjoyed the mutual Gal taxis adventure, and he invited me in 1970 to give a lecture at CalTech, Biology Division; this happened also to be the first 'Jean Weigle Lecture', which was later published in *Science* [36].

Preparing for U.S. Citizenship

In the early summer of 1958, I collected papers and instructions to apply for my U.S. citizenship. Being a resident of Montgomery County, MD, my hearing would eventually take place in Rockville, MD.

I now spent an hour or two studying the version of U.S. history rendered to me from the brochure. The Rockville brochure obviously also contained a more or less colourful history of the constitution of Maryland.

My main attention, especially under the Eisenhower administration, tended to center around the U.S. Constitution. Constitutional rights on behalf of many scholarly colleagues had been violated in several cases by the State Department. However, I soon realized, partly through rumours, that the flunking frequency in Maryland seemed more a function of spurious replies to examinations on the constitution and history of the State of Maryland than of the U.S. Constitution. This seemed rather unsettling.

During the early part of the summer of 1958, I came to Woods Hole for teaching and research and I was lucky enough to share a room with François Jacob. Obviously, it was a great privilege for me to listen and exchange scientific thoughts with him. Besides, François has a genuine sense of humour and he generously offered, as a sort of evening prayers, to initiate a brief but succinct dialogue with me on the U.S. Constitution. When he sensed my extra worries about the possibility of some 'sadistic' Rockville sessions which might center mainly around the constitution of Maryland, François in a masterly way switched to a stern examination of this somewhat limited, less philosophical topic. Before we moved away from the room and went our ways, François deemed it proper to start me on a last brisk 'dry run' on this prosaic topic.

I received my U.S. citizenship in autumn, 1958, in Rockville, MD. My examination was conducted in a civilized and fair way by a woman. The 'domestic' state laws were scarcely discussed. In the discussion of the New England states I had forgotten the history of the origin of the smallest state of the union, Rhode Island. Its foundation about 350 years ago was apparently attributed to lack of religious freedom in the puritanical state of Massachusetts. Perhaps I suppressed this charge against one of my favourite states, which later would become my home state for well over a quarter of a century.

In any case, I became a U.S. citizen, vintage 1958, and still a relative newcomer to The Johns Hopkins University, 'Homewood' campus, Biology Department. The university president was the wise and popular Milton S. Eisenhower, 'Ike's' brother.

A number of unexpected and interesting events shaped my plans for the near future. During the last week of April, 1959, I received a cable of congratulations from Milton S. E. on behalf of the Johns

Hopkins University; I had been elected member of the National Academy of Sciences (NAS). My old friend from CalTech '39, David Bonner, had been elected in the same session. This was just the beginning of a more exotic story.

Just at that time, the NAS and its president, Detlev Bronk (Milton S. E.'s predecessor as president of Hopkins), had just initiated a scholarly exchange programme between the NAS members and members of the USSR Academy of Sciences, 'Akademi NAYK'. Dave Bonner and I immediately applied to the foreign secretary of NAS for the first exchange in 1959 and listed Moscow, Leningrad, Sverdlovsk, Novosibirsk, Kiev, and Yerevan as scientific centers to visit. Dave Bonner, a geneticist, was particularly interested in the Ural and Siberia academies. Besides Moscow, Leningrad and Kiev, I wanted also to go to Yerevan, in Armenia. My interests in Kiev and Yerevan were not merely touristic, but as I will explain later scientific, to meet V.A. Belitser (see p.159) and the excellent pharmacologist G.K. Bunyatyan in Yerevan.

The U-2 Incident

May–June also happened to be the time when 'Ike' was supposed to meet his enthusiastic host Nikita Krushchev and in the Eisenhower entourage was also his brother, Milton S. E. He was of course very interested in hearing about Bonner's and my plans and, in his generous way, offered any help he was able to render us.

On May Day 1960, dramatic events erupted, or rather descended upon us. This was the day of the U-2 incident, when the Russians shot down a secret U.S. spyplane over the Ural mountains and caught the pilot Gary Powers and his parachute, as well as most of the hardware and the sophisticated U-2 plane. The situation was awkward to say the least and a dialogue between the two chiefs of state had been ruined. The 'cold war' had started.

What about the academic exchange? A week or two into May, we were relieved to learn that the academic exchange was essentially unaffected by the U-2 incident, albeit with some reservations. Visits to the regions of Ural as well as Novosibirsk were not permitted. In spite of the cancellation of the Eisenhower visit, Milton S. E. showed a warm interest for my planned trip. He wrote a very nice letter to Detlev Bronk (Dear Det, . . .) asking for his help and advice to make it possible for me to visit as many scholarly institutions in the U.S.S.R. as possible during the month of June 1960.

I had been looking forward to join David Bonner on this trip. However, at the last moment, he was forced to cancel his trip because of a serious health problem, diagnosed as a beginning Hodgkin's disease, which required intense treatment. Hence, I became a solitary visitor in that first informal academic exchange to Russia.

My solo visit to the U.S.S.R., June 1959

On June 1, I flew to Paris and from there with 'Aeroflot' directly to Moscow; the jet was not more than half filled. In Moscow I was met by a young Scandinavian-looking student from 'Akademi NAYK.' His name was Michael Kritsky (let us call him Mike). Most of Moscow, including the hotel, I found rather depressing. However, Mike turned out to be an interesting and attractive travel companion.

Mike had learned most of his English from his father who was an academy Professor of Hydrobiology and had travelled in India, Pakistan and Egypt, as a consultant in planning the construction of river dams. Mike was a 24-year-old graduate student in biochemistry at one of the academy institutes in Moscow under Prof. Belozersky, a highly regarded biochemist in the field of nucleic acids. I met him later.

In these autobiographical notes I will pass relatively briefly over scientific reports of my visits to the many labs for obvious reasons. I handed NAS my scientific report some time in the fall of 1960, a delegation of leading American biochemists had visited Russia in 1959 and since the 1961 International Congress of Biochemistry was held in Moscow with heavy American participation (including myself), it seems more natural here for me to try to give a little personal flavour to my lone, but not lonesome, journey from the Finlandian border to the Eastern slopes of Mount Ararat within the borders of Turkey.

Mike first saw to it that I paid my respects to the Vice-President of the Akademi NAYK in Moscow, Prof. Topchief, who is an organic chemist. He was also a member of the Pugwash International Peace Conferences. Topchief also expressed interest in my article about strontium fallout. Bentley Glass from Johns Hopkins University had told him of my article about the milk teeth census.

Prof. Topchief asked me whether the US National Academy of Sciences had elected any Foreign Associates from the U.S.S.R. I was able to tell him that in 1959 they had just elected Victor Ambartsumanian, astronomer from University of Yerevan, Armenia. Physics in the U.S.S.R. was highly respected at that time. The Royal Danish Academy of Sciences had elected a Russian pupil of Niels Bohr, Lev Landau, to Foreign Member already in 1951, and in 1960 Landau was elected as foreign associate by NAS.

Since the main purpose of my trip was to meet Russian biochemists, we visited the A.N. Bach Institute of Biochemistry at Leninsky Prospect in Moscow. It houses some of the most creative biologists and biochemists in the Soviet Union. The history of its

center, is well described by W. L. Kretovich in Chapter 11 of *Comprehensive Biochemistry*, Volume 35.

At the Bach Institute, I had the opportunity to meet quite a number of distinguished and active biochemists in fields more related to that of my own.

W. A. Engelhardt, a grand old friend and scholar, I had recently seen when he came to visit the Biology Department of Johns Hopkins University, where he spent quite a few minutes in the darkroom, admiring a new manifestation of, as I recall 'ATP-power', Bill McElroy's luciferin-luciferase-ATP fireflies. At the time, Engelhardt was involved in the study of membrane ATPases and cation transport. In a few months he was due to move to a new academy institute. His wife, M. N. Lyubimova was studying ATPases in various filamentous proteins, spermosin, mimosin (from mimosa plants) and filaments in trypanosomes; all these filamentous proteins seem to contract when exposed to ATP. Academician A.L.

Oparin had not lost his interest in his own speculations on the origin of life, the 'coascervate' bubbles, an artificially made 'cell' formed through a stochastic event from a mix of protein with another polymer of opposite charge, for instance a nucleic acid.

It had been arranged that the lecture which they expected me to deliver should take place in Oparin's division.

Let me first try to reconstruct the background for my decision to lecture on a topic like metabolic basis for human hereditary enzyme defects. I was encouraged by many colleagues, among whom DeWitt (Hans) Stetten from NIH and Bentley Glass from Hopkins to talk about the work on hereditary galactosaemia in which we had just been involved. It might still be a bit adventurous to lecture about genetic diseases while Lysenko was still alive, although his reputation in the Krushchev era was fortunately going down, so why not?

I needed of course lantern slides in Russian, and for this purpose I needed help and got it in a particularly nice way.

Preparations for my lecture in Moscow

The text was first translated into Russian with the help of a Russian immigrant, Mrs. R. Ernsberger from NIH Medical Library. Since most of my text was biochemically oriented, additional professional help was desirable. This came from an American biochemist, my special friend Dr. Simon (Si) Black. Si had made outstanding contributions to basic biochemistry; he was codiscoverer of the reductive pathway of aspartyl phosphate to homoserine and also one of the early pioneers in the Lipmann research group in postulating adenylyl formation from ATP. His surprising knowledge of Russian, which he had developed well before his travels in the U.S.S.R. in 1957, was partly selftaught by subscription to *Isvestia.* He was particularly well received on the many traintrips (3rd class). When asked why he did not subscribe to *Pravda,* Si said with his dry wit, 'because it costs a penny more.' This created warm smiles and laughter, and created a congenial spirit, especially on the modest local trains out on the countryside. I wished that he might one day write about his solo trip in the U.S.S.R.

In any case, the Russian slides he handed to me were in various sizes, not only 2 by 2 inches but also 4 by 4.1/2 inches. The latter old-fashioned format of slides turned out to be the size which fitted in Oparin's projector.

It is common knowledge that during the Stalin years, Mendelian genetics was denounced by Trofim Lysenko, a propagandist who managed to get all professional geneticists in the U.S.S.R. silenced, often by exile in Siberia.

Mendelian genetics is of course also at the heart of biochemical genetics. How would my talk be received right now at the onset of the 'cold war' prompted by the U-2 incident?

I decided first to center my lecture on our own work on hereditary galactosaemia in humans (the prevailing type being the marked defect in the enzyme PGalUT; see page 144). Subsequently, I would then give a brief account of a hereditary human enzyme defect, linked to the X chromosome, described about the same time as ours from two hospitals, one at the university of Chicago and one at the Telomer hospital in Tel Aviv. This enzyme defect is that of G6P dehydrogenase in the red blood cells, a defect which manifests itself as haemolytic anaemia resulting from exposure to agents such as antimalarial drugs or to certain natural plant material such as the Fava bean. The first-mentioned is relatively common among black males; the second, 'Favism', is common among sephardic male Jews living in the Middle East including several southern provinces of the U.S.S.R., bordering Iran and Afghanistan.

Hereditary galactosaemia

My lecture in Moscow entitled *Nasledstvennaya Galacktotsemia* took place in a lecture hall in the Bach Institute. Oparin presided and Dr. V. Gorkin translated paragraph by paragraph into Russian. In this way almost two hours elapsed, in spite of the Russian slides which I duly presented. The audience included junior as well as senior biologists and biochemists (Wladimir Engelhardt, among the latter). The topic of inborn errors of metabolism in man seemed to interest the audience greatly.

I had arranged for a 20-min question period, if necessary. Five minutes turned out to suffice since no questions were raised, either philosophical or physiological! I then decided to give a brief summary about the X-linked Glc-6-P dehydrogenase defects.

The audience was surprised to learn about the high incidence in their own country of Favism (especially in oriental Jews), this sex-linked hereditary Glc-6-P dehydrogenase defect in red blood cells, promoting haemolytic anaemia.

Chairman Oparin expressed his thanks in a friendly speech in English as well as in Russian and he and Engelhardt hoped to hear more about biochemical genetics next year during the International Congress in Moscow. It is well-known now that this was the time when Marshall Nirenberg broke the genetic code by presenting the data for the code for biosynthesis of polyphenylalanine which he found to be UUU ('the U-3 incident,' as my friend Rollin Hotchkiss calls it), a more friendly event than the U-2 incident of 1960.

Returning to the U-2 incident for a moment, I joined the Moscovites together with staff members of the Danish embassy to take a look at the well-retrieved U-2 plane on exhibit in Gorky Park. The coat-of-arms of the Connecticut manufacturers 'Whitney & Pratt' was well-polished, as were Gary Powers' saucy magazines. Gary Powers, the pilot, apparently received only a brief prison sentence.

My accompanying student, Mike, was silent on this occasion, but expressed warm appreciation of my lecture. Human genetics was not taught at that time.

At that time, when I was in Moscow in 1960, I had the pleasure to meet the great biochemist and polymer chemist, Ephraim K. Katchalski, from Israel. This started a friendship which also resulted in several invitations to visit The Weizmann Institute of Sciences in Israel. I did not participate actively in the polymer symposium in Moscow. I learned, however, that this discipline was considered important in the U.S.S.R. and that during the Krushchev regime at the time, Jewish scientists were held in high esteem and several of them such as N.N. Semenov occupied leading positions.

Mike and I travelled by train to Leningrad (the night express was called 'Red Arrow' and was comfortable), but since most of the chemists were at the meeting in Moscow we met only a few of interest at the polymer institute. One of them was W. Volkenstein studying the biophysical aspects of ribonucleic acids.

The light nights in Leningrad encouraged us to long evening walks near the Neva. At the Hermitage there was an exhibition of the ancient Crimean Gold. Yet, for me the unforgettable sights were the French paintings, culminating with the magnificent painting by Henri Matisse 'Les Danseuses.'

From Leningrad a jet carried us non-stop to Kiev. At the airport in Leningrad we had met a young flutist from the Leningrad Philharmonic who was on his way to a summer job in Kiev. He did not speak English, but, like me, spoke a bit of German. We talked about Mozart and especially about Shostakowitch. I tried to hum a motiv from the so called 'Leningrad Symphony.' He said this demonic motiv symbolizes the Fascists who nearly killed all of us. The word 'Nazi's' was not used. Perhaps he felt that 'Sozialist' should not be used for the Hitler hordes. The young flutist had survived the siege of Leningrad. He had lost most of his teeth, at least the front ones, perhaps as a school child during the Leningrad siege. In any case, he seemed to go for a career as a flutist. As many other artists from Leningrad, Mozart was high on his list of great composers, not the least his flute concerto. He and Mike also taught me how to pronounce SPOTNIK: do not say 'Spotnik,' which means 'sweat' in Russian; say 'Spootnik.'

Kiev is an old city, with old churches, now museums. My scientific host was another of those old commizar-type Russians, Palladin, who did not speak English but some German. Much like with a local politician, several photoshops carried photographs of The Academician Palladin.

My main interest was not so much to meet the chairman, but V.A. Belitser, who in 1939 first described in detail a reliable system which generated oxidative phosphorylation. I later received permission from the USSR reprint office to have his 1939 article which Tsibakowa translated into English and reprinted in my monograph of 1969. My immediate problem in 1960, however, was to find Belitser, before or after my lecture, but before anybody took a photograph of me. I politely refused, therefore, to be photographed with the chairman Palladin. After a good deal of searching after my lecture, Belitser was finally found in the basement. He had not been informed about my visit, nor my lecture (an abbreviated 'German' version of my talk in Moscow), probably because his present work was in the field of blood coagulation (as I understand it, a continuation of his work assigned during the war years). I found Belitser very cordial and seemingly enjoying to meet me and to be photographed with me. We spent about half an hour together when most of the staff and audience had left. Palladin seemed pleased with the lucky search. After the usual thanks I left together with Mike Kritsky.

To Yerevan, the Armenian Akademi NAYK

Returning from Kiev to Moscow, Mike and I were planning our long trip to Yerevan, Armenia. It seemed to me that the poor Armenians were treated somewhat stepmotherly by the Muscovites, at least when it came to 'Aeroflot' traffic. From Leningrad to Kiev we were flown non-stop in a big jet. In contrast, no jet was available for the long journey from Moscow to Yerevan, supposedly via Kharkov, Rostov (at the Don river), Tbilisi, to Yerevan. This long route sounded bad enough, but it turned out to be a good deal worse than just a lot of stops. We were supposed to take off from Moscow by 7 pm. A lousy little 'DC-3'-looking propeller plane appeared around 7:30 with one of its engines seemingly on fire. It was fortunately decided not to board the plane in its present condition. After three to four hours waiting at the not too comfortable airport, another twin-engine plane appeared. It turned out to be the transporter to Yerevan, and to our dismay, it was the same little aircraft which seemed to be afire three hours earlier. Hopefully it had been fixed properly. I felt like a prisoner during the first two hours of the flight and I believed that Mike felt much the same. Additional complications ensued. The plane which was supposed to make a stop in Kharkov was denied landing there and had to land at Voronezh where we were sitting for several hours without permission to leave the plane. A mother with children ranging from 2 to 8 years had counted on reaching Kharkov, Mike told me. After an hour or so, I told Mike that he should have one of the pilots intervene with the authorities to get the mother and children to their destination. Mike first hesitated, but showed his genuine human qualities by asking for help for the little Kharkov family and getting action going. Our morale went up after this little rescue operation.

We finally reached Yerevan, via Tbilisi, next day in the afternoon. The hospitable Armenians, not being familiar with any of the details of our strenuous trip, had arranged that we attend one of the so-called historic Armenian 'operas' that same evening. In spite of some strong 'turkish' coffee, I am afraid that I slept in my seat through more than half of the performance.

Our scientific as well as cultural host was the delightful Prof. G.K. Bunyatyan, head of the Institute of Physiology of Akademi NAYK, Armenia. He spoke both English and German and had travelled in Great Britain as well as in the U.S.A. He invited me to stay in his house. Like Engelhardt, Bunyatyan was familiar with European literature and in his attractive home the book shelves were filled with classics in German and English. One of the starting points of his present research was Eugene Roberts' discovery of aminobutyric acid, which was properly quoted in Bunyatyan's papers, written in Russian, with Armenian and English summaries. His institute seemed well equipped. At that time they were interested in the stimulation of glucose uptake into skeletal muscle by γ-aminobutyric acid, an insulin-like effect which they found to be effectively inhibited by β-alanine.

At that time they were building profusely on the campus of the Armenian academy and it was good to see that they had abandoned the ugly 'Stalinist' architecture which dominated the Moscow University building. Instead, they used Armenian sandstone for the new buildings on the campus.

I also met Prof. Camelian, head of the Department of Pharmacology, and Prof. A.L. Mndzhoyan, Chemist and Vice President of the Armenian Academy of Sciences. The President of the Academy, Prof. V.A. Ambartsumyan, astronomer, one of the Foreign Associates of our own Academy, was at that time occupied at the Armenian Observatory which is located in the mountains.

Some social events in Armenia

Prof. Bunyatyan and his wife were kind enough to arrange a fine dinner party for me in their home. The party counted several scientists of that area. It was, of course, a genuine Armenian dinner.

I was introduced to the head of the Academy Museum of Historic Culture and was presented with a fine personal gift - an old clay vase found in an ancient fortress destroyed in the year 59 A.D. The vase, therefore, was considered older than that date. Later the ancient vase was dated by Prof. George M. Hanfman of Harvard University to 7–9 B.C., as a fine specimen of Urartu (Ararat) culture long since extinct.

I was taken on an excursion to the old fortress and to excavations at various locations in the beautiful countryside with a view of Mt. Ararat, now to the sorrow of my host on Turkish territory. A visit to a collective farm was improvised and the foreman kindly invited me to an elaborate lunch in his home; the food, as well as the cognac and wine, were Armenian products. The farms, which mainly raise silk worms, cotton, and grapes, used mostly manual women labour, and the labourers lived in very modest ancient-style, mainly clay, huts.

Yerevan is a city of 400 000 inhabitants. It has, however, a good deal of chemical industry (plastics, synthetic rubber). There are few tourists and transportation, even by air, remains relatively primitive as compared with facilities in Tbilisi and along the Black Sea coast.

The evening before leaving Yerevan, I stuffed the precious old Ararat water jug with newspaper, i.e., *Pravda*, and the next morning Mike and I were driven to the small Yerevan airport. It was breakfast time for the workers, and it was the custom to consume a couple of glasses of domestic cognac. My host, the professor, was surprised when I declined, although I had joined the customary cognac toasting the previous afternoon. After embracing my dear host, Prof. Bunyatyan, we boarded the ridiculously small twin-engine plane the long way back to Moscow. We were flying in the clouds which grew heavier when we approached Tbilisi, where we stopped. No chance of getting on a regular Aeroflot jet flight back to Moscow. Coming from Soviet Armenia we were treated by Aeroflot as if we came from a leper colony. Our little plane now flew under the clouds next to the Black Sea, which was black indeed, with twisters here and there. From there our little craft followed the rivers between the mountains. Mike looked very pale; I tried to cheer him up, joking about my cognac day at the foot of Mount Ararat.

Arriving in Moscow, I soon realized that I had lost my connection to Helsinki. As at prior occasions, Engelhardt rendered splendid hospitality. He also gave me a ticket to the next night gala ballet evening, in honour of the visiting President of India. Mr. Krushchev, I was told, does not care for ballet; instead, he usually sends a ceremonially looking guy with bushy eyebrows, named Breshnev, and true enough the guy was sitting in the imperial loge. At gala occasions the Bolshoi Ballet usually presents, 'Swan Lake' in a competent rendition, although I did not find it inspiring. Next day, Mike and I hugged each other farewell. I was on my way to my hometown, Copenhagen, and then back to Johns Hopkins University. Mike even ventured the idea of coming to Johns Hopkins with me. Returning there, I had, however, to review some new perspectives.

My future in the United States

In 1960 I was faced with some very interesting offers. The late Leo Szilard had encouraged me to join a small committee chaired by Jonas Salk, the 'architect' of the first polio vaccine, and including eminent molecular geneticists, like Seymour Benzer, Melvin Cohn, Renato Dulbecco, Matthew Meselson and others, planning an Institute of Molecular Biology in La Jolla, CA. I was in a way flattered to be invited to become a member of this group, but also worried being the only biochemist so far acceptable to the molecular geneticists. My old and happy affiliations with Max Delbrück seemed not sufficiently strong to convert me to an active molecular geneticist, as Jim Watson had come to realize already ten years earlier (cf. [19a]).

Although some of my closest friends had left Johns Hopkins, I felt happy there, except for my long commutating home to Barbara and the three children in Kensington, MD. This commutating in the long run became more and more exhausting. Barbara grew up at the Pacific Coast; having a home in the sporty La Jolla would make sense in this regard.

When I returned from my travels, a new attractive offer from Boston had arrived.

Biochemical research at Massachusetts General Hospital (MGH)

Let me try to narrate the events which made it possible for me to become a principal investigator and laboratory chief at MGH in Boston and, as we shall see, leave my 'nomadic' existence as a scientist to become a more stable Bostonian.

In 1960 I was approached by Dr. Paul Zamecnik, MGH, Harvard and Eugene P. Kennedy, Professor of Biological Chemistry, Harvard Medical School, whether I would consider a distinguished type of a 'hybrid' research position left open, since Fritz Lipmann, who had occupied this position for 17 years (and winning a Nobel Prize in 1954) had left for Rockefeller University, three years earlier.

This proposition sounded interesting to me. I came by night train from Washington to Boston.

What about Paul Zamecnik's offer? Barbara and I both felt that we should make a decision soon. The decision in favour of MGH was once more catalysed by Paul Zamecnik who had not overlooked Barbara's distinguished credentials in microbiology and offered her an independent research position too. This generous attitude made the new choice unbeatable and it took us less than a week to decide in favour of the MGH-Harvard position. I was also aware that the new research enterprise was planned on a broad basis.

I returned in early 1961 for some very interesting negotiations. The Biochemical Research Laboratory was revitalized by the following funding: the main capital supporting the new research lab came from The Wellcome Trust in London. This bold plan was initiated by Sir Henry Dale, one of the great pharmacologists of the century, and President of The Welcome Trust. His close and trusted friend was Dr. Walter Bauer, Professor of Medicine, Harvard Medical School and Chief of Medical Services, MGH.

My own position would be Professor of Biological Chemistry, Harvard Medical School and Chief, Biochemical Research Department, MGH. The main funding of my salary came from The Wellcome Trust in London, but Harvard University would pay 25% of my salary. The Dean of Harvard Medical School, George P. Berry and the other 'statesmen' now tried to put some of their wit together, tayloring in a proper way my official stationary. After a few months, the following 'gracious' type of 'collage' was agreed upon. On the left upper corner, my name and title, 'Professor of Biological Chemistry, Harvard Medical School', followed by Harvard's 'VE RI TAS, should appear. On the hopefully equally 'creative' right side of the stationary appeared first a portrait of the crisp and wise Indian in full alert with his arrows, the coat-of-arms of MGH. Then followed a discreet 'Henry S.Wellcome Biochemist'. This double seal turned out to solve my scientific career for 1/8 of a century.

I cannot list all the names of the extensive research and office staff except in certain contexts. Thanks to Paul Z.'s quiet but effective efforts, my office became very stately with panels of blond cherry wood. This setting was adorned by magnificent curtains with cellular organelles, a cherished gift from Freda Lipmann. When Christmas time approached we (Fran, Betsy and I) invited Dr. Zamecnik and his group to join us. As Santa Claus, I was supposed to read a story. Since Paul Z. and I shared our common memories about Kaj Linderstrøm-Lang of The Carlsberg Lab, my choice, apparently a successful one, was the humoresque: *The Thermodynamic Activity of The Male Housefly* (Muscus domesticus *L-L*) by F. FitzLoony and K. Linderstrøm-Lang'. The theoretical part comprised 'The Uncertainty Equation', and the 'appalling' idea of invoking Pauling's helices; this story has two faces, intriguing and witty at the same time.

A new type of microbiology, 'Ektobiology'

I have already mentioned (on page 149) that microbial genetics was pursued in our research lab at MGH by Henry Wu and Winfried Boos. However, I have not mentioned my involvement in another type of microbial physiology which was initiated in Japan, by Hiroshi Nikaido and Toshio Fukasawa in 1960. They described a *Salmonella typhimurium* mutant, unable to convert UDPG to UDPGal, which showed resistance to a bacterial virus (P-22), apparently because the outer cell wall of this mutant grown in media without addition of galactose, loses the receptors for this virus. I named this new type of approach 'Ektobiology'.

I invited Nikaido to join our new research lab and develop his own research shop. Shortly before his arrival, two researchers in my own group were able to describe another bacterial mutant of a related type this mutant was unable to synthesize UDPG. The two researchers were T. Sundararajan from New Delhi and Anette Rapin from Lausanne, Switzerland. Annette Rapin had joined us on the recommendation of her mentor, Prof. Roger Jeanloz, at MGH (and also from Switzerland). The UDPG-defective mutant did also show ektobiological changes; in this case it became sensitive to another phage, C21 [37,38].

During the same year a related case was reported by Kiyoshi Kurahashi and his coworkers (see Nikaido's review [38]). I used the homespun word 'Ektobiology' in an attempt to classify these types of events which depend on a variety of surface receptors. In his comprehensive review, Nikaido adds his interesting reflections about the use of the word 'Ektobiology' beyond events depending on specific virus receptors [38]. In any case, the terminal monosaccharide in the outer membrane of many Gram$^-$ bacteria seems to play a role for the recognition between the bacterial host and the bacterial virus [38].

Cellular regulation of hexose transport

At MGH I became very interested in tumour biology and hexose transport regulation. A strong factor stimulating me in this direction was Kurt Isselbacher's return to MGH from a year of sabbatical research at The Imperial Cancer Research Institute in London. During this year Kurt had clearly demonstrated that amino acid as well as hexose transport are greatly enhanced in mammalian cell cultures, transformed by oncogenic viruses [39]. Kurt's presence at MGH stimulated my active interest in this type of deregulation of hexose and amino acid transport in mammalian fibroblast cultures. Admittedly, my own contributions in this field got a slower start than those by Kurt or by Arthur Pardee.

Before continuing my report on some of our lab activities, it behooves me to report briefly about my personal life during the Boston years. My relations with Barbara and our children, Sonja, Nina and Niels were active and happy until the late Fifties. At that time my protracted and compulsory long-distance travelling, changed me too much; I became too much of an absentee. No wonder that Barbara turned more and more to her own skill and inventiveness, especially in water sports. After all, she grew up at the Pacific coast where she became a bold swimmer and diver. In contrast my participation in our weekly kayaking did not extend to white-water kayaking. I had, however, kept my enthusiasm for horseback riding, including hurdle jumping (occasionally getting myself into 'hairy' situations). I did not expose the kids to such kinds of situations. Admittedly, my 'Don Quixote'-like activities were kind of dead-end compared with Barbara's white-water Kayak lessons which I mostly attended from a 6-feet safe distance. My admiration of Barbara's skill as a mentor for children as well as for youngsters, slowly dwindled to a vicarious enjoyment from a distance.

Coming to Boston I met two old friends from The Carlsberg Lab in Copenhagen, Mahlon Hoagland and Paul Plesner and their guest, an exciting beautiful lady with a serene poise. She came from Copenhagen. Thirty years ago I had met and admired Agnete Fridericia. Agnete lost her husband, Svend Laursen in early 1960, but friends and colleagues more than ever respected and admired Agnete's courage in face of the agony and her ability to get through singlehandedly with her two young children. Agnete's and my cheerful conversations during the 1960s, mostly in Danish, changed my life again. After several years of indecision, I left Barbara and married Agnete in 1968. Agnete also became a friend of my lab and office workers.

In the early 1970s, I changed my research field again. A series of studies on chick fibroblast cultures by Harold Amos and his coworkers from the Department of Microbiology of Harvard Medical School, caught my attention. Amos and his associates had discovered that the mere withdrawal of glucose from normal chick fibroblast cultures, elicited a large and consistent enhancement of the hexose transport system. It became my good fortune that Dr. Amos as well as his excellent associate, Dr. C.W. (Bill) Christopher selected me as a mentor in transposing part of their basic programme to transport regulation in mammalian fibroblast cultures. I had the additional good fortune to add Donna B. Ullrey to our research group. Donna was very well trained in biochemistry by the cell biologist Donald F.H. Wallach. Moreover, Bill as well as Donna gathered additional experience from personal visits to the laboratory of Sen-itiroh Hakomori and his associates, and so did I.

Among our joint contributions in this new field, I am trying to single out what I feel as a few more personal touches. I was taking a liking to give new names to observations of unanticipated types of biological responses. For instance, in avoiding the name 'repression' of hexose transport which might imply participation of RNA or DNA, I chose a noncommittal name for the downregulation of the hexose transport system of cell cultures, a response which ensues after a protracted exposure to D-glucose or D-allose; I called it, 'the transport curb'. Since we also found that the transport curb can be released by inhibitors of oxidative phosphorylation, like dinitrophenol, oligomycin or malonate [40] we are trying to understand why a transport curb should require oxidative energy.

It seems indeed ironic that an event like a downregulation of the hexose transport systems in mammalian fibroblast cultures, should require oxidative energy.

My first realization of the requirement of respiration for phosphorylation and ATP formation in cell preparations goes back to my reading of an extensive and lucid article by Ejnar Lundsgaard in early 1932, in an obscure Danish medical journal. Eventually, I translated this article into English, with a few notes from Lundsgaard himself (see [41]). About the same time, i.e., 1932, W.A. Engelhardt described the stimulation of pyrophosphate formation in nucleated pigeon erythrocytes by addition of redox dyes in air, and later John Runnström in Sweden did related work.

All this brings me back to my early years with my great mentor, Fritz Lipmann, and his exciting discoveries.

My deep attachments to Copenhagen, 'The Athens of the North'

Although I was a capricious pupil in school and occasionally felt quite miserable, I had many interesting and rewarding memories from that old-fashioned school which, according to modern standards, especially for athletic activities, must be considered very modest. Its name was 'Østre Borgerdyd Skole', which meant literally: the school for civic merits in the Eastern (Østre) borough of Copenhagen. I suppose that 'merits' were meant to be applied to a larger part of the world than that quiet and attractive quarter of the Danish capital in which the school was located (the three Kalckar sons, of which I was the middle one, lived at easy walking distance from the school). The 'Athenian' flavour of 'Borgerdyd Skolen' could not help making itself felt. The retired rector (headmaster), J.L. Heiberg was a world-renowned scholar in Greek and a member of numerous academies.

Our physics teacher, H.C. Christiansen (a stubborn Jutlandian from the austere the heather-covered region of Jutland), was a formidable and passionate teacher especially of the Ørsted-Ampere rules. He had also written a fine concise textbook on physics for the gymnasium; calculus played a prominent part in this text. Admittedly, we were a little scared of him and his passionate temper, but never depressed. His temper would flare up if some of us tried to cut corners, by trying to memorize the text and formulas from his textbook. Hans Christian Christiansen insisted in no uncertain terms that physics, calculus and vector analysis can only be grasped by using pencil and paper over and over again, in order to grasp the right 'polarity' in the network of a scientific creative argument. The teaching went as far as Faraday and Maxwell. Einstein and Bohr were left for science teaching at Copenhagen University.

Here, my brother Fritz soon had the privilege to be associated with Niels Bohr and the enthusiastic spirit which pervaded young scholars from all corners of the world.

In my own case, my interest in biology and physiology was not awakened by our ordinary school teachers in these disciplines. But luckily the world-famous 'Jutlandian' zoophysiologist August Krogh, professor at the University of Copenhagen (and Nobel Laureate for his description of the blood capillary bed), brightened our belief in physiology by his stimulating school demonstrations of human physiology. To my knowledge he was the only physiologist in Copenhagen who took an active interest in introducing high-school pupils into the principles of human physiology. At these demonstrations, Krogh was accompanied by one of his young pupils, striding on an ergometer bicycle while breathing into a 'Krogh respirometer', consuming oxygen. Krogh was the original inventor of the microtonometer which rapidly registers the oxygen tension in microsamples of blood. This made it possible to detect deviations in the reaction of haemoglobin with oxygen at different pHs, or different carbon dioxide tensions. This effect is known as the 'Bohr effect', after Christian Bohr, the great mentor of the exquisite Danish school of physiology, which included not only Krogh but also Valdemar Henriques and his illustrious pupil, Ejnar Lundsgaard who directly inspired me to go into physiological research, as I mentioned at the very beginning of this memoir.

How did I fare as a medical student at the University of Copenhagen? Thanks to the teachings by Henriques and Lundsgaard, my interest was sufficiently well focussed to obtain an A in physiology and a B+ in Anatomy in spite of a pedantic old teacher who became notorious for trying to flunk students by the dozens. My grades in the clinical disciplines were more up and down. Erik Warburg, Professor of Clinical Medicine, did inspire me by his brilliant lectures. At that time, I also became aware of the pioneering work by Otto Warburg in Berlin and his 'magical' manometric methods for studying the cell physiology of mammalian and avian tumours. This passionate interest of mine arose already in 1927, only one year after the famous book, *Über den Stoffwechsel der Tumoren*, had appeared. This interest for Warburg's work must have been on my own initiative since it arose five years before Fritz Lipmann came to Copenhagen and Lundsgaard wrote his review about glycolysis in *Annual Review of Biochemistry*.

My home and family

In spite of my monomaniac traits, nobody else in my happy family showed this kind of disposition. The relatively modest Kalckar home did in fact create an oasis for me. My parents developed a sense of humanistic disciplines in me and my two brothers. Both insisted that we treat the Danish language with care and reverence. We belonged to a middle-class Danish-Jewish family, over several generations. We were largely free-thinkers, however, our mother did feel attached to some traditions in the Jewish faith and thought that it behoved us to learn a minimum of Hebrew to prepare us for Bar Mitzvah. My father found most of this rather boring, but he liked to read Danish-Yiddish humorists. All of us responded to Copenhagen humour, cartoons, limericks and especially to the theater. In a way, both my older brother, Emil, and my younger brother, Fritz, were more mature than I was. In their lovable way they coached me to smile a little more at the world.

My father was 'vintage 1860'; already as a young man he loved the theater. He attended the 'Urpremiere' of Ibsen's 'A Doll's House' at The Royal Theater in Copenhagen, November 1879, and wrote a delightful and enthusiastic report on the bold play. I still possess his spirited report as a little treasure. He continued to observe the fate of Ibsen's strong heroine women with an amazing foresight. Another Norwegian, Ludvig Holberg, who became the 'Danish Molière', became my father's hero, especially because of his immortal comedies written more than 200 years ago. Holberg, who never married, also became one of the most efficient chancellors of the University of Copenhagen.

My mother was especially fond of French literature (Proust, Flaubert and others). She spoke French well from her visits to Paris during the colourful 'nineties'.

Music life in Copenhagen at The Royal Theatre was well developed. Mozart's 'Don Giovanni' was actually first recorded at that theater, and Søren Kierkegaard had gone to this theater again and again before he wrote his hymn to Mozart. My father and Fritz loved the Mozart operas. Beethoven was my mother's main hero and also mine. I remember in 1927 that our fine 'Royal Opera Kapel' was very proud to receive, as conductor of two Beethoven symphonies, the legendary conductor, Karl Muck, the main builder of the elite Boston Symphony before World War I. Unlike the Bostonians, the Copenhageners disregarded Muck's old affiliations with the 'Kaisertum' and felt like the music circles in Boston that he was a profound musician. I was fortunate enough to hear Dr. Muck once more as conductor of Brahms' old orchestra, the Hamburg Philharmonic, in 1929.

From school I had two friends who managed to get me active, at least at the piano, playing four hands. When alone I preferred to play some of Bach's easier pieces composed for Anna Magdalena, his wife; some of these Gavottes became, after much rehearsing, my favourite pieces on the piano.

Denmark during and after World War I

Contrasting World War II, Denmark was rather lucky during and after the First World War, almost too lucky in terms of gluttony, be it in profiteering at the stock market, or in the wild export of milk products. My father was a genuine liberal. Although a businessman, by necessity rather than by choice, he carried on his profession as a consultant and broker most conscientiously. Investment savings interested him but he stubbornly refused to deal with the stock market. During these years, he could easily have earned 10 to 20 times more, like many other brokers. We admired him for his genuine unassuming modesty. He muttered in his low-key way, that he did not care to earn 'Kroner' in droves on all the ships sunk with loss of life. This did not mean that he felt in any way like distancing himself from friends and family who played on the stock market. Many of them had later to sell their Daimlers or other luxury cars when the stockmarket dipped.

Interest in the history of Denmark's relations with its neighbours, especially Germany, was vivid in our home. Germany with the Kaiser was not felt as a trustworthy neighbour and neither of my parents, as so many other Danes, liked him. My father had an extensive library on the history of Schleswig and Holstein, the disputed border provinces. In 1918–1920, the popular and nationalistic Danish King Christian X wanted both provinces for Denmark, as an additional humiliation to the beaten Germany. My Francophile mother fully supported these ideas of Christian the Tenth. In contrast, my father felt that restraint was more important than blind nationalistic emotions. He was all for the return of North Schleswig, but felt unable to make a judgment in regard to the southern provinces. Fortunately, Denmark was blessed with a statesman, Danish by birth and by sentiment, yet very familiar with the populations of both North and South Schleswig. His name was Hans Peter Hanssen-Nørremølle and he and his equally wise friend, H.V. Claussen, used their firsthand knowledge to approach people as human beings, regardless of whether they were from North or South Schleswig.

The verdict of this fair and patient enquiry was that Denmark should concentrate its efforts on North Schleswig, where the pro-Danes were in the majority, and not on South Schleswig, where the majority of the population remained pro-German.

The liberal Danish government at that time respected this verdict and bravely sacrificed their political future for a number of years. I believe that in general the liberal Danes at that time showed great and constructive foresight. Imagine the abysmal conditions in Denmark under the Nazi occupation, 1940–45 if it had annexed the Schleswig–Holstein provinces! As I have written, the real Nazi terror in Denmark started only three years after the invasion. I may be wrong in my analysis of these vectors; after all, why should the brave Norwegians be exposed over five years to Nazi terror? Both countries were lucky to have their staunch Kings at that time, Christian X and the heroic Haakon VII of Norway.

Let me close with a few personal experiences. After my medical exams, I was drafted into the King's Army as a medical officer. After a dreary army training, which I wished I could have dodged (and where I incidentally was exposed to the only antisemitic slur by a fat and stupid lieutenant), I hastened to apply to the Danish cavalry school. After some 'wild' riding lessons, I was transferred to a wholesome Jutlandic cavalry-artillery unit and the colonel there did apportion me a gentle and handsome horse, Nana by name. I had permission to trot and canter over the distant stretchers of heather-covered purple meadows as they look on a Danish September autumn day. There were but few medical problems with the riders. But the nice and friendly veterinarian had to try to heal scores of sores on the horses, once we started to ride across Jutland. Ten to twelve hours of riding were often the term of the day. I volunteered to try to help our Vet with some of these 'horsehold' problems.

Acknowledgements

In order to blow some life into this autobiography I was so lucky to have the help and inspiration from my old friend and colleague, Rollin D. Hotchkiss, Professor Emeritus, Rockefeller University. I first want to express my gratitude for his warm interest and creative comments. My coworker Donna B. Ullrey, M.A., for her great help not only in word processing but also for her interest and awareness during proofreading. Other members of the Boston University Chemistry Department were also very helpful, as was my wife Agnete Kalckar. Finally I want to thank Prof. Giorgio Semenza, ETH Zurich for his active interest and for his patience.

REFERENCES

1 H.M. Kalckar, Enzymologia 2 (1937) 47-52.
2 H.M. Kalckar, Nature 142 (1938) 871.
3 H.M. Kalckar, Biochem. J. 33 (1939) 631-641.
4 F.E. Hunter, in Phosphorus Metabolism, Vol. 1, The Johns Hopkins Press, Baltimore, 1951, pp. 297-330.
5 V.A. Belitser and E.T. Tsibakowa, Biokhimiya 4 (1939) 516-535 (translated into English by H.M. Kalckar; see [41] pp. 211-227).
6 H.M. Kalckar, Chem. Rev. 28 (1941) 71-178.
7 S.P. Colowick and H.M. Kalckar, J. Biol. Chem. 148 (1943) 117-126.
8 F. Lipmann, Adv. Enzymol. 1 (1941) 99-162.
9 H.M. Kalckar, J. Biol. Chem. 148 (1943) 127-137.
10 H.M. Kalckar, J. Dehlinger and A. Mehler, J. Biol. Chem. 154 (1944) 275-291.
11 W. Klein, Z. Physiol. Chem. 231 (1935) 125-148.
12 H.M. Kalckar, Fed. Proc. 4 (1945) 248-252.
13 O.H. Lowry and J.A. Lopez, J. Biol. Chem. 162 (1946) 421-428.
14 H.M. Kalckar, N.O. Kjeldgaard and H. Klenow, Biochim. Biophys. Acta 5 (1950) 575-585.
15 M. Friedkin, H.M. Kalckar and E. Hoff-Joergensen, J. Biol. Chem. 178 (1949) 527-528.
16 M. Friedkin and H.M. Kalckar, J. Biol. Chem. 184 (1950) 437-448.
17 M. Clark and H.M. Kalckar, J. Chem. Soc. (1950) 1029-1030.
18 H.M. Kalckar, Enzymatic Reactions in Purine Metabolism, The Harvey Lectures, Series 45, Thomas, Springfield, IL, 1952, pp. 11-39.
19 J.D. Watson and O. Maaloe, Biochim. Biophys. Acta 10 (1953) 432-442.
19a J.D. Watson, in J. Cairnes, G.S. Stent and J.D. Watson (Eds.), Phage and the Origin of Molecular Biology, Cold Spring Harbor Laboratory, Cold Spring Harbor, NY, 1966, p. 423.
20 H.M. Kalckar, M.S. MacNutt and E. Hoff-Joergensen, Biochem. J. 50 (1951) 387-400.
21 H. Klenow, Arch. Biochem. Biophys. 48 (1953) 186-200.
22 M. Saffran and E. Scarano, Nature 173 (1953) 249-253.
23 E. Scarano, Nature 172 (1953) 254-258.
24 A. Kornberg, in A. Kornberg, B.L. Horecker, L. Cornudella and J. Oro (Eds.), Reflections on Biochemistry, In Honour of Severo Ochoa, p. 249.
25 L.F. Leloir, The metabolism of hexosephosphates, in W.D. McElroy and B. Glass (Eds.), Symposium on Phosphate Compounds, Vol. 1, The Johns Hopkins Press, Baltimore, MD, 1951, pp. 67-93.
26 A. Munch-Petersen, H.M. Kalckar, E. Cutolo and E.E.B. Smith, Nature 172 (1953) 1036-1039.
27 H.M. Kalckar, B. Bragança and A. Munch-Petersen, Nature 172 (1953) 1039-1041.

28 H.M. Kalckar, E.P. Anderson and K.J. Isselbacher, Biochim. Biophys. Acta 20 (1956) 262–268.
29 E.S. Maxwell, J. Biol. Chem. 229 (1957) 139–151.
30 H.M. Kalckar, H. de Robichon-Szulmajster and K. Kurahashi, Galactose metabolism in mutants of man and microorganisms, in Proceedings International Symposium on Enzyme Chemistry, Maruzen, Tokyo, 1958, pp. 53–56.
30a H.M. Kalckar, Nature 182 (1958) 283–284.
30b L.Z. Reiss, Science 134 (1961) 1669–1673.
31 E. Jordan, M.B. Yarmolinsky and H.M. Kalckar, Proc. Natl. Acad. Sci. USA 48 (1962) 32–40.
32 W. Boos, Eur. J. Biochem. 10 (1969) 66–73.
33 W. Boos, in E. Haber (Ed.), The Cell Membrane, from Galactose Recapture to Galactose Chemoreceptors, Plenum, New York, 1983, pp. 101–107.
34 J. Adler, Sci. Amer. 234 (1978) 40–47.
35 F.G. Hopkins, 1936, quoted from [41] p. 317.
36 H.M. Kalckar, Science 174 (1971) 557–565.
37 T.A. Sundararajan, A.M. Rapin and H.M. Kalckar, Proc. Natl. Acad. Sci. USA 48 (1962) 2187–2193.
38 H. Nikaido, Bacterial cell wall, in B.D. Davis and L. Warren (Eds.), The Specificity of Cell Surfaces, Prentice Hall, Englewood Cliffs, NJ, 1962, pp. 67–71.
39 K.J. Isselbacher, Proc. Natl. Acad. Sci. USA 69 (1972) 585–589.
40 H.M. Kalckar, C.W. Christopher and D.B. Ullrey, Proc. Natl. Acad. Sci. USA 77 (1979) 5958–5961.
41 H.M. Kalckar, Biological Phosphorylations, Prentice Hall, Englewood Cliffs, NJ, 1969, pp. 344–354.

G. Semenza and R. Jaenicke (Eds.) Selected Topics in the History of Biochemistry: Personal Recollections, III.
(Comprehensive Biochemistry Vol. 37)

Chapter 5

Enzyme Regulation: from Allosteric Sites to Intracellular Behavior

ALBERTO SOLS

Instituto de Investigaciones Biomédicas C.S.I.C. and Departamento de Bioquímica, Facultad de Medicina U.A.M., Arzobispo Morcillo 4, 28029 Madrid (Spain)

Alberto Sols died unexpectedly while this autobiographical chapter was being written. We went thoroughly through his notes trying to put together in a logical order what he left. We hope that what we have collected may give a glimpse of the personality and scientific work of A. Sols. Unfortunately, we have been unable to find anything concerning his establishment in Spain when back from Cori's laboratory in St. Louis, MO (U.S.A.). We have heard him, however, tell how he arrived in Spain. He had spent all his savings buying a Klett photometer and some reagents and, having used his last money to clear the customs, had nothing left to pay the shuttle from the ship to the harbour, not to mention the train ticket to Madrid. Very likely this part was not written. The work of Sols was of paramount importance for the creation of a high-quality Spanish biochemistry. His emphasis on excellence was constant and permeated other laboratories. No wonder that he was able to attract enthusiastic young people who respected him for his dedication and intellectual rigor. Indeed it would be difficult to find some established biochemist in Spain who has not been in one way or another influenced by Sols. Generous with his ideas he was always ready to give advice on how to reorient a research or how to stress important points in a paper. The creation of the Spanish Society of Biochemistry was mainly due to his enthusiasm and the standards of quality he set for it have guided other Spanish scientific societies.

Some of his enzymological work has become classical and his accomplishments in this field brought him a wide international audience. Those of us who were his students and remained his friends will miss him deeply.

C. GANCEDO
J.J. ARAGÓN

Plate 7. Alberto Sols.

Facts to be of my service must be for or against some view.
Charles Darwin

Allosteric . . . proteins . . . as molecular receivers and transducers of chemical signals [are] the most characteristic and essential components of cellular control systems.
Jacques Monod, 1963

As a biochemist my life started in Carl Cori's laboratory in St. Louis, rather late in chronological age: I was already 34 years old. Why I started so late in what is both my vocation and my profession is briefly told in the following paragraph.

I was born in Sax (Alicante, Span) in 1917. My early education was in Sax primary school and with my father, who had a rather broad culture, including sciences; it was my father who early instilled in me the first seeds of curiosity for nature's problems. My secondary education was in Valencia, first in the Jesuits' 'San José' College and later in the public institute 'Luis Vives' while being a resident in the Colegio Mayor 'Ribera'. At the Institute I was fortunate in having an enthusiastic professor of natural sciences, Antimo Boscá; with him I was enthralled by the beauty and the mysteries of the workings of life; at the same time I enjoyed the opportunity of playing a little in a chemical laboratory at the Ribera College. Afterwards, having decided to go for a career in biology, I entered Medical School, as the best approximation available at the time. With the three years' parenthesis of the civil war in Spain I obtained an M.D. from the 'Complutense' University of Madrid in 1943. Biochemistry was still in its infancy as a science, and its teaching was very poor in Spanish medical schools, with the most unfortunate consequence of provoking in me a profound distaste for what much later I came slowly to realize was my true vocation with biological research. This slow awakening took place during my postgraduate studies in Barcelona as I will tell in the following section.

Physiological prelude in Barcelona

For graduate work towards a Ph.D. in Medical Sciences I went to the Institute of Physiology in Barcelona Medical School, where I worked on the intestinal absorption of sugars with Francisco Ponz, who had spent some time in Switzerland with L. Laszt, a former student of F. Verzár who had postulated the theory that glucose and galactose were actively absorbed because they were phosphorylated in the intestine. Within the poor intellectual framework of the metabolic enzymology of the time - mid 1940s - I studied the alkaline phosphatase of the intestinal mucosa trying to correlate it with the absorption of glucose. And I found what was expected: an apparent positive correlation. (I'll come back to it later.) As a by-product we developed a new method for the study of intestinal absorption based on the infusion of intestinal segments in vivo, which fulfilled my wish of making comparisons by successive observations in a single animal. The method [1] was used for several decades, and was described in detail in the *Handbook of Physiology of the American Physiological Society* (Vol. 3, 1968, pp. 1193–1194). I presented this method for successive absorptions at the International Congress of Physiology held in Oxford in 1947; this was my first experience outside Spain. As a complement to this work, later, in 1950, I spent three months at Oxford, in the Department of Physiology, visiting R.B. ('David') Fisher and Donald Parsons, who had developed a more elaborate method for the study of intestinal absorption in perfused isolated segments of intestine; with a side trip to Birmingham to visit Alastair C. Fraser, who was studying the intestinal absorption of lipids, and a visit to Laszt in Fribourg (Switzerland) on my way back to Spain.

Meanwhile I had started a professional career in clinical chemistry, to make a living (there were no fellowships in Spain at the time), did some work on phosphatases mixing a little medicine and a little physiology, and invented an original method of 'two-standards colorimetry', whose potential usefulness was killed by the abandoning of colorimeters which were then substituted by photometers. Finally, I developed a very simple method for the determination of cholesterol in serum [2] that was used for several decades in many laboratories; in work towards a method to estimate free cholesterol I was able to obtain a linear relationship between the amount of saponin and the amount of haemolysis [3]. I did develop and publish some other analytical methods, but they were less successful or not at all.

Meanwhile, I became part-time assistant professor of Physiological Chemistry at Barcelona Medical School. It was a very boring experience because of my ignorance and the lack of scientific atmosphere in Barcelona. Eventually, in 1950, I decided to leave everything I had in Barcelona (my 'burning of the ships'), leaving behind for good both the Physiology assistant professorship and the Clinical Chemistry practice to go and learn about modern metabolic enzymology with Carl F. Cori in St. Louis, MO (U.S.A.)*

In Carl Cori's laboratory: St. Louis 1951–1953

In 1950 I wrote a long letter to Carl Cori asking to be admitted to his laboratory to learn about the then flourishing enzymology and to fulfill my ambition of discovering the real enzymatic basis of the active absorption of glucose. The laboratory of Carl and Gerty Cori was at the time the mecca of metabolic enzymology, and Carl had begun his scientific career working on the intestinal absorption of glucose. Once I had his admission I obtained a fellowship and after bypassing difficulties concerning visa to enter the U.S.A., I arrived at St. Louis in early September, 1951. In my first contact with Cori I reminded him of my aim to unravel the enzymatic basis of glucose absorption, but, sensing that he did not warm up to the project, I inquired if he was not interested in it. He answered: 'No, but you can work on it here if you wish'. And I immediately took a critical decision, telling him that the really important thing for me was to learn from him about enzymes, and that for this I preferred to work on any project of his current interest, adding that, if I really learned about enzymes, I could work later on my intestinal enzymes pet

* For a biographical chapter on Carl and Gerty Cori, see [41].

project. Carl then proposed that I joined another postdoctoral, Robert (Bob) K. Crane*, just arrived from Fritz Lipmann's laboratory, in studying the queer strong inhibition by glucose-6-phosphate (glucose-6-P) of brain hexokinase, just reported by Weil-Malherbe [4]. We would work together in his personal laboratory. The immediate objective he proposed to us was to purify brain hexokinase up to a preparation free of enzymes of glucose-6-P metabolism, particularly phosphofructokinase. And curiously enough for me, he recommended that we use as starting material acetone powder extracts of brain as they had been shown to be a suitable preparation for the study of brain hexokinase by Severo Ochoa – my countryman – when he was in his laboratory a decade earlier.

Carl Cori's real interest was a consequence of his postulate several years before that hormones were likely to affect the activity of certain enzymes; and preliminary and as yet unconfirmed evidence from his laboratory suggested than insulin could activate hexokinase by overcoming some inhibition [5]. So when he learned of the observation of Weil-Malherbe in Great Britain on the strong inhibition of brain hexokinase by glucose-6-P he thought that there could be the key to his long-standing dear theory.

My close association with Bob Crane, sharing not only a lab but also a desk for two and a half years of full-time dedication to the project of both of us, except for his moderate teaching commitments as an Instructor, was a very good experience and quite profitable; each of us learned from the other, always under the constant interest and availability for advice of Carl Cori. Plus the added very important benefit of living in the bright and enthusiastic atmosphere of his department, with frequent contacts with people as good as Gerty Cori, Mildred Cohn, Earl Sutherland and others. It was indeed a decisive opportunity for me. After eight months I was granted a research fellowship from Washington University.

Crane and I began making active powder extracts from calf brains – a procedure as popular then as forgotten now – and fractionating them assaying both the hexokinase we were interested in and the

---

* For an autobiographical chapter on R.K. Crane, see [42].

phosphofructokinase we wanted to get rid of. After almost two months we still were unable to purify the hexokinase; we were first baffled and finally worried. Then, one day in the second week of November, I wondered and asked to Carl and other experienced people in the Department: how much hexokinase has a fresh brain? To my surprise nobody had any idea: the good material for assay was supposed to be the acetone powder extract. And so, on November 15 – my patron saint's day – I killed a rat, got its brain, homogenized it, kept a sample, centrifuged the rest, took a sample of the supernatant, and assayed hexokinase in both the homogenate and the supernatant. The homogenate had an activity some tenfold greater, per gram of original tissue, than that of the acetone powders! And the supernatant had little activity. Then I remembered that the sediment left in the cold room had two layers: a coarse one at the bottom and a layer of fine material on top. I removed the latter, dispersed it, and tested it, alone and together with the initial supernatant. The result was quite exciting: the top layer of the sediment had also rather little activity, but the mixture of sediment and supernatant gave several times more activity than the sum of both tested separately! The whole thing ended near midnight. It was a happy patron saint's day indeed. We had been on the wrong track, and now we had the key to our goal: acetone treatment was very bad for hexokinase, and differential centrifugation offered a simple way to separate hexokinase from the interfering enzymes that prevented the accumulation of the inhibitory glucose-6-P! From this finding we moved quickly to attempt to solubilize the hexokinase from the particles, already free of glucose phosphate isomerase and phosphofructokinase. In this we failed; but one of the treatments, with lipase, led to a severalfold increase in specific activity. And so we were able to submit a good abstract for the 1956 FASEB Meeting, before the deadline of December 31, and after nine versions: written or amended by Crane, myself and Carl Cori (three each!). We later published our first paper [6] centered on the association of brain hexokinase to particulate material, its partial purification and general kinetic properties. We reported that more than 90% of the hexokinase activity of brain homogenate is associated with the particulate fraction that sediments after moderate centrifugation;

from this material we were able to obtain an approx. 50-fold purified preparation, still particulate, free of interfering enzymes. Significant proportions of the hexokinase activity of an homogenate of other tissues examined, except erythrocytes, could also be easily sedimented. Thirty years later, after it had been identified that the particulate hexokinase was bound to mitochondria, and that within physiological conditions there were changes on the bound and free enzyme, brain hexokinase was proposed by John E. Wilson as the prototype for his proposed concept of 'ambiquitous' enzymes [7]. Meanwhile a curious, but unproven (and unlikely) 'hexokinase acceptor theory' of insulin action, with the later interlinking hexokinase with mitochondria was proposed by Samuel P. Bessmann [8].

During the next one and a half years we carried out a systematic characterization of the substrate specificity of brain hexokinase, probably the most systematic one carried out with any enzyme so far, introducing what we called the 'phosphorylation coefficient', later generalized by others to k_{cat}/K_m. We discovered that brain hexokinase could phosphorylate quite a variety of sugars, some easily, some poorly. Among the practical consequences of our specificity work on hexokinase and on enzymes acting on glucose-6-P it deserves mention that our conclusion that 'the use of 2-deoxyglucose isolates the hexokinase reaction' [9], served as the basis for the elegant methods developed by Louis Sokoloff at the NIH for the study of energy metabolism in brain of intact animals first, and later, using the analogue 2-deoxy-2-fluoroglucose, also in humans by positron emission tomography (PET), a method of considerable diagnostic importance. And we compared the sugar substrate specificity with a similarly systematic study of the effect of hexosephosphates as potential inhibitors of the enzyme, in the noncompetitive way peculiar to glucose-6-P. This study included not only phosphorylated products of the hexokinase reaction itself on a variety of substrates, but also other potential analogues whose unphosphorylated precursors were not phosphorylatable by hexokinase itself, or only with very small affinity, including the L-sorbose-l-phosphate found to be a strong inhibitor by Henry Lardy, who pointed out its close structural similarity to glucose-6-P [10]. We were lucky, very

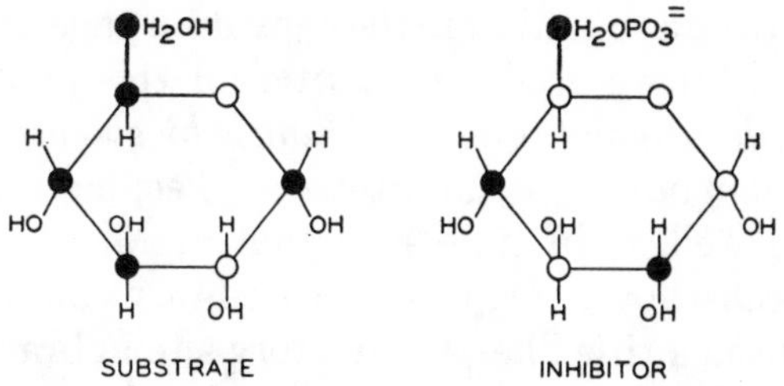

Fig. 4. Comparison of the substrate and inhibitor specificities of brain hexokinase. The Haworth formulas for glucose and glucose-6-P are drawn with each atom of the ring and carbon 6 represented by a circle. The solid circle is used if the evidence indicates that the hydroxyl group on that atom is an attaching group in the enzyme-substrate or enzyme-inhibitor complex or that the glucose configuration at that carbon is essential; the open circle if it does not. The importance of the pyranosidic oxygen atom is unknown. [11]

lucky: the specificity for noncompetitive inhibition by hexosephosphate was dramatically different from that for free hexoses as substrates. What led to the last and most important paper in the series of four in the *Journal of Biological Chemistry*: the one reporting the discovery that 'brain hexokinase possesses, in addition to the binding sites for substrates and adenosine triphosphate, *a third specific site for glucose-6-phosphate*', as '*part of an intrinsic cellular mechanism for the control of the hexokinase reaction*', postulating that 'the mechanism of inhibition must involve, after formation of the ester, the release of the ester from its site of formation followed by a reattachment to the inhibitor site where its presence interferes with the transfer of phosphate to the substrate' [11] (italics added). The specificities for substrate and inhibitor are graphically summarized in Fig. 4, taken from the 1954 publication. The picture is quite valid today; the only important addition was the finding, many years later, that 5-thioglucose-6-phosphate is a much more powerful allosteric inhibitor than glucose-6-P, with a K_i of about 1 μM, although 5-thioglucose is a very poor substrate [12, 13].

We had thus fully accomplished our main goal, with some significant additional findings along the way. But we – myself and Bob Crane – also made four important mistakes, namely: (1) we did not ask Carl Cori to be a co-author in the regulation paper, as I believe he fully deserved; (2) we used a poor title for this paper, *The non-competitive inhibition of brain hexokinase by glucose-6-phosphate and related compounds*, as unfortunate - keeping the distance in magnitude as well as in time - as that of Mendel for his first discovery of rules governing certain hereditary characteristics; we should have used a title like 'A regulatory site in brain hexokinase for specific inhibition by glucose-6-phosphate'; (3) the discussion was too short, rather timid, and on the whole unexciting; and (4) both Crane and I abandoned the subject of specific regulation of enzyme activity, Crane definitely, and I temporarily. If one adds to these mistakes in the presentation of our discovery the confusing fact that glucose-6-P is the main physiological primary product of animal hexokinases (for a discussion of the concept of the 'glucose phosphorylation pathway' see [14]), it is not surprising that our discovery of the first specific - later called allosteric - regulatory site in an enzyme (or, in general, in a biologically active protein) passed largely unnoticed for well over a decade. Later on, to make things more difficult for us and for science, hexokinases are monomeric enzymes, and thus supposedly unfit for regulation according to the almost mythical belief, developed in the 1960s, that oligomerism as basis for regulatory enzymes is a virtual necessity (I shall come back to this later). Nevertheless, meanwhile some progress was made by Herbert Fromm's finding that the inhibition by glucose-6-P was competitive with respect to ATP [15] (although Fromm did not believe at the time in an allosteric effect) and the demonstration by Irwin (Ernie) A. Rose that there was an inverse correlation between glucose phosphorylation in erythrocytes and the level of glucose-6-P [16].

In telling the story of the discovery of the first regulatory site in an enzyme I wish to mention two antecedents. Zacharias Dische published in 1941 a paper on the possible 'régulation automatique' of different glycolytic enzymes by mono- and particularly di-phosphoglyceric acid [17]; the approach was very crude, with the limitations of the time and circumstances (in France during the Second World War), but the results pointed to inhibition of glucose phosphorylation, leading Dische to forward the concept later known as feedback inhibition, as he insisted in an assay published in 1976 [18]. Nevertheless, the possibility of hexokinase being inhibited by 2,3-bisphosphoglycerate, although confirmed by several labortories including mine, cannot be physiologically significant because it is unspecific (all other kinases examined, including yeast hexokinase, can also be inhibited) and because the inhibitory concentrations far exceed the levels of 2,3-bisphosphoglycerate not bound to haemoglobin. In a letter written in 1975 Dische told me the story of a hypothesis of him to account for the inhibition through interaction of a phosphoglycerate-ATP complex with the hexokinase, but that the editor (Jean Roche) asked for its removal from the discussion because it was too speculative. On the whole, I have no doubt that Dische was a precursor, the first known precursor, of the concept of feed-back regulation.

The second antecedent is the well-known activation of phosphorylase *b* by AMP discovered in Cori's laboratory in the 1940s. Within the framework of the enzymology of the time it was thought to be related to a prosthetic group required by the enzyme, presumably present in phosphorylase *a* but lost in phosphorylase *b*.

During the 1980s there has been full confirmation of the physical reality of the allosteric site of glucose-6-P on hexokinases of higher animals, from my own and other laboratories. With John E. Wilson and Pedro A. Lazo I produced evidence that 'brain hexokinase has two spatially discrete sites for binding of glucose-6-phosphate' [19]. With Tito Ureta and Lazo I published a kinetic study of the reverse reaction in which brain hexokinase (but not yeast hexokinase) is strongly inhibited by excess substrate when it is glucose-6-P, but not when it is 2-deoxyglucose-6-phosphate [20]. And very recently, Wilson has obtained the complete amino acid sequence of rat brain hexokinase, giving direct support for the hypothesis that the 100-kDa mammalian hexokinases evolved by a process of duplication and fusion of a gene encoding an ancestral hexokinase similar to the yeast enzyme of 50 kDa; with the remaining active site in the C-terminal region and an evolved regulatory site in the N-terminal region [21], which in addition can be partially blocked by certain monoclonal antibodies (J.E. Wilson, personal communication). Wilson's sequencing was paralleled in part for human hexokinase,

Moreover, Tito Ureta and myself, in a transatlantic and interhemispheric collaboration (Spain–Chile) – with exchanges of visits of each of us to the other's laboratory – have obtained preliminary evidence that in the evolutionary origin of the allosteric site in animal hexokinases, the transformation of an active site into a regulatory site has happened independently at least twice, one of them within part of the vast world of the insects, leading to allosteric sites that while having in common the binding of glucose-6-P as allosteric effector, are markedly different in the structural requirements for efficient analogues.

Going back to Carl Cori's hope of identifying an activation of hexokinase by insulin, I feel it is appropriate to tell here the last chapter of a story that went on for half a dozen years. Early in 1953, after Crane and myself had purified brain hexokinase, Joe Bornstein, an Australian on sabbatical in St. Louis, and Mike Krahl, the expert in hormones in Cori's Department, proposed to Carl to try insulin on the purified enzyme. He was of course eagerly interested. Crane and I supplied the enzyme and gave advice on how to handle and assay it. And they began obtaining positive results! For several days there was expectation in the afternoon on the results of the experiment started each morning. Carl Cori planned a draft of a communication with the positive finding, but said that nothing would be published unless Crane and I confirmed the much cherished positive results of Bornstein and Krahl. We discussed weaknesses because of deficiencies in the handling of the enzyme. And finally it was aggreed that Crane would repeat exactly their experimental protocol, while I would do it with all the appropriate controls. Crane's result gave a trace of possible activation. When I was going to read at the photometer the results of my critical experiment, both Carl and Gerty and some other people came to our lab. And while I was getting the photometer ready Carl anxiously had a look at the coloured tubes and said 'I think there is an activation'; then Gerty took the rack from Carl, had an intent look, and said 'I don't think there is any effect'. A few minutes later I completed the photometer readings: no effect at all. To my knowledge this was the end of the affair for Carl Cori: I believe that right then he abandoned for good his long-cherished hypothesis. He never mentioned it to me, neither during the rest of my stay in St. Louis nor in later years, when we became very closely attached personally to each other, up to an almost father–son relationship, particularly during his last decade.*

....

....

More on sugar metabolism

Wide curiosity without strong ambition easily leads to too much dispersion. During much of my scientific career I have investigated in extension rather than in depth. On retrospect, I regret it. The deep enjoyment of reaching a real bottom far surpasses the combined minor enjoyments of skimming a variety of surfaces. Seldom have I aimed for big game. An even when pursuing ordinary prey, I often got distracted shooting at any small fry that came across my way.

Before going back to the pursuit of my major long-term enterprise of physiological regulation of enzyme activity I will just mention some minor miscellaneous achievements.

The accomplishment of many an enzymologist's dream: the discovery of my first enzyme, the first truly specific fructokinase (EC 2.7.1.4) in 1956, found by chance in peas in collaboration with a young plant biochemist, Antonia M. Medina, who came to my laboratory to learn something about enzymes [23]. This enzyme is highly specific for fructose giving fructose-6-P, in contrast with the

---

* Cori's interest in science never abated, as shown in the last letter (June 7, 1984) which I received shortly before he passed away. He wrote:

'Yesterday came your card with lovely wishes and signed by friends I value most. Included were also two photos of just about a year ago when we met in Madrid for Ochoa's lecture at the Universidad Autónoma. I see myself still in very good shape a year ago, but now I am only a shadow and still have to use oxygen most of the time. No matter, one must bear it. Actually I am quite cheerful, read the scientific literature which amazes me every day anew and have visits from a lot of our friends.'

liver enzyme, which is really a ketohexokinase (EC 2.7.1.3). We had no hypothesis whatever on its significance in peas, where we found also a common hexokinase. And to my knowledge, 30 years later, the physiological significance of plant fructokinase is still unknown. I found this enzyme by serendipity. What makes me remember a comment by Carl Cori, during lunch in his laboratory, in answer to a young fellow who asked for advice on where one could find a new enzyme; Carl shrugged his shoulders and said simply: 'new enzymes, like oil, are where you find them'. Well, this was the case for my first new enzyme.

To compensate the load for the biochemists by the discovery of a new enzyme, I soon erased one from the literature: with Carlos Villar-Palasí, then a postdoctoral fellow in my laboratory we identified the muscle 'fructokinase' of the literature as merely a marginal activity of phosphofructokinase [24]. Later on, with Claudio F. Heredia, I discovered a new enzyme with unknown biological significance: a phosphatase apparently specific for 2-deoxyglucose-6-phosphate that enables yeast mutants to become resistant to the toxicity of 2-deoxyglucose [25]. Understandably, the Enzyme Commission has refused to recognize this new enzyme. In later work Heredia has confirmed the high specificity of the enzyme, although it is also able to hydrolyse fructose-1-phosphate. Presumably it is an enzyme for something else, normally present in small amounts, but its gene can easily be hyperexpressed in mutants to which it confers resistance to deoxyglucose.

When Albert Keston invented the glucose oxidase-peroxidase colorimetric method for the specific estimation of glucose in biological samples, I decided that the popular belief that glucose oxidase was very specific had a large but not firm basis. And in a few week's work with Gertrudis de la Fuente we found two additional good substrates, 2-deoxyglucose and glucosone [26], of which the former became of great interest for certain experimental purposes.

Direct utilization of glucose-1-P had been claimed for muscle by E.B. Chain and co-workers, in 1954, and possibly for yeast. We were able to demonstrate that both claims were in error, while confirming the experimental observations but showing the artefact involved in the first case: escaped phosphoglucomutase that converted acid-labile glucose-1-P into acid-stable glucose-6-P [27], and by the finding that the ectophosphatase in yeast is repressible by phosphate and is essential for the utilization of glucose-1-P as carbon and energy source [28].

A long time after the entomologist Karl von Frisch - of bee-language fame - unexpectedly found that mannose was strongly toxic to bees, I discovered its enzymatic basis as a deficiency in mannosephosphate isomerase with a normal hexokinase [29]; the presumed mechanism: mannose-6-P accumulation with concomitant ATP depletion was later confirmed [30].

In the course of an attempt to eliminate glucosephosphate isomerase as a potential regulatory step in glycolysis I opened, with my students Margarita Salas and Eladio Viñuela, the field of the spontaneous and enzymatically catalysed anomerization of glucose-6-P and of the anomeric specificity of related enzymes [31]. This opening led to useful work in several laboratories.

In an attempt to extend the induced fit effect found in yeast hexokinase to other enzymes I failed my objective but found that arsenate could enable a variety of enzymes that have phosphorylated compounds as substrate to act also on the corresponding nonphosphorylated compound [32]. This was an amusing and unexpectedly profitable experience, that led to a series of studies by Rosario Lagunas in Madrid and by others in other laboratories on arsenate esters in a number of enzymological contexts.

As part of the doctoral dissertation of Carlos Gancedo we unraveled the structural requirements for substrate of two key enzymes of glycerol metabolism, glycerol kinase and the highly specific glycerophosphatase [33].

Another particularly successful thesis under my direction was that of Ignacio V. Sandoval. Contrary to conclusions based on isotopic tracers, we proved unequivocally at the enzyme level that in rat liver the major pathway of serine as a gluconeogenic precursor goes by way of serine dehydratase [34], an enzyme studied in my laboratory by Angel Pestaña. In addition to the ingenuity and efficient work of Sandoval we benefited from the advice of Sorab P. Mistry, while on a sabbatical in my laboratory, with his expertise on the biosynthesis of pyruvate carboxylase.

To finish this brief account of miscellaneous work not primarily related to regulation, I have to mention the invention of a non-invasive method for the assay of intestinal lactase applicable to infants and already tested in suckling rats. It is based on the synthesis of 3-methyllactose that, when administered orally, leads to the appearance of 3-methylglucose in urine [35].

....

....

Allosteric regulation of phosphofructokinase, fructosebisphosphatases and pyruvate kinases

Several years after the discovery of a specific mechanism involving a regulatory site in animal hexokinases I became involved through the sixties with three other regulatory enzymes.

That phosphofructokinase was likely to be a regulated enzyme was beginning to be suspected by some in the late 1950s. Its queer behaviour could be related to some regulatory mechanism. Already in 1951, when Bob Crane and I wanted to separate phosphofructokinase from hexokinase I, fully ignorant about the former, asked around in Cori's laboratory and got the rather unhelpful information: 'it is an *unreliable* enzyme'. In 1962 I decided to study the possibilities of regulation of yeast phosphofructokinase, which I proposed, as subject for his Ph. D. thesis, to a bright graduate student: Eladio Viñuela. We had considerable difficulty in just assaying it; fearing that phosphatases could destroy the substrate ATP we added more and more of it without success, until we realized that the increase in ATP concentration quite surprisingly made things worse. Right then, in 1962, appeared two papers on phosphofructokinase from animal tissues, one by Tag Mansour [36] and the other by Oliver Lowry [37], briefly reporting that they were affected by several metabolites, including inhibition by excess ATP. We soon made it clear that it was not excess substrate: it was specific inhibition by ATP, not occurring with another good substrate, such as GTP. In 1963 we published the feedback inhibition of yeast phosphofructokinase by the end product of glycolysis, ATP. From allosteric inhibition by a primary product (hexokinase and glucose-6-P) I had come across to my second major regulatory finding, an allosteric inhibition by a substrate! At that time, during a visit to the U.S.A. I was a guest of Carl and Anne Cori in St. Louis and reported this work at the Biochemistry (Cori)-Pharmacology (Lowry) seminar. After my seminar, while Carl and I were leaving for his home, he muttered to himself: 'it looks like a *neurotic* enzyme'. To which we joked: 'may be it is an *all-hysteric* enzyme!'

Some time later, I was able, with my students Marisa and José Salas, to desensitize yeast phosphofructokinase to the specific ATP inhibition by brief treatment with trypsin; this fact, and the differential protection against desensitization by various adenine nucleotides led to the conclusion that there should be in this enzyme one inhibitory site for ATP and a different site for AMP, able to counteract the inhibition by ATP. All this was just the beginning of the work in my laboratory on the physiological regulation of phosphofructokinase. (I shall come back to it later on.)

....

....

Multimodulation of enzyme activity

In the winter of 1978 I arrived in Santiago, Chile, to participate in an advanced course on Enzyme Regulation for Latin American young scientists, organized by Tito Ureta. On arrival I met - we were going to stay in the same appartment - Dan Atkinson and Joe Katz, coming from Los Angeles for the same purpose. Dan handed me a letter inviting me to lecture in a UCLA Symposium on Enzyme Regulation he was planning to have in Keystone, CO, up in the Rockies. The title for the Symposium was 'Covalent and Non-covalent Modulation of Protein Function'. When I read the letter I immediately decided that I would not accept, mainly because I had nothing interesting to report since the last of the periodical Symposia on Metabolic Interconversion of Enzymes; but since I dislike to say no to a good friend, I decided to delay the answer, at least until the following day. The following day, July 20, at about 5 a.m. I awoke with an internal pressure in my mind: it was a concept, in one word, coming from nowhere: 'multimodulation'! And I quickly wrote the outline of a lecture considering facts and problems of enzymes that have *several* regulatory mechanisms, not infrequently *both covalent and noncovalent*. I presume the idea was cooked in my subconscious, out of my long first-hand experience with the multiply regulated phosphofructokinases and certain pyruvate kinases, plus my second-hand long experience concerning the complex regulation of glycogen phosphorylase. By the time we met for breakfast I had changed my mind and told Dan that I would gladly participate in his symposium speaking on 'Multimodulation of enzyme activity'.

Back home in Madrid a month later I had to do considerable homework on the subject: reading, pondering, and eventually reaching conclusions. Finally, I wrote in a hurry the text of the lecture that I had been asked to bring along to Keystone for publication in a UCLA volume devoted to this Symposium. The Meeting, February 25 to March 2, was very good, with plenty of opportunities for informal personal discussions. Particularly interesting was a long talk with Dan Koshland, the two of us pondering the origin and biological significance of the oligomeric nature prevalent in intracellular enzymes, including many nonregulatory ones.

Some time afterwards, in July, Bernie Horecker happened to pass through Madrid and I told him about my article in press, poorly written since a complex subject had to be treated in a short time because of a deadline, but which I felt was basically good, and I gave him a preprint. A couple of days later he phoned me from New York:

> 'I read your manuscript in the plane home. I like it very much. Take your time to revise and complete it as you wanted and send the improved version to me: I'll publish it in the *Current Topics (in Cellular Regulation)*.'

And so it eventually came to appear in a form [38] much better than the 'preliminary' one (printed in the Symposium volume by Academic Press in 1979) because, in addition to my unhurried further work, the final text was carefully polished by Bernie as expert editor friend.

The conclusion was the unequivocal demonstration that multimodulation of enzymatic activity in metabolic regulation is a significant and widespread fact, even if frequently blurred by uncritical claims. This fact, and the variety of possible interactions among the different modulators for a given enzyme, has an heuristic value. Exploratory search on possible regulatory effectors for an enzyme should be carried out in assay conditions with good sensitivity for the detection of effects on the affinity for the substrate (by working below saturation, preferably at concentrations around or below the $S_{0.5}$) or of modulation by other effectors (by adding them at concentrations around their K_a or K_i). This particular condition should be superimposed on the general rule of avoiding any marked deviation from physiological conditions in pH, ionic environment (K^+!), temperature and, as far as possible, enzyme concentration. Finally, as a consequence of the fact that multimodulation is common (and sometimes quite complex) among regulatory enzymes, care should be taken to avoid dogmatic conclusions in trying to account for the metabolic regulation of an enzyme on the basis of one (or even several!) modulatory effect(s), even if it (they) is (are) likely to be of physiological significance. There is always the possibility of some additional as yet undiscovered modulatory effect that could be not only important, but even more important for metabolic regulation than what was already known. An outstanding case was the discovery of fructose-2,6-P_2 with its powerful activation of the multimodulated phosphofructokinase.

In my 1981 paper I speculated on the possibilities for the evolutionary origin of regulatory sites, particularly in multimodulated enzymes, considering the *why*, the *how* and the *when*. As for the how I emphasized that multimodulated enzymes – in contrast to multifunctional enzymes – have monomeric or subunit sizes not markedly greater than common enzymes, which led me to postulate the likely frequent occurrence of mutational development of a new specific site. Nevertheless, recent work has given strong evidence for gene duplication and fusion with posterior evolution of one of the active sites to a regulatory site as an important factor. I will emphasize that in the two cases thus identified, mammalian phosphofructokinase and hexokinase, the monomeric molecular weights are about twice the average molecular weight of the subunits of most enzymes [39].

Multimodulated enzymes are marvellous integrators of metabolic signals. They are the ultimate in the miniaturization of computers. Integration of signals in multimodulated enzymes is, on a microscale, an average in time of the fluctuating activity of each enzyme molecule, or, on a macroscale, the average activity of a population of enzyme molecules in a given compartment; the possibility of integration at the single molecule level is compounded in populations of molecules, where the multiplicity of identical molecules efficiently buffers the response to the variety of signals.

Criteria to authenticate regulatory mechanisms of modulation of enzyme activity can be outlined as follows:

(a) Allosteric effects. (1) must have *specificity* (among metabolites); (2) must be *genuine* (not due to a contaminant); (3) must have *appropriate affinity* (respect to physiological concentrations; (4) must have *significant magnitude*; and (5) must make *physiological sense* (even then it could be an analogue of a true physiological effector).

(b) Cooperative effects. (1) must have *significant magnitude* $n_H >$ 1.5; and (2) must make *physiological sense.*

(c) Covalent modification. (1) must significantly affect some *kinetic parameter* within the physiological range; (2) must be *physiologically reversible* (except for the irreversible activation of zymogens or specific proteolytic inactivation); and (3) its operation must be *related to metabolism.*

The appropriate requirements should be shown to be fulfilled by a new presumptive regulatory mechanism for it to be acceptable by the scientific community. In the absence of this evidence a claim could be accepted only as suggestive in original publications, should be identified as such if quoted in reviews, and should not be collected in textbooks.

Ultimate proof will include physical identification of the regulatory sites and observation of functional and, if possible, biological disturbances caused by spontaneous mutational change or directed mutagenesis or chemical specific modifications of the enzyme.

Multiphosphorylation is still a baffling problem. In addition to the probability of simple noise being a factor in many cases there is recent evidence that in certain extreme cases the significance is not linked to specific residues but rather to a large change in the charge of the molecule, as seems to be the case for eukaryotic RNA polymerase II which can be phosphorylated at dozens of residues [40].

I wish to emphasize that the hindrance introduced in the mid-1960s by the use of 'allosteric' to cover the supposedly linked allosteric effects and cooperativity, unfortunately still lingers, as can be seen in the proceedings of the Jacques Monod Conference: *Twenty-five Years of Allostery*, organized by the CNRS on March 1989.

It has been said that for every Sherlock Holmes there tends to be some 20 Lestrades messing up the data, and hence hindering the reality rather than contributing to its unraveling. So has happened in metabolic regulation in what could be referred to as the post-allosteric period with fancy speculations under labels such as 'biochemical systems theory', 'metabolic control theory', 'molecular democracy', or plain applications of 'Occam's razor to brush aside any wide significance of allosterisms'. I have taken expressions from the literature, but do not wish to give references or even quote names. Ever since the recognition of highly efficient specific regulatory mechanisms in certain enzymes, metabolic regulation at the enzyme level is aristocratic, not democratic. Regulatory enzymes are well fitted to be true pacemakers of metabolic pathways.*

....

....

* The symbols for graphical representation of enzyme regulation recommended by the CBN (see e.g., [43]) are: ⊖ for inhibition and ⊕ for stimulation; ⊟ for repression and ⊞ for induction; o $\rightleftharpoons$ m, for interconvertible enzymes. These symbols were adopted by an ad hoc subcommittee of Nomenclature of Interconvertible Enzyme that met in Seattle in 1976; the first four were proposed by myself, the last one by Earl Stadtman.

REFERENCES

1 A. Sols and F. Ponz, Rev. Esp. Fisiol., 3 (1947) 207-211.

2 A. Sols, Rev. Esp. Fisiol., 3 (1947) 225-241; Can. J. Med. Technol., 10 (1948) 163-165.

3 A. Sols, Nature, 164 (1949) 111.

4 H. Weil-Malherbe and A.D. Bone, Biochem. J., 49 (1951) 339-347.

5 C.F. Cori, Influence of hormones on enzymatic reactions, in Proceedings of the First International Congress of Biochemistry, 1949, pp. 9-18.

6 R.K. Crane and A. Sols, J. Biol. Chem., 203 (1953) 273.

7 J.E. Wilson, Curr. Top. Cell. Regul., 16 (1980) 1-44.

8 S.P. Bessmann, J. Pediatr. 56 (1960) 191-203; N. Pal and S.P. Bessmann, Biochem. Biophys. Res. Commun. 154 (1988) 450-454.

9 A. Sols and R.K. Crane, J. Biol. Chem., 210 (1954) 581-595.

10 H.A. Lardy, V.D. Wiebelhaus and K.M. Mann, J. Biol. Chem., 187 (1950) 325.

11 R.K. Crane and A. Sols, J. Biol. Chem., 210 (1954) 597-606.

12 E. Machado de Domenech and A. Sols, FEBS Lett., 119 (1980) 174-176.

13 J.E. Wilson and V. Chung, Arch. Biochem. Biophys., 269 (1989) 517-525.

14 A. Sols, in F. Dickens, P.J. Randle and W.J. Whelan (Eds.), Carbohydrate Metabolism and its Disorders, Academic Press, London, 1968, pp. 53-86.

15 H.J. Fromm and V. Zewe, J. Biol. Chem., 237 (1962) 1661-1667.

16 I.A. Rose and E.L. O'Connell, J. Biol. Chem. 239 (1964) 12-17.

17 Z. Dische, Bull. Soc. Chim. Biol., 23 (1941) 1140-1148.

18 Z. Dische, in A. Kornberg et al. (Eds.), Reflections on Biochemistry, Pergamon, Oxford, 1976, pp. 215-225.

19 P.A. Lazo, A. Sols and J.E. Wilson, J. Biol. Chem., 255 (1980) 7548-7551.

20 T. Ureta, P.A. Lazo and A. Sols, Arch. Biochem. Biophys., 239 (1985) 315-319.

21 D.A. Schwab and J.E. Wilson, Proc. Natl. Acad. Sci. USA, 86 (1989) 2563-2567.

22 S. Nishi, S. Seino and G.I. Bell, Biochem. Biophys. Res. Commun., 157 (1988) 937-943.

23 A.M. Medina and A. Sols, Biochim. Biophys. Acta, 19 (1956) 378-379.

24 C. Villar-Palasi and A. Sols, Bull. Soc. Chim. Biol. 39, suppl. II (1957) 71-75.

25 C.F. Heredia and A. Sols, Biochim. Biophys. Acta, 86 (1964) 224-228.

26 A. Sols and G. de la Fuente, Biochim. Biophys. Acta, 24 (1957) 206-207.

27 R. R-Candela, A. Sols, F. Alvarado, E. Santiago, C. Villar-Palasí and J.L. R-Candela, Bull. Soc. Chim. Biol. 39, suppl. II (1957) 71-75.

28 C.F. Heredia, F. Yen and A. Sols, Biochem. Biophys. Res. Commun., 10 (1963) 14-18.

29 A. Sols, E. Cadenas and F. Alvarado, Science, 131 (1960) 297-298.

30 M. de la Fuente, P.F. Peñas and A. Sols, Biochim. Biophys. Res. Commun., 140 (1986) 51-55.

31 M. Salas, E. Viñuela and A. Sols, J. Biol. Chem., 240 (1965) 561-568.

32 R. Lagunas and A. Sols, FEBS Lett., 1 (1968) 32-34.

33 C. Gancedo, J.M. Gancedo and A. Sols, Eur. J. Biochem., 5 (1968) 165–172.
34 Sandoval, I.V. and A. Sols, Eur. J. Biochem., 43 (1974) 609-616.
35 M. Martinez-Pardo, P.G. Montes, M. Martín-Lomas and A. Sols, FEBS Lett., 98 (1979) 99–102.
36 T.E. Mansour and J.M. Mansour, J. Biol. Chem., 237 (1962) 629.
37 J.V. Passonneau and O.H. Lowry, Biochem. Biophys. Res. Commun., 7 (1962) 10.
38 A. Sols, in Current Topics in Cellular Regulation, Vol. 19, Academic Press, New York, 1981, pp. 77–101.
39 A. Sols and C. Gancedo. Primary regulatory enzymes and related proteins, in E. Kun and S. Grisolía (Eds.), Biochemical Regulatory Mechanisms in Eukaryotic Cells, Wiley, New York, 1972, pp. 85–114.
40 L.J. Cisek and J.L. Corden, Nature, 339 (1989) 679–684.
41 H.M. Kalckar, in G. Semenza (Ed.), A History of Biochemistry, Personal Recollections. I, Comprehensive Biochemistry, Vol. 35, Elsevier Amsterdam, 1983, pp. 1–24.
42 R.K. Crane, in G. Semenza (Ed.), A History of Biochemistry, Personal Recollections. I, Comprehensive Biochemistry, Vol. 35, Elsevier, Amsterdam, 1983, pp. 43–69.
43 IUPAC-IUB Commission on Biochemical Nomenclature. J. Biol. Chem. 252 (1977) 5939-5941.

G. Semenza and R. Jaenicke (Eds.) Selected Topics in the History of Biochemistry: Personal Recollections, III.
(Comprehensive Biochemistry Vol. 37) © 1990 Elsevier Science Publishers BV (Biomedical Division)

Chapter 6

Jean Brachet (1909–1988)

H. CHANTRENNE

82, Chaussée de Tervuren, 1160 Brussels (Belgium)

In July 1988, Prof. Jean Brachet, invited to write a chapter on his scientific life for *Comprehensive Biochemistry*, had offered to translate into English a lecture he gave in November 1987 at the Belgian Royal Academy, and to complete it somewhat. But he died in August. The Editors asked me to do what Brachet had promised, an almost impossible task, indeed.

Let me first quote* the relevant part of the lecture [1] he was referring to:

> 'In 1927, as a young medical student, I decided to enter the laboratory of Embryology at the Brussels Faculty of Medicine. I had been attracted to research by the first lesson of Pol Gérard's histology course: he had told us that a fragment of cell devoid of a nucleus can survive for hours, or even days without losing its ability to move. This looked incredible to me and I wanted to find out why removing the nucleus does not kill the cell right away. P. Gérard advised me to work in the laboratory of Albert Brachet, my father. I could not bear to think of it. My father found a way out in directing me to Albert Dalcq (who eventually became his successor).
>
> After a few weeks learning techniques, I had to choose between two research subjects proposed by Dalcq. I decided to study the localization of 'thymonucleic acid' during oogenesis (i.e., the growth of the oocyte within the maternal organism). Thymonucleic acid was what we now call DNA. The lucky choice of this subject thus introduced the young student that I was to nucleic acids.

---

* Translated from the French.

Plate 8. Jean Brachet (1909–1988).

Oocyte chromosomes become very tenuous during oogenesis; it was then believed that they loose their nucleic acid at that moment; if so, thymonucleic acid would not be a constant constituent of chromosomes and could not have any genetic function; it was indeed assumed at that time that genes are proteins.

Detection of DNA in cells was made possible by a recently developed cytochemical method, the 'Feulgen reaction'. My work with this technique led to two conclusions: (1) DNA is a constant constituent of chromosomes and might therefore possibly play a role in heredity; (2) the cytoplasm of oocytes does not contain any storage of DNA detectable by the Feulgen reaction.

These conclusions led me to address, in 1931–1932 a related question: is there any DNA synthesis during egg cleavage, a period during which the fertilized egg divides into thousands of cells? The matter was controversial in spite of the little interest that most biochemists and biologists attached then to nucleic acids. Whereas Jacques Loeb thought that during development a complete synthesis of DNA occurs at the expense of simple constituents, such as phosphoric acid, sugars, purines and pyrimidines, the embryologist Godlewski believed that there is no net DNA synthesis and that DNA stored in the egg cytoplasm migrates to the nucleus during oogenesis. My own results with the Feulgen reaction were incompatible with the idea of migration; on the other hand, published data on phosphorus and purine determinations indicated that unfertilized eggs do contain a considerable store of nucleic acid material sufficient for providing nucleic acid during egg cleavage into numerous cells.

This contradiction between cytochemical and biochemical data led me to determine nucleic acids during the development of sea-urchin eggs; for that purpose, I spent two summers at the Zoological Station of Roscoff, in France. To my great surprise, it turned out that both contenders seemed to be right in some way: net synthesis of DNA indeed occurs, I was able to confirm it using a newly established method of deoxyribose determination, but the total amount of nucleic acid (as measured by purine assay) did not change to any significant extent.

Nothing is more stimulating for a research worker than to be trapped into contradiction. I had carefully read and reread the few pages concerning nucleic acids one could find in biochemistry textbooks of the time; it was stated in all of them that there are two kinds of nucleic acids, thymonucleic acid in animals and zymonucleic acid in plants and yeast. The former contains a strange unidentified sugar, the latter a pentose.

The only way that came to my mind to explain the contradiction I mentioned was to suppose that sea-urchin eggs contain a large amount of 'plant nucleic acid', i.e., pentose nucleic acid in their cytoplasm. Could a 22-year-old student dare to break the dogma accepted by all biochemists, namely that one kind of nucleic acid belongs to animals and the other to plants? Before being such an iconoclast, I had the wisdom of writing for advice to Joseph Needham, the pope of 'chemical embryology'. My assumption looked so bold to him that he asked the opinion of his boss, F.G. Hopkins; here is the advice 'Hoppy' gave: 'Tell that young man he should not believe all that is in the books, they are full of errors. Let him do experiments to check his interpretation'.

And that I did; right after passing my summer examinations I returned to Roscoff, taking with me an apparatus made for the determination of pentoses in straw; my French friends, Ephrussi, Lwoff and Monod, who were also spending the summer in Roscoff, often teased me about my straw chemistry equipment. The assays soon showed that I was on the right track: sea-urchin eggs indeed contain large amounts of pentoses, mainly in the form of pentose nucleic acid. They also showed that the amount of pentoses decreases during development, and this led me to the incorrect assumption that part of the RNA is converted to DNA. At present we know that deoxyribose ultimately comes from ribose but that there is no conversion at the level of macromolecules.

Back in Brussels, I rushed into the office of my master, A. Dalcq, to tell him my story. He listened in a placid mood and told me that he would not believe in the presence of plant nucleic acids in sea-urchin eggs unless I could show it to him under the microscope. After a few discouraging attempts, I eventually devised in 1939 a simple cytochemical method for detecting RNA on histological preparations using the recently isolated enzyme, pancreatic ribonuclease, for reliable identification of RNA. To my delight, I found that RNA, like DNA, is a universal constituent of all examined cells (bacteria, plant and animal cells).The location of the two nucleic acids, however, is different: while DNA is found in chromatin and chromosomes, RNA is found in cytoplasm and nucleoli. Moreover, while the amount of DNA per nucleus is constant in any given species (except that it doubles when the cell is about to divide), the amount of RNA per cell differs considerably between tissues of the same organism. I noticed a completely unexpected, if striking, correlation between the amount of RNA in a cell and its capacity to make proteins. This led me to another iconoclastic proposition, namely that proteins are not synthesized under the action of proteolytic enzymes working backwards, but rather by some unknown mechanism in which RNA must be involved. The same idea was expressed independently by T. Caspersson who, in Stockholm, was using a cytophotometric method for detecting nucleic acids on histological preparations.

My ideas (net synthesis of DNA at cell division, presence of plant nucleic acids in animal cells and involvement of this nucleic acid in protein synthesis) raised much criticism, and it took many experiments to refute it.

During the war, together with my colleague, R. Jeener, we tried to approach my working hypothesis with biochemical methods; our results supported the hypothesis but we were unable to get a final demonstration because the indispensable tools (radioactive amino acids) were not available to us. In the early fifties, a few American laboratories showed that the integrity of RNA is required for labelled amino acids to be incorporated into nascent proteins. When I read their papers, I was as happy as if I had done the experiments myself. At that time, nobody could understand how DNA or RNA are made, nor how RNA might be involved in protein synthesis. It is only in the sixties that the solution was found.

Nevertheless, as early as 1942, essential aspects of the 'central dogma' of molecular biology were emerging in my laboratory thanks to research on eggs and oocytes. I did not know in the period 1930–1940 that I was doing molecular biology; this name, indeed, had not yet been coined.

Later, I returned to the first problem that I had hoped to approach as a young student, namely, the interactions between nucleus and cytoplasm. The opportunity came in 1950–1955, when I had the possibility of using large amoebae as well as the giant unicellular alga *Acetabularia mediterranea*, that it is easy to divide in two parts (nucleate and anucleate). These experiments showed that RNAs are synthesized in the nucleus and migrate to the cytoplasm, whereas protein synthesis and energy production occur in the cytoplasm.

Acetabularia was of special interest: anucleate fragments indeed survive for weeks and are able to develop a complex structure, the 'umbrella'. Hämmerling had shown as early as 1934 by interspecific grafting experiments that the shape of the umbrella corresponds to the species of the fragment containing the nucleus. My own experiments showed that anucleate fragments can develop a perfect umbrella typical of their species. This was an interesting paradox. Besides, the biochemical experiments I did in collaboration with H. Chantrenne showed that amino acids are incorporated into proteins of anucleate fragments at the normal rate for at least two weeks, after which protein synthesis stops. My interpretation was that nuclear DNA releases into the cytoplasm some RNA which carries genetic instructions for organizing the umbrella and that it survives for a long time and plays its part in protein synthesis.

A little later, Jacob and Monod, working with bacteria, arrived at a similar conclusion and gave to the information carrier the striking name of 'messenger'; a notable difference was that bacterial messengers are short-lived.'[1].

That is the lively manner in which Brachet presented his personal recollections of that part of his scientific life concerned with nucleic acids and nucleocytoplasmic interactions. Some additional matter can be found in three short publications [2–4].

I might complement Brachet's text by mentioning the conditions under which this early work was done and how Brachet's laboratory developed. At the end of 1938, when he had just obtained the first indication of the presence of RNA in animal tissues, Brachet left the Faculty of Medicine and was appointed Associate Professor of Animal Morphology at the Faculty of Science. But no laboratory space was made available to him. He eventually associated with a colleague, R. Jeener, who had arranged, mostly at his own expense, a small laboratory of Animal Physiology in an empty shack in a corner of the botanical garden of the University, in the outskirts of Brussels.

Pooling all their equipment, they had altogether a table centrifuge, a Pulfrich optical colorimeter, a home-made water thermostate, a microscope, a microtome and a clumsy home-made pH meter located in a dark corner because of the unwieldy second-hand flying spot galvanometer. Brachet had, however, an excellent zoologist collaborator for the histological work, Mrs. De Saedeleer, and Jeener had a mechanic. I believe it is during that year 1939 that Brachet developed his ribonuclease method for histological localization of RNA, and with it harvested in a short time a considerable amount of data. The next year, conscription and the war stopped every activity.

Both Brachet and Jeener were fortunate enough to return to their miserable laboratory at the beginning of 1941 in occupied Belgium; I had the privilege to join their small group at the end of that year. Brachet and Jeener were then studying the association of RNA and several enzymes with 'granules' (i.e., microsomes and ribosomes) by fractional centrifugation of tissue extracts, using an air-driven ultracentrifuge invented by Henriot and Huguenard. Henriot was Professor of Physics at the Faculty of Science, and the air-driven centrifuge was in the basement of the Physics Institute, a few miles from the laboratory. If rudimentary, it was a remarkable machine, the actual prototype of the present-day air-driven ultracentrifuge distributed by Beckman. There were two rotors with tiny plastic tubes, spinning at 45 000 and 75 000 rpm, respectively; we had a stroboscope for monitoring the run. The only protection against possible rotor rupture was an old wooden box filled with sand; there was no braking system whatsoever, the only way to slow down the rotor was reducing the air flow, but if it was reduced too much, the rotor would touch the basis, jump out and run on the floor like a mad spinning top, bumping into everything, and the experiment was spoiled. In order to avoid such incidents, we used to maintain the slowing-down rotor with our left hand protected by a thick leather glove until it came to a standstill; it was great fun.

By the end of November 1941, the German administration required that Jewish professors be expelled from the University; their colleagues, almost unanimously, refused to continue teaching if this should happen, and all teaching indeed stopped. For some time, the existence of a small laboratory in the slummy shelters of the botanical garden passed unnoticed, and we were able to continue using the ultracentrifuge in the cellar of the Physics Institute with the complicity of the doorkeeper; the work went on for six months. In May 1942, the University was definitively closed down by the occupation power; our group was dismantled. A little later, Brachet was taken as hostage by the Nazis and sent to a fortress where he was kept for several months in the company of high magistrates, artists, professors and other intellectuals. When released, having no laboratory in which to work, he wrote his first book, *Embryologie Chimique* [5], which appeared in 1944 and was later translated into English and Russian. Both this book and his proposition that RNA is in some manner linked with protein synthesis impressed young biologists and quite a few biology-oriented biochemists.

At present, 50 years later, it is difficult to recapture the atmosphere and the state of knowledge of biochemistry in that immediate post war period. The only macromolecules that were reasonably well known were starch and cellulose; a few proteins had been crystallized but many biochemists still doubted whether they were defined chemical species; their structure was unknown, even their polypeptidic nature was in doubt; nucleic acids were described as tetranucleotides in biochemistry textbooks, and biologists regarded them as phosphorus storage compounds. No biochemical synthesis whatsoever was understood (except urea formation, which is actually part of catabolism). The nature of genes and the way they influence enzyme activity and morphogenesis were complete mysteries. All the known biochemical processes were desperately catabolic.

The bold proposition that RNA was involved in protein synthesis, sounded like nonsense to biochemists, but it was exciting to young biologists less reluctant to adopt new ideas in a field where so little was understood anyway; they were probably struck by the conceivable analogy between plant viruses and cytoplasmic particles containing RNA.

To cytologists, Brachet's ribonuclease method was of unusual interest: it showed that most of the basophilic structures familiar to them bind basic dyes because they contain RNA, thus providing a chemical basis to an otherwise empirical technique. Brachet's method was accessible to any cytological laboratory, it made use of a known histological staining technique (Unna's classical methyl green-pyronine recipe) combined with a very simple biochemical manipulation (dipping one slice out of two in a solution of pancreatic ribonuclease, an enzyme that had been purified by Kunitz). The beauty of the method was that the enzyme insured reliable identification of RNA, while the histological staining gave its fine localization down to subcellular structures.

In 1947, Brachet spent a year in Philadelphia and gave seminars or lectures in several American universities and research centers. He was a good lecturer and he communicated to many a young researcher his excitement about RNA. When Brachet returned to Brussels, the small laboratory that he shared with R. Jeener was somewhat enlarged and a pseudo-Flemish front wall had been built to hide its shabby appearance. Brachet received a grant from the Rockefeller Foundation for a good microscope, some microanalytical equipment and the marvel of the time, a Beckman DU ultraviolet spectrophotometer, ideally suited for nucleic acid research.

The dynamism, determination and boldness that prevailed in the group were amazing. Let me give an example: Brachet's colleague, Jeener, wanted to study RNA synthesis, but radioactive tracers were not available in Belgium; unabashed, Jeener decided to make radioactive ^{32}P himself. He obtained from the Union Minière a decent source of neutrons (1 g of radium with beryllium). A shelter was rapidly arranged in the garden, where the neutron source was dipped into a 10-liter carbondisulfide bottle, surrounded by a double shield of lead home-poured from old pipes and paraffin. The ^{32}P resulting from sulfur transmutation was collected every week by nitric acid extraction and purified, providing enough labelled phosphate for a week's work. Experiments with that home-made ^{32}P established that there are several classes of metabolically different RNAs in a cell, an essential point indeed.

Postdoctoral fellows from Europe and the U.S.A. came to the laboratory, and the number of applications increased so much that the University reluctantly agreed to successively add a number of small buildings, the recurrent process being as follows: Brachet and Jeener would ask the University for a seminar room (any cheap prefabricated hut would do). As soon as it was built, the hut was invaded step by step by laboratory equipment, frogs, mice and researchers. As a result, there was no seminar room anymore, and we needed one badly (just a cheap prefabricated hut would do). No sooner was it built than it was invaded . . . etc. The laboratory was thus expanding like budding yeast. Fortunately, resources for research were improving and the laboratory acquired equipment to match its growth. By 1950, it was full of American postdocs. Each Spring we would welcome Dr. Pomerat from the Rockefeller Foundation, who was very helpful, and eventually gave a bulky grant for a Spinco ultracentrifuge, counters of radioactivity and some other expensive instruments.

By 1960, the laboratory had been considerably enlarged by the addition of a brick building for housing the groups of Biochemistry and of Genetics which were offsprings from Brachet's laboratory. For by that time a few of his first students, who had been abroad as postdoctoral fellows, were coming back. So the laboratories of Animal Morphology (J. Brachet) and Animal Physiology (R. Jeener) were soon flanked by groups of Microbiology (J. Wiame), Biochemistry (H. Chantrenne), Biophysics (M. Errera) and Genetics (R. Thomas), a tightly united group of good friends forming a multidisciplinary center of modern biology. The group was selected by Euratom with three other European institutes as a training center for Radiation Biology, and moved to new quarters built by the University at Rhode St. Genèse, eventually to become the Department of Molecular Biology. This achievement was due primarily to Brachet's prestige and his determination; he fought relentlessly for obtaining the support of the University, of the Belgian State and of Foundations.

Nucleic acids and nucleocytoplasmic interactions were not Brachet's only interests. Before anything else, he was an embryologist. The optimal period for obtaining good embryological material is short in our climate; it is early Spring when amphibians are about to lay their eggs. Spring was a hectic period in Brachet's laboratory. Several of his collaborators would first go to the countryside, often as far as the Ardennes, to catch frogs and newts. For several weeks everyone would interrupt his or her own research and work on eggs under the direct supervision of Brachet: removing the jelly coat, fertilizing eggs, treating them with various drugs, cutting eggs, doing microchemical determinations, fixing material for histology, and usually sacrificing the Easter holiday; but it was an exciting time.

During the past 25 years, Brachet used to spend part of the year in Naples where sea-urchin eggs are available; he had a part-time arrangement with the Laboratorio Internazionale di Biofisica e Genetica and later with the Laboratorio di Embriologia Molecolare.

Brachet never was a biochemist at heart, he was a biologist; his real fields were cytology and embryology. He published more than 300 papers between 1929 and 1988; most of them were on aspects of early embryonic development or cell structure. But he deplored the separation he had experienced as a student between morphology and physiology. Let me quote him again:

> 'There was sharp opposition between "anatomists" and "physiologists" when I was a medical student in the University of Brussels, some 60 years ago. This split was exemplified by the presence of two separate buildings, called respectively Institutes of Anatomy and of Physiology in the newly erected Medical School. The first housed embryologists, histologists and pathologists, the second physiologists, biochemists, bacteriologists and pharmacologists. Biochemistry was a recent outgrowth of the older and larger Physiology laboratory; the young professor, E.J. Bigwood, was at that time mainly interested in redox potentials. There was no inner communication between the two buildings except a long dark underground corridor; we called it the "tunnel". Students used it, but in general, senior "anatomists" and "physiologists" were not much interested in meeting each other.
>
> I became an "anatomist" in 1927, although I had a much greater interest in organic chemistry than in human bones.'[4]

Brachet's constant endeavour was indeed to bridge the gap between morphology and biochemistry. For he had very early realized that the key to complex biological phenomena like cell division or morphogenesis could only be found in simpler underlying processes. He used to watch the current literature closely, all the way from biochemistry and fundamental genetics to cytology and embryology. Each progress, each new biochemical discovery prompted him to find out what it might mean for cellular structures or processes. He would, for instance, try on cells or embryos the newly discovered selective inhibitor of an enzyme hoping for a clue as to the function of the relevant metabolic chain in some complex biological process.

And he encouraged the use of all possible micromethods for adapting the biochemical approach to the scale of single cells, for example microinjections, ^{14}C autoradiography on microscopic slides, the development of which in his laboratory was largely due to A. Ficq. At the end of his life he found comfort in the fusion of the different approaches that he had so much contributed to bring about. I quote:

> 'Already in 1940, I had learned a lesson that I shall never forget: one should always try to combine the biochemial and cytochemical approaches if both biochemists and morphologists (as well as yourself) are to be convinced. I tried to persuade fellow scientists of this truth in two books. The title of the second (*Biochemical Cytology*) [6] led to sharp adverse reactions from a few Anatomy professors as late as 1960; I was openly accused by one of them to have produced a lethal, unviable hybrid between Cytology and Biochemistry. Today, thanks to the development of new and powerful methods (electron microscopy and differential centrifugation of homogenates first developed by Albert Claude, autoradiography, immunocytochemistry, in situ hybridization of specific DNA sequences), the old battle has been won. Far from being lethal, the hybrid has been exceedingly fertile. There are few papers published today in the leading journals of cell biology and developmental biology where electrophoresis gels are not found next to micrographs depicting the intracellular localization of the substance of interest. Cytochemistry and Biochemistry are no longer enemies: they help each other on the long, arduous way of scientific discovery.'[4]

A true reductionist, Brachet disliked vague holistic or vitalist concepts. He was convinced only by simple direct experimental data. Nevertheless, his interpretations rested more on the general impression he gathered from a set of results than on rigorous reasoning; in that, he was more intuitive than cartesian. He hated all forms of dogmatism and preconceived ideas. Witness his reaction to Lysenko's genetics. Immediately after the war, quite a few intellectuals joined the communist party; Brachet was among them, but he did not stay long. For ideological reasons the European communist parties were trying to enforce the 'Progressive Genetics' of Lysenko, which rejected classical genetics and insisted on the dominant effect of the surroundings. Pressed to lecturing to propagate that new genetics, Brachet said he needed first some more objective information about its experimental basis. He went to the U.S.S.R. for that purpose, met Lysenko and realized that he was a quack supported for political reasons only. Brachet bluntly refused to advocate

Lysenko's theories, and cut all relations with the communist party to the dismay of friends, the more so since some respected biologists in Belgium and in France were supporting the dogmatic Progressive Genetics.

Brachet was a good lecturer and a stimulating professor. At the end of his lessons or lectures, he would usually underline something that was not well understood and make suggestions as to conceivable experimental approaches to solve the puzzle. This awakened the curiosity of students and may explain in part why so many brilliant students were attracted to his laboratory.

Brachet was a prolific author; between 1944 and 1985, he published at least nine monographs [5–13] on the biochemical aspects of cytology or on early embryonic development. He wrote most of them during the summer vacations in his cottage near Saint Raphael. Each of these books reflects the state of knowledge at the time it was written, so that the set gives a striking picture of 40 years during which fundamental biology and biochemistry were deeply renewed, changing from chemistry of natural compounds and morphology to present-day molecular biology.

For those who worked in his laboratory, Brachet was a charismatic figure. All his pupils respected, admired and liked him. The door of his study was always wide open, anyone was welcome at any time for advice, encouragement or help. Entirely devoted to disinterested fundamental research, Brachet was a man of perfect integrity and great generosity.

REFERENCES

1 J. Brachet, Bull. Acad. Roy. Belg. (Classe Sciences) 73 (1987) 441.
2 J. Brachet, Acta Embryol. Morphol. Exp. 4 (1983) 169.
3 J. Brachet, in R. Peters and M. Trendelenburg (Eds.) Nucleocytoplasmic Transport. Springer, Berlin, 1986.
4 J. Brachet, Trends Biochem. Sci. 12 (1987) 244.
5 J. Brachet, Embryologie Chimique. Desoer, Liège, 1944, p. 509.
6 J. Brachet, Biochemical Cytology. Academic Press, New York, 1957, 516 pp.
7 J. Brachet, Chemical Embryology. Interscience, New York, 1950.
8 J. Brachet, Le Rôle des Acides Nucléiques dans la Vie de la Cellule et de l'Embryon, Actualités Biochimiques No. 16, Desoer, Liège, 1952, 122 pp.
9 J. Brachet, The Biochemistry of Development, Pergamon Press, London, 1960, 320 pp.
10 J. Brachet, Embriologia Molecolare, Mondadori, Milan, 1973, 209 pp.
11 J. Brachet, Introduction to Molecular Embryology, Springer-Verlag, New York, 1974, 176 pp.
12 J. Brachet, Molecular Cytology, Vol. 2, Academic Press, New York, 1985, 935 pp.
13 J. Brachet and H. Alexandre, Introduction to Molecular Embryology, 2nd ed., Springer, Berlin, 1986, 229 pp.

G. Semenza and R. Jaenicke (Eds.) Selected Topics in the History of Biochemistry: Personal Recollections, III.
(Comprehensive Biochemistry Vol. 37)

Chapter 7

My Love Affair with Membranes

ASER ROTHSTEIN

Research Institute, The Hospital for Sick Children, Toronto (Canada)

Introduction

When I was younger I almost always looked ahead, anticipating the excitements of tomorrow; yesterday's events were soon stored away as past history, only rarely to be examined. With the passage of time my future potential has inexorably diminished, whereas my past has continuously expanded. My swollen memory banks can no longer be ignored. I find myself more frequently looking backward, curious about what actually happened in the course of my personal history compared to my original expectations at the time.

Science is supposed to be a rational activity; as a scientist I always prided myself on being a rational person. At the time of each decision I perceived myself as assembling information, weighing the pros and cons and making intelligent decisions. Alas, in retrospect, I have to admit that nonlogical events played a sometimes crucial role. I was not always as rational (or as smart) as I thought. Many of my decisions may even have bordered on the irrational. Fortunately most of my decisions, whether rational or not, turned out well because the fates were kind to me. A case in point was my decision to major in Biology rather than Physics at the University of British Columbia.

Plate 9. Aser Rothstein.

Early times – University of British Columbia

I was born in Vancouver of parents who had emigrated from Eastern Europe. They had not had time or opportunity for much formal education. My father was smuggled out of Russia at the age of 16 to join his brother, a mature 18-year-old, in New York. Survival required an independent spirit and hard work at many menial jobs. They somehow migrated to Vancouver, Canada, where they settled down and ultimately established relatively successful business enterprises. Given the Jewish tradition, with its emphasis on learning, there was no question that the first choice for me was to proceed through university and into a profession. The same was true for my brother. The end result was, in a sense, tragic for my father inasmuch as both sons became successful academics, so that there was no one left to succeed him in his business. He ultimately, on retirement, sold out to his brother's company, run by my cousin who had no academic ambitions. In turning my back on a business career to become a Professor, it turned out ironically, that many years later as Director of a large Research Institute, I ended up running a research business that was considerably larger than my father's commercial business. It required just as much business-like activity except that its success rewarded me with satisfaction rather than monetary profit.

At the time that I entered the University of British Columbia (1934) a university education was not the goal for most young people. It was the time of the great depression and less than 10% of high school graduates continued to a university. Entrance standards were relatively high, and poor grades could result in removal from the university by the end of the first semester, a situation given the title 'BAC' (Bounced at Christmas). In my first Physics course I became the target of the professor, probably because I was often late (it was an early morning class). He took pride in inventing ever more sarcastic comments denigrating my potential for academic advancement. He assured me in front of the class that I would surely receive the BAC degree. Much to his surprise but not mine I wrote a perfect exam. He would, however, only report my grade as 99%, because he claimed that no one was perfect. During the second semester he ignored me. On the final exam I scored 94%, and because he had given everyone a bonus of 10% (the average score was far too low), I of course demanded a grade of 104% but received only 99%. When I came to register for second year there was a note from my friendly physics professor asking me to meet with him. He asked me if I would major in Physics. Although I liked Physics, largely to spite him, I announced in triumph that I planned to major in Biology. Thus my major career decision was made in about 30 seconds, on impulse, to spite a professor. To tell the truth, I was already 'leaning' toward Biology but my conversation with the physics professor 'tipped the scale'. The professor proceeded to a distinguished career. He founded and was the first president of Simon Fraser University in Vancouver. It became clear to me that talent and success do not necessarily correlate with 'niceness', perhaps the contrary.

During the next decade my somewhat irrational decision appeared to be a serious misjudgement. It was the beginning of the Atomic Age so that young physicists were rapidly promoted to Full Professorships, whereas biologists were lucky to have any position at all, even at low rank and low pay. Of course, in the long run Physics became a quieter field, but Biology expanded enormously; I was swept along with the tide of its growth and became a professor, chairman of a department, director of a large Atomic Energy Project, and director of a large biomedical Research Institute. I became relatively well known for my research in Membrane Biology. Fate had determined that my decision to major in Biology was, in the long run, both rational and clever.

As it happened, Biology at the University of British Columbia was, at that time (1934–1938) taught in a classical descriptive mode, not at all a first-rate training for my ultimate research career. I learned the names of all the phyla and species, Mendelian genetics, embryological development, comparative anatomy, histology, and evolution, but no biochemistry, cell biology, or physiology. In other words, I was taught the biology of the past, but little about the biology of the future. I was behind the times even before I started on my career.

Graduate School – Berkeley and Rochester

After graduating from the University of British Columbia in 1938 I had decided that I wanted to become a professor and do research. The little I knew about research had been gleaned from romantic novels describing 'magic bullets' and 'miracle cures'. All I knew about professors was that they gave a few lectures and seemed to have a lot of time for interesting activities. In retrospect, my perception of what my choice entailed was entirely naive. In any case it was clear that I had to attend Graduate School and earn a Ph.D. This degree was not then available in the sciences at British Columbia. In fact, the only Ph.D. programs in Biological Sciences in Canada were then given at McGill in Montreal and at the University of Toronto. The main track for aspiring students from Vancouver was south, usually to the University of California at Berkeley. Without much thought and with virtually no guidance I ended up there as a graduate student in the Department of Zoology. There was little that could be called rational or informed about this decision, but at Berkeley chance favored me once again. A few stray students unsuited for Classical Zoology (the Head of the Department was well known for his book on the Elasmobranch Fishes) were passed down from the fourth floor of the enormous Life Sciences Building (the location of Zoology) to a small enclave in the basement presided over by Prof. S.C. Brooks. He was a maverick, in charge of a small, almost invisible, interdepartmental program called Physico-Chemico-Biology (PCB). In it Brooks attempted to demonstrate that physicochemical processes underlay many biological phenomena at the cellular level. It was Biophysics long before Biophysics was recognized as a discipline. In this effort Brooks was a pioneer. He emphasized membrane transport phenomena at a time when few paid attention even to the existence of a limiting membrane. He was a pioneer not only in a conceptual sense, but also in terms of experimentation. For example, soon after the world's first cyclotron became functional at Berkeley, Brooks arranged to obtain some of its metal targets. From them he managed to extract small amounts of $^{42}K^{+}$ and with a primitive home-built Geiger counter full of large radio tubes spread across a wide bench top he did the first isotopic transport experiments, measuring K^+ uptake in the algal cell, *Nitella*. Brooks opened my eyes to a new world, exciting my imagination. I also took my first biochemistry course on amino acids and proteins, another 'eye opener', from a then young professor, David Greenberg, who subsequently became a leader in that field. Brooks spent his summers at Woods Hole and took me with him as a lab-assistant. There, with exposure to Woods Hole 'lore', my education in Cell Physiology was further extended. In 1940, I performed the first study, to my knowledge, of biological phosphate uptake (with fish eggs) using isotope techniques with ^{32}P. I was enormously influenced by my two years with Brooks. I thereafter maintained a continuing interest in membrane phenomena, especially in ion transport.

Unfortunately, during my second year at Berkeley, Brooks suffered a heart attack and was unable to carry on his academic duties. I consequently lost both my advisor and my teaching assistantship. So I decided to continue my studies elsewhere, although I did not have complete freedom of choice in selecting a new location. I wanted to continue research in cell physiology particularly related to membrane phenomena, but I was serious about a music student from Calgary who was also studying at Berkeley. In selecting a new graduate school I was, therefore, restricted to the strange combination of music and cell physiology. The answer turned out to be the University of Rochester, the home of the famous Eastman School of music. In the University of Rochester catalog an interdisciplinary program in Biological Sciences was described that looked attractive to me. One of the listed faculty members in the program, whom I recognized, was W.O. Fenn whose work on K^+ metabolism Dr. Brooks had mentioned. My wife to be was accepted as a student at the Eastman School. I was offered a teaching assistantship in the Biology Department with a stipend of $70 per month. My next best offer was $55, so, partly influenced by the sum of $15 per month we decided on Rochester. Today $70 will pay for a moderately good dinner for two, but in 1940 we lived, or rather survived, on that sum. With our new location determined, Evelyn, my musical friend, and I proceeded to marry and departed for Rochester. Thus an important step in my career was initiated by a professor's illness, and was strongly influenced by a romance and the location of a music school. Rational choice played only a minor role.

We arrived in Rochester in the Fall of 1940. The first winter we were appalled and surprised by the amount of snow. We had not taken climate into account in our choice of location. On the academic side we were surprised to find that the University was spread over four campuses. The Eastman School of Music was downtown, and almost completely distinct from the University. The Biology Department, in which I was to work, was on the new River Campus, but my teaching duties in Comparative Anatomy were at the old campus, downtown (which served as the Women's Campus until a number of years later when sexual integration occurred at the River Campus), and Dr. Fenn was in the Physiology Department of the Medical School, located about a mile from the River Campus. Furthermore, the attractive interdisciplinary program in Biological Sciences turned out to be the figment of imagination of some unknown person who had prepared the University catalog. It simply did not exist. We rented an inexpensive apartment downtown near Eastman School. It is hard to believe today but the cost was only $25 per month. We found out later that the low cost was partly influenced by a location in Rochester's 'red light' district. Evelyn never understood, at the time, why the men around were so friendly.

I was finally able to arrange to get my degree in Biology but to use Prof. Fenn as my thesis supervisor. For reasons that I cannot remember I had started to examine phosphate uptake and changes in phosphorylation that occurred after sugar was fed to yeast cells. Fenn showed me a short abstract that had recently appeared, demonstrating that, on addition of sugar, yeast cells took up K^+. He suggested that the K^+ uptake might be related to PO_4 uptake or a phosphorylation process, and steered me into the problem of K^+ transport. To set the background, knowledge of electrolyte metabolism in 1940 was most primitive, to say the least. Isotopes had not been available so that the concepts of ion exchanges, unidirectional fluxes and steady states were as yet to evolve. It was known that cells maintained large Na^+ and K^+ gradients in opposite directions, but how was not known. By the criteria available cells appeared to be relatively impermeable to ions, so the simplest explanation was that membrane impermeability prevented dissipation of pre-existing gradients. On the other hand, cells had to retain K^+ and exclude Na^+ for a long time. Furthermore, it was known that when cells were cooled they lost K^+ which was regained on rewarming, and that muscle lost K^+ on contraction, which was regained during the recovery period. Fenn believed that some kind of 'active' process must be involved that allowed accumulation of K^+ and exclusion of Na^+. The observation demonstrating uptake of K^+ by yeast seemed to present an opportunity to define mechanisms that might be important in its accumulation and in the maintenance of ion gradients.

It became obvious from my early experiments that the amount of K^+ taken up by yeast greatly exceeded the amount of PO_4 and that the movements of the two ions were not directly related [1,2]. I ended up, therefore, with a two-theme thesis. In the case of K^+, I was able to demonstrate that the inflow involved a stoichiometric one-for-one exchange for H^+, the first description of a cation exchanger and a proton-linked transport system. The process was dependent on specific modes of metabolism, and K^+ and H^+ were transported against rather enormous concentration gradients [2].

During the course of my thesis research, Fenn was encouraging but he never imposed his own ideas. In fact, he left me almost entirely on my own resources in terms of both experimental approach and development of concepts. Typically, I would nervously bring in some data; he would look at them and after an appropriate pause would exclaim 'that's very interesting'. After a few minutes of silence, I would pick up my graphs and depart assuming that everything was probably alright, but was never quite certain. After a few such 'consultations', I did not come back to see him until I was able to hand him the first draft of my thesis. His approach, by necessity, quickly induced in me a high degree of independence. Fenn's influence was, nevertheless, considerable. To some degree I am still following the research track on which he started me. I have continued my interest in cation transport, first with yeast, and later with red blood cells, lymphocytes, platelets, and various cultured cells. With my own students, I also have followed Fenn's philosophy to some degree: encourage, but do not unduly get in the way of their independent development.

At the time of my thesis research, only one other laboratory was concerned with yeast transport, that of E.J. Conway, an eminent Irish biochemist. Conway was a pioneer in examining electrolyte metabolism in muscle, especially its capacity to maintain K^+-Na^+ gradients. Unknown to me he had also turned to yeast as a model system for studying cation transport. When I published my thesis in 1946, demonstrating that K^+ uptake involves a stoichiometric exchange for H^+, I was surprised to receive, soon after, a rather vituperative letter from Conway. He had devised a 'redox theory' to account for the energy requirement for K^+ transport. Basically, he felt that iron-containing enzymes (cytochromes) provided oxidative energy for transport. To support his theory he published findings similar to my own on K^+-H^+ exchange [3], except that my paper preceded his by about three months. He was very upset that I had 'preempted' his work, and berated me for not, at least, giving him priority by referring to an abstract he had published in the *Irish Journal of Science* (of which I had never heard). I was rather perturbed, as a young scientist publishing his first paper, to be unjustly 'attacked' by such an established scientist. After I cooled off I replied to Conway suggesting that arguments over priority were not appropriate, that we had mutually confirmed our results and, besides, that I had published an abstract in *Federation Proceedings* preceding his abstract. I later learned that I was not alone in incurring Conway's wrath. He, sooner or later, became involved in acrimonious debates with almost all of his 'rivals'. There was growing support for the concept that phosphorylation processes directly provided the necessary energy maintenance of cellular Na/K gradients, a concept that was soon confirmed by the isolation of Na^+-K^+ ATPase [4]. In 1959, a symposium was sponsored by the CIBA Foundation in London in Conway's honor. I was privileged to be invited [5] along with about 25 others, the transport establishment of the time. It was rather amusing. Conway was still clinging to his redox pump theory, whereas the 25 individuals who were honoring him disagreed with him. For a change, however, the exchanges were good-natured, with smiles and some laughter.

Conway was a complex individual. In his laboratory he was the 'king'. In Ireland, he was one of the few recognized scientists, and was held in high esteem. Although I was much younger, I became one of his only scientific friends. In 1950, I had the honor of being a recipient of a travel award for promising young scientists to attend a physiological congress in Copenhagen. It was my first international overseas meeting. Conway gave one of the plenary lectures. After the meeting I flew to Amsterdam and by chance Conway was on the same plane and also planned to spend a day in that city. We went sight-seeing and had dinner together. After a drink and some wine, this shy crusty man who appeared to be at odds with everyone, turned into a charming, cultured, erudite individual with many interests. Two years afterwards, at a meeting in Bangor, Wales, to be mentioned later, my wife Evelyn sat next to Conway on a sight-seeing bus and conquered him. He invited us to visit him in Dublin, which we did. We had a delightful time and were friends ever after.

Desert physiology – Rochester

Toward the end of my thesis work my career was pushed in an unexpected direction by an outside event of considerable magnitude, World War II. Although I was a Canadian national, I had been deferred from the Canadian Draft because I was a graduate student. As a foreigner in the U.S.A., I had been exempted from the American Draft. Then the U.S. rules changed encouraging me to participate in war research rather than be drafted. Prof. Edward Adolph, an eminent physiologist had been asked to investigate the physiology of man in the desert as a prelude to the American invasion of North Africa. I joined the effort. We operated a 'hot' lab in which we tortured poor medical students by paying them poorly to pedal a bicycle in a room at 49°C, while wired up for all sorts of measurements. We spent the summer at an Air Force base in the California desert (Blythe). It was a fascinating experience. A platoon of soldiers was assigned to us for experimentation. We were trying to determine water requirements for varied activities and also to determine physiological limits with and without water. We, of course, also had to expose ourselves to all tests and to outperform the soldiers. The incentive for them was an afternoon off at the swimming pool. A sweaty good time was had by all. We measured the highest rates of sweating ever reported. We carried out 'model' survival marches with a limited water supply. I well remember coming in with a group after an overnight forced march with water losses of up to 5 kilos, over 7% of body weight, almost the limit of endurance. Our findings were used to determine the logistics of water supply for the US Army and Air Force in Africa. Transport of water was usually the limiting factor in any advance. Our overall results were published in a volume called *The Physiology of Man in the Desert* [6].

I was involved in some special projects. For example, I discovered a phenomenon that we called 'voluntary dehydration'. Man's thirst mechanism can be overridden by involvement in 'interesting' activities. As a result, serious dehydration can occur even if cool water is readily available. The Blythe Airport was an Air Force training and test base. I was assigned the interesting task of evaluating the stresses on aviators of flying in the desert. Eggs could be fried on metal surfaces of the wings of planes parked in the desert sun. The initial temperature in the cockpits was over 60°C. By the time a cooler altitude was attained dehydration was often severe and the stresses on fliers were substantial. It was quite an adventure to fly in B17 and B26 bombers. I had a feeling of power, for by simply signing a small slip of paper I could requisition a plane and its crew to fly at a series of altitudes while I took their temperatures and body weights. Every weekend a plane would be flown to San Francisco ostensibly to test the bomb-bay doors, but really to visit the big city.

Manhattan Project– Rochester

My experiences in Desert Physiology, though most interesting, did not induce me to change research fields, but my next involvement in war research was to have an enormous influence on my future career. I was approached by a young faculty member, Harold Hodge (of whom more later) to join an important but secret research activity. In fact, it was so secret I could not even be informed of its nature before accepting the job. On the other hand, I would be deferred from army service and I would receive the magnificent salary of $3200 per year, a substantial increase over the $1800 I was paid to sweat in the desert. As I write, these amounts seem minuscule, less than 3% of my most recent salary. Yet, at the time, $3200 was a relatively high starting academic salary, a sad comment on inflation.

The secret program turned out to be a part of the Manhattan Project, later to be known as the Atom Bomb Project. At that time, not much was known about radiation or heavy metal toxicology. One of the few professors of Radiology involved in research was Stafford Warren, Chairman of Rochester's Department (after the war he founded and became the first Dean of the Medical School at the University of California at Los Angeles). When the decision was made to develop the 'bomb' it was recognized that additional information was quickly needed with respect to the potential hazards to personnel associated with the rapidly expanding effort. Dr. Warren was fitted with a shiny Army uniform in the rank of Colonel and appointed Chief Medical Officer. He set up several crash research programs related to health protection, one of them at Rochester. Because of this series of world events, I was soon to become an instant 'expert' in heavy metal toxicology. At the time (end of 1942), I was still a graduate student just finishing thesis research. On reporting to work I was informed that I was to study the toxicology of uranium compounds. The main concern was inhalation toxicity of dusts. In preliminary experiments, exposure of animals to dusts of uranium compounds resulted in severe contamination of their fur. In the process of cleaning themselves the animals ingested large amounts of uranium, thereby complicating the interpretation of 'inhalation' toxicity. My particular job was to expose animals in such a way that fur-contamination was controlled and minimized. To place the situation in perspective, at that time the inhalation of toxic particles had not been investigated in any serious manner at all. Technology was primitive and had to be developed as experiments proceeded. At first, dust clouds were produced by a simple contraption, a large salt shaker filled with ground uranium compound, that was intermittently banged on a hard surface by a primitive mechanical device. Each noisy bang produced a small cloud of dust to which rats were exposed. We were able to determine the total amount of dust in the air by drawing a fixed volume through a filter that could be weighed, but we had no capacity to characterize the particle-size distribution, a key parameter in evaluation of retention in various parts of the respiratory system. In time we developed sophisticated methods for preparing materials of known particle-size distribution, for maintaining well-characterized suspensions in the atmosphere, and for evaluating the amounts inhaled by the animals and retained in various parts of the respiratory tract. Ultimately we learned about the disposition of the inhaled materials in the body and about the various toxic manifestations. In other words, we developed the essential technology for a whole new sub-field of toxicology. With it we quickly generated a body of knowledge that was used to set industrial health standards in the atomic energy industry. These studies were ultimately 'cleared' and published in a series of volumes in the *National Nuclear Energy Series*, in which I was author of a number of chapters [7,8].

Nothing in my previous experience had prepared me for the task I had been given. Nothing was available in the library. I was told to build a chamber in which animals could be exposed to uranium dusts without contamination of their fur. I was assigned a carpenter to do the construction and had access to a machine shop. We did not bother with diagrams or blueprints. After we made some preliminary decisions about the kinds of species and number of animals, the carpenter and I started the construction. I would hold my hands a distance apart and tell the carpenter, 'let's make it this big'. He would saw a piece of wood and nail it in place. If it did not work we would dismantle and start again. We decided to try to build a unit in which only the animals' heads were exposed to the dusty atmosphere. To do so we developed, by trial and error, a series of restraint cages, one for each species, in which the animals were partially immobilized but were reasonably comfortable. We placed a soft rubber collar mounted on a circle of sponge rubber around each animals's neck. The cage was placed in a special rack so the animal's head would protrude through a hole in the side of the exposure chamber with the rubber collar forming an air-tight seal. Thus the animals could be 'plugged in'. Those holes in the exposure chamber that were not in use were sealed by a trap door. Within the chamber we learned to maintain uniform, controlled, concentrations of dusts of characterized particle sizes. The two ends of the chamber were of plexiglass so the animals could be observed. It was quite a sight to observe all those animal heads protruding through the side walls. The unit achieved a certain amount of fame and we had many visitors from the higher echelons of the Manhattan Project on 'sight-seeing' tours. Much of the existing information concerning inhalation toxicity of insoluble uranium compounds derived from studies using this unit. These and other results produced by the Rochester Project were the data on which the Atomic Energy Commission's (AEC's) industrial health standards were initially based.

My subsequent career was enormously influenced by my unchosen war-time experiences. Soon after I joined the Manhattan Project, although still a graduate student, I found myself in charge of six assistants, running the inhalation toxicology program described above on a two-shift basis (16 hours per day). Later, the number of assistants grew to ten. The assistants were assigned by the army, based on any kind of scientific background. They were a diverse group, including an M.Sc. in Chemistry, another in Biochemistry, a Ph.D. in Plant Physiology, and several kinds of Engineers in various stages of training. For me it meant a crash course in administration, which has served me well over the years. I learned how to get science done despite all the red tape that a secret army-run program could engender. Coping with civilian administrations, later in my career, proved to be much easier.

The Rochester Atomic Energy Project – my initiation into administration

When the war was over, the pressure for immediate production of toxicological data was diminished. The AEC succeeded the army-run Manhattan Project, assuming responsibility for our program. The Rochester Manhattan Project became the University of Rochester Atomic Energy Project (AEP), devoted to basic, as well as applied research, and also to training in Toxicology, Radiation Biology, and ultimately, Biophysics. It was also mandated, for a time, to train health physicists who were in short supply for AEC installations, hospitals, universities etc. I was destined to play a role in the AEP's development and its conversion from a secret, wartime, non-academic, applied research operation into what became one of the largest and, in my view, one of the best departments in the Medical School and a major academic unit of the University and the Graduate School. In the course of this involvement administration as well as research became a major part of my activity. My career was, from this time onward, split into two virtually separate activities, administration and research, with which I shall deal separately in this narrative.

Initially, the Rochester AEP simply assumed the existing Manhattan Project organization and personnel, which it operated under government contract. It had, at that time, no direct academic role although some individuals were appointed into Medical School departments (I was made an Assistant Professor of Pharmacology, of which more later). I became the Head of Physiology Section of the Toxicology Division. The other Divisions were Radiobiology and Medicine. In 1961 I became Associate Director and in 1965, together with Dr. W. Neuman, Co-Director. It was during the 1960s that major changes were to occur in the status of the project. Some of the applied programs related to inhalation toxicity had been continued, but we also moved into more fundamental kinds of research, to an increasing degree. An entity, called the Department of Radiation Biology and Biophysics, had been created to administer the project. At first it was a department in name only, but in the course of time it became a functional department with its own appointments and its own educational programs. When Dr. Neuman and I were appointed Co-Directors of the AEC project we were also appointed Co-Chairmen of this University Department, so we held a dual responsibility, reporting on the one hand to the AEC and on the other to the university. We completed the transition. While adding a number of excellent young staff members, we weeded out some of the older, nonproductive staff; we abolished the line organization (Divisions and Sections); we expanded into Ph.D. programs in Radiation Biology, Biophysics, and Toxicology, reaching a peak of some 100 graduate students in residence; we developed an interdisciplinary graduate curriculum (with the cooperation of other departments); we generated a series of training grants to support the graduate programs; we added a building to the two existing ones previously built for the Manhattan and AEC Projects; we diversified our sources of funding so that we were not unilaterally supported by the AEC, and we talked the University into providing tenure for the staff (who were all paid out of the AEC contract or grant and training funds). By the time I left for Toronto in 1971, the Project was one of the largest departments on Campus. Within the Medical School it had about half of all the research funds and graduate students. By any criteria it was a great success. For the University it was a real bargain. A major component came into existence without appreciable cost (although they did assume risks by granting tenure to the staff). For me it was a great adventure in setting up an organization. The administrative activity was less frustrating than usual because we had virtual control of money and space resources through our contract with the AEC. We did not have to come hat in hand to the Dean or the President. We were blessed with an extraordinary degree of autonomy which we exploited to the full and protected with our lives. It was a type of administration that one can enjoy, which I did. Bill Neuman and I took great pride in the conversion of a war-time project with a 'pick up' staff into a first-class academic department.

There were a few specific incidents worth recording. The first relates to opportunities for entrepreneurship. We needed space badly. The AEC was sympathetic but after several years of discussions and negotiations we were told that it was not politically feasible to put the capital costs of a new building into the budget request because it would open the door to similar requests from other universities from States with powerful lobbies and powerful Senators. We did find out, however, that it would be possible to request a 'use charge' if the University were to construct a building for us. At the time, it was also possible for universities to obtain matching funds from other units of the Federal Government for privately funded Capital Projects. I suggested to the University that they build a building for us, requesting 'use charges' from the AEC, and that we also apply for matching funds. Everyone laughed at this idea because in essence the Government would be asked to donate matching funds against a building for which they were already paying through 'use charges'. To humor me I was given permission to obtain a legal opinion from University and AEC lawyers. Surprise, surprise, the lawyers could find nothing illegal or immoral about the proposition. The situation, perhaps, defied common sense, but since when are legal matters sensible? Unfortunately, this positive opinion arrived just two and a half weeks before the deadline for the matching fund program. As the President of the University and the Dean of the Medical School were away to some unattainable location, perhaps the Gobi Desert, I frantically tried to reach the only University official who happened to be in Rochester, the Treasurer, and the Chairman of the Long-Range Planning Committee (of which I was a member). The potential bargain was too good to miss, so we decided to submit a proposal and to explain later. The Treasurer, being also an adventurer, pledged one million dollars toward the project so it would not appear that the University was getting a completely 'free ride'. The application forms were terribly complicated, requiring enormous amounts of information about the University and the Medical School and plans for future growth and development. Fortunately, much of this was available in various places in the Office of Research Administration, to be regurgitated for our application. More difficult was the request for Floor Plans with layouts of laboratories and facilities, as well as descriptions of the programs that would be housed in the new building, and why they were needed. Here, imagination proved its value to a scientist. We resuscitated some old floor plans for a mythical building that had never received funding and proceeded to carve up the space for specific laboratories and program expansions, more or less invented on the spot. We finished the whole thing in about ten days. An enormous application emerged, as if by a miracle. To everyone's surprise we were awarded the largest (at that time) matching fund allowance. Once we received the award, the floor plans were redrawn, and space was reallocated. The process took a year and endless hours of committee meetings. The result, however, was not too much different from our more spontaneous concoction. The AEC Project and the Medical School got badly needed space and, for a time at least, I was a hero. The incident demonstrates that opportunism is extremely important in resource management; that applications for funding only need to be plausible, but not necessarily 'real' (no one really checks in any detail to see if proposals and execution are an exact match); and that action can be quick and effective if carried out by individuals uninhibited by the necessity of endless debates in committees and approvals by layers of bureaucracy. The success of the building proposal provided me with a sense of accomplishment much the same as I might enjoy after an important discovery in the laboratory.

A continuing Government Contract can be a treasured asset, but it also comes with some 'strings' attached. The contract, for example, specifies in some detail all manner of things that the institution is supposed to do to keep the government bureaucrats happy. Our contract was to support AEC-relevant research (heavy metal and radiation toxicology) and training. Fortunately, the AEC managers in the Division of Biology and Medicine were sympathetic to a very liberal interpretation of relevance (they were ex-academics). They accepted the importance of basic research and training in domains of their interest, as long as the programs were of high quality and as long as we devoted some fraction of the effort to more obviously relevant and practical studies. I will illustrate, in discussing my own research, how I was able to exploit heavy metals as probes of membrane function to elicit basic information, while at the same time providing relevant information on toxicological mechanisms. We had no serious problem in carying the dual responsibility of being a Toxicological Center for the AEC and a first-class academic unit for the university.

While at the program level we had no serious problems, at the operating level we had to learn how to cope. Government auditors do not understand much about research and they always seem willing to spend dollars to track down pennies. They hate flexibility in the use of funds. Research, on the other hand, is full of surprises and requires fiscal flexibility. The auditors appeared quarterly and sometimes dropped in unexpectedly. Each year-end they had a field day with our books. Although research programs did not run from July 1 to June 30, the contract did. On July 1 all unspent money had to disappear and overspending had to be covered by the University. Money could not be moved from one budget category to another even though it made sense from the point of view of the goals of the contract. Despite the auditors' best efforts we managed to outwit them without doing anything illegal or immoral, while keeping them happy. In fact we were cited several times as the best managed (from a budgetary point of view) AEC Contractee, and were given Awards of E for excellence. We almost always managed to bring our multimillion dollar contract to year-end within a few hundred dollars of budget. We gained flexibility mainly by using two loopholes, animal food and heating and cooling. We ran a very large vivarium and bought animal food, in bulk, as much as a year or two ahead, arranging for its warehousing. If it appeared that we were going to be short of funds we stopped buying food for that year; if we had a budget surplus we would buy an extra year or two's supply. We purchased power from the University. Each year we had to estimate our needs and on that basis we would make monthly payments during the year (to maintain an even, predictable dollar flow). After the year had ended and the real costs were calculated, the university would refund us or charge us for the shortfall. But because the new fiscal year had already started, the readjustments had the effect of moving funds from one fiscal year to another. If, in any fiscal year, we judged that we would underspend, we would predict an exceptionally cold winter and hot summer, thereby overpaying the University for power and achieving transfer of the surplus into the next year when we might need it. I outline this story to point out that creativity is necessary in administration as well as in research. Without the flexibility achieved by this creativity, the research output would have been diminished, or some young scientist might not have been hired.

I have pointed out that I was appointed Co-Director and Co-Chairman with Dr. W. Neuman. He and I had both joined the Manhattan Project as graduate students in 1943. While I was doing the uranium-inhalation studies, he was running the Biochemistry Laboratory (blood and urine analyses on experimental animals). After the AEC took over the operation, I became Head of the Physiology Section and he of the Biochemistry Section (of the Toxicology Division). While I was making a name in membrane research, he was doing the same in bone and mineral metabolism (his group were known affectionately as 'boneheads'). We played tennis and squash together; we were bosom companions; we were the young 'Turks' of the Project, the troublemakers. When I became Associate Director he became Head of the largest Division (Toxicology). Because the Director had been ill with a heart attack, we were, de facto, in charge of the whole operation. With the Director's pending retirement it became clear that we were in competition for his job, the two logical internal candidates. Given the circumstances it was a serious dilemma for us, about which we talked frankly. We both sort of wanted the job, but we realized that whoever was appointed would have a difficult time maintaining a viable research program. We both enjoyed research very much and were quite good at it. Neither wanted to give it up to become a full-time administrator, so we reached this solution: the only acceptable circumstance would be our dual appointment as Co-Directors. We would alternate as Director on a two-year cycle, with the other serving as backup and carrying a minor administrative load. In essence we would alternate between 75% and 25% involvement in administrative matters. The University finally accepted our unorthodox proposal. It worked very well from our point of view (we both maintained exciting research programs), and as I have already pointed out the Project flourished. It was a partnership that drew us even closer together in a personal as well as a professional sense. It was a sad day for me several years ago when Bill Neuman suffered a fatal heart attack.

Academic matters – Rochester

My first academic appointment at Rochester was as Assistant Professor of Pharmacology (1948). The appointment was not based on my expertise in pharmacology; the closest I came was my newly acquired knowledge of heavy metal toxicology. Indeed, my appointment well illustrates the sometimes strange ways in which universities do business and the unexpected impact on career development. At Rochester, the first Dean was Dr. G. Whipple, who received a Nobel Prize for his work on anemia. He was a STRONG DEAN. In those days the Dean's office comprised the Dean himself (only part-time, because he was also Head of the Pathology Department) and one secretary. Compared to the size of Deans' offices today, there was virtually no bureaucracy. The Dean did not happen to believe that Pharmacology was a discipline that deserved departmental status, so by his fiat there was no Pharmacology Department. Nevertheless Accreditation demanded that pharmacology be taught to medical students. Whipple's solution was to assign a very junior member of the Biochemistry Department to give the course. The 'victim' was H.C. Hodge, who was not even a biochemist, but an inorganic chemist by training. He had become interested in bone and tooth structure (and ultimately was at the cutting edge of research on fluoride, which led to fluoridation for prevention of caries). Soon after, when someone was needed to develop the Toxicology Program in the soon to be started Manhattan Project, Hodge, with his newly acquired expertise in pharmacology and with his knowledge of bone biochemistry (heavy metals were known to deposit in bony structures), was a logical choice. As I pointed out above it was Hodge who recruited me into the Manhattan Project.

After Whipple retired, the Medical Faculty decided to legitimize Pharmacology and appointed Hodge Chairman of the new department. To save money, however, they did not at first give him any new positions, but had him appoint the Toxicology unit of the AEP including myself, as the faculty. That is how I became a Professor of Pharmacology. As neophyte pharmacologists, I and my newly appointed colleagues, learned on the job and managed to stay one step ahead of the medical students. Hodge was an excellent teacher and spent much time and effort in mounting a first-rate pharmacology course. He also tried to infect the rest of us with his enthusiasm with somewhat mixed results. I remember one year that he decided to develop the best possible laboratory program by distilling the essence of the experiences of all other medical schools. He wrote away and received the pharmacology laboratory manuals of the other 90 medical schools. We tore them up and proceeded to make large piles of similar laboratory exercises from which could be synthesized the essence, the best and most dramatic way of illustrating each pharmacological principle. I'm not at all certain that this 'essence' was much of an improvement. Another year Hodge decided that lectures had little impact unless accompanied by dramatic demonstrations. Some of the staff developed several imaginative and complicated activities directly demonstrated on TV screens and tape recorders, and with animal preparations scattered about the lecture hall, somewhat like a three-ring circus. I was not overwhelmed by the concept. My first lecture happened to be on the effects of caffeine. I entered the lecture hall carrying a cup of coffee which I carefully placed on the lectern. After 30 min of lecturing I stopped and drank the lukewarm coffee, and asked the students to take careful note of the pending improvement in the second half of my lecture. The students appeared to appreciate my 'demonstration'. Fortunately Hodge was tolerant, good-natured, and optimistic about my conversion to pharmacology.

My active career as a pharmacologist ended when I assumed the role of Co-Chairman of the Department of Radiation Biology and Biophysics. I was, of course, appointed Professor of those disciplines. I could certainly not claim any expertise in Radiation Biology. As a membrane biologist, however, I was supposed to know something about Biophysics. I find it somewhat amusing that I carried out biophysical types of research, became a Biophysics Professor and Chairman of a large Biophysics Department, supervising Ph.D. students, and was on the Council of the Biophysical Society without ever being exposed to appropriate Biophysics training. In fact, I had never taken a course in Calculus and only a Freshman course in Physics. So much for the importance of academic bonafides and labels. The greatest mismatch was to come after I moved to the Hospital for Sick Children in Toronto with my appointment as Professor of Pediatrics. Based on my research activities I should probably have been classified as a biochemist, but no one ever offered me an appointment in that discipline.

Research – Rochester

After the war was over and the AEC had superceded the Manhattan Project, the pressure for immediate practical results diminished. Although I still supervised some continuing inhalation toxicology studies I began to look at a more interesting aspect of uranium toxicity, especially at the cellular level. Soluble uranium compounds were known to be nephrotoxic agents that damaged the luminal cells of proximal tubules. The earliest effects, after a few minutes of perfusion, were inhibitions of sugar and amino acid resorption in the kidney. Based on my thesis experience I turned to the yeast cell as a model. The results were immediately exciting. At micromolar concentrations the soluble form, uranyl ion (UO_2^{2+}), inhibited sugar-dependent respiration or fermentation [9–13]. Yeast was a lucky choice for several reasons. At physiological pH, UO_2^{2+} forms insoluble complexes with OH^- or $H_2PO_4^-$, and stable but soluble complexes with HCO_3^- and organic anions; it does not, therefore exert an appreciable toxic action. Yeast, however, acidifies the medium during the fermentation of sugars. The UO_2^{2+} is, in acidic solutions (below pH 6) in uncomplexed form, available for inhibition. For this reason yeast turned out to be an ideal organism with which to elucidate the basic mechanisms of uranium toxicity at the cellular level. The pH dependence also provided the explanation for its specific nephrotoxic action. The proximal tubule is the first site of acidification in the kidney. Uranium is carried in the blood plasma primarily as a filterable, soluble, UO_2-HCO_3 complex. After glomerular filtration, on exposure to acidification in the proximal tubule, UO_2^{2+} is released, allowing binding to luminal cells and producing the toxic response [14]. Thus, in the case of UO_2^{2+}, yeast mimics the proximal tubule. Fortuitously, the yeast-cell studies also brought me right back to my major research interest, the cell membrane, and more particularly, the mechanism of sugar permeation.

In order to place the UO_2^{2+} studies in context, I will briefly describe the state of membrane research at the time (about 1946–1950). The post-World War II period marked a major turning point in membrane research. In earlier times, dating back to the mid 1800s there had been an interest first in osmotic phenomena and later in the permeation of non-electrolytes. From studies on the shrinking of plant protoplasts (plasmolysis) and later, on the swelling and shrinking of red blood cells and marine eggs exposed to media of different osmotic pressures, the concept emerged that cells were surrounded by semi-permeable membranes (permeable to water but relatively impermeable to solutes). Thus, water would move across the membranes until osmotic equilibrium was attained, accounting for the observed shrinking or swelling. It was on this basis that the existence of a cell membrane was first proposed. Somewhat later the permeability to non-electrolytes was explored, also by measuring the osmotic volume changes that accompanied the entry or exit of solutes. By the turn of the century Overton would note that the rate of permeation was directly related to lipid solubility and inversely to the size of the penetrating molecule. He proposed that the membrane was essentially a lipid membrane, but with aqueous patches [15]. Thus, he anticipated the lipid-mosaic membrane structure [16] by some 70 years. Permeation behavior continued to be explored in greater and greater detail, with quantitative definition based on diffusion equations. This period of research culminated in a book entitled *The Permeability of Natural Membranes* by Davson and Danielli published in 1943 [17]. The membrane, at this time, was an invention to explain osmotic and permeability data. It was presumed to be a lipid bilayer based on the relationship of lipid solubility and permeation rates, and on the finding of Gortner and Grendel [18] that the 'stroma' (membrane fraction or ghost) of the red blood cell had just sufficient lipid to form a bilayer. The membrane would not be actually 'seen' by electron microscopy until many years later. Diffusion gradients were considered to be the driving forces; the concept of active transport had not yet evolved. There was little understanding of the permeation of electrolytes. As noted earlier, the techniques available could only measure net movements of salts, which occurred only under certain limited circumstances. The molecular basis of membrane potentials and the role of electrical driving forces were still obscure. There was little work on the permeability to metabolites, because they did not seem to obey diffusion and osmotic derivations so their transport was not readily measurable in terms of predictable volume changes. Most permeability studies were, therefore, carried out with 'off-the-shelf' organic chemicals.

The late 1940s were an exciting period of almost revolutionary change. The impetus derived largely from technological developments in electrophysiology (improved electrodes, and sophisticated applications such as measurements of ion currents under short-circuit and later, voltage-clamp conditions) and the availability of isotopes, allowing for the first time measurements of unidirectional fluxes. There followed rapid conceptual developments: the role of electro-chemical driving forces; the concept of flux ratios; the recognition of the ionic basis of membrane potentials, conduction and current flows; the development of transport kinetics and the carrier concept; and the concept of active, or 'uphill', transport supported by metabolism. These important developments were discussed at the first International Symposium on Membrane Transport held in Bangor, Wales, in 1953. Membrane research was not a particularly popular field in the U.S.A. in those days. I was fortunate to be one of the three Americans to be invited and I summarized our studies on inhibition of sugar transport in yeast by UO_2^{2+} [19]. As a young man, I was duly impressed and excited by the presence of so many 'classical' figures in Membraneology including: Ussing, Hodgkin, Danielli, Davson, Wilbrandt, Lundegardh, Conway, Theorell, Keynes, etc. I had, however, one serious and unanticipated obstacle to overcome. At that time 35-mm slides had not yet come into use. For presentations one used 3×4 inch glass photographic plates. A box of 20 such plates weighed several pounds. Because my wife and I had planned to travel before the meeting we did not wish to be burdened by the weight, so I arranged to mail the plates to the conference center, in advance. They were not there when I arrived. Some frantic telephoning by my hosts elicited the fact that they had been impounded by British Customs at Liverpool. I was guilty of trying to import photographic materials without a license. There was no way of getting them in time for my talk. Thus I had to perform at my first major international conference without slides. I gained the sympathy of the audience by pointing out that whether or not I was a 'great' scientist I was, at least, considered to be a 'threat' to the British film industry.

We had demonstrated that UO_2^{2+} blocks the entry of sugar into the cell and that it does so without entering the cell [9]. It, therefore, must act on the outside of the membrane. I concluded that the limiting factor in sugar metabolism was a specific permeation step through the cell membrane. It was not diffusion-limited, but involved a ‘saturable’ process driven by metabolism and involving specific membrane ligands [10–13,19]. My unplanned interest in uranium had placed me in a position to identify and exploit the first non-penetrating transport inhibitor. If left to my own resources I would never have given a thought to the possible use of heavy metals as membrane probes.

During the early stages of the uranium studies a lab ‘accident’ was enormously helpful in reaching appropriate conclusions. The effects of UO_2^{2+} had been amazingly consistent and predictable until one day, when inconsistency took over. Inhibitions became minimal and variations maximal. We went through a frustrating period of remaking all solutions, washing and rewashing all glassware, cooking everything in nitric acid etc. Nothing helped. As a result a graduate student’s (Leon Hurwitz) thesis was in peril. My confidence in ‘rational’ science was badly shaken. Finally, just before the graduate student and I developed simultaneous nervous breakdowns, the world was set straight. One morning, when walking into the laboratory, I noticed two boxes of detergent on the glass-washing sink. They were both labeled ‘Dreft’, but one of them had a large red star on which was printed ‘NEW IMPROVED DREFT’. The light dawned. We had been using Dreft as a glass-washing detergent because it was virtually pure sodium dodecyl sulfate. In NEW IMPROVED DREFT ‘hexametaphosphate’ had been added as a water softener. This addition may have been a great improvement for the housewife, but it was an abomination for us, because hexametaphosphate is an undefined mixture of phosphate polymers that have an exceedingly high affinity for polyvalent cations such as UO_2^{2+}. It also binds to glass and cannot be easily desorbed. It was coating our glassware and binding the low concentrations of UO_2^{2+} we added; thus the expected inhibitory effect was substantially diminished. The solution was simple. We found a new supply of detergent and tested each batch for polyphosphate. The experience was, however, useful in a conceptual way. My neurones were titillated. I reasoned that the surface of the yeast cell, like the surface of the glass vessels, must also contain ligands capable of interaction with UO_2^{2+} and that this interaction must account for the inhibitory effect. On this premise we proceeded to titrate the UO_2^{2+}-binding sites of the yeast cell surface [10,11]. There were a fixed number, about 10 million per cell, of high affinity, and the percentage of the sites occupied by the metal was directly proportional to the inhibition. This was, to my knowledge, the first demonstration of the presence of a defined population of specific-transport related membrane ligands. The concept of transport mediation by special entities (called 'carriers' at that time) was a relatively new idea derived largely from the kinetic behavior of sugar tranport by red blood cells [20]. We were able to demonstrate the presence of real, quantifiable membrane-transport entities. Fortuitously we were also able to determine the chemical nature of the transport ligands from binding properties [12]. They turned out to be polyphosphate groups that, as J. van Steveninck (a visitor from Leiden) and I were later to demonstrate, were transferred to the transported sugar [21,22]. The sugar transport systems of yeast appear to be similar to the phosphotransferase system of bacteria, described some years after our UO_2^{2+}-binding studies [23]. Thus we were able to turn a frustrating problem into a scientific advance. Ironically, the chemical nature of the membrane-binding sites turned out to be the same as the contaminant in the detergent, a polyphosphate. It was at this point in my career that I began to understand the important role of chance and serendipity in research. I was to be favored many more times by 'unexpected' results. As it turned out, I did not have long to wait for the next important incident.

During our first UO_2^{2+} studies described above [9] we tested a number of UO_2^{2+}-binding compounds for their ability to reverse the inhibitory effect. With most of the test substances the system was well behaved, but with ATP we had trouble with reproducibility. We finally realized that the effects of ATP were time-dependent, diminishing to a low value after about 15 min. It finally became apparent that ATP was destroyed by the cells. Because the products of its breakdown, inorganic phosphate and adenosine could be quantitatively recovered in the medium, it appeared that the ATP was being hydrolysed by an enzyme (ATPase) located on the outer surface of the membrane. This conclusion was, at that time, highly unorthodox. As I have pointed out above, the plasma membrane was thought to be an inert diffusion barrier. How was I to 'prove' that the membrane contained enzymes? ATP was a rare chemical not simply ordered out of a catalog. It was made in the laboratory from rabbit muscle by a tedious process. I had recently collaborated with Dr. Alex Dounce of Rochester's Biochemistry Department on an improved procedure for the preparation of ATP. I decided to make ^{32}P-labeled ATP to test my hypothesis. We injected a very large rabbit with a very large amount of [^{32}P]phosphate. To insure that the label was incorporated into ATP, we provided the rabbit with a large tub of water and allowed him to swim for about 15 min (rabbits, I found, are excellent natural swimmers). We then killed the rabbit and prepared the world's first [^{32}P]ATP [24]. When added to a yeast suspension it was hydrolysed and the ^{32}P was recovered quantitatively in the medium. No change in specific activity was evident even though the cytoplasm contained much more phosphate than the medium. We could, therefore, conclude that no mixing of the ATP-derived phosphate with cytoplasmic phosphate had occurred and that ATP must have been hydrolysed at the outside of the membrane barrier; because no hydrolytic activity could be found in the absence of cells, ATPase must be present on the membrane surface [25]. I submitted an abstract of this study to the *Federation of American Societies for Experimental Biology* and it was placed in a biochemistry section concerned with metabolic phosphorylation, a very active field. It was my first national meeting and the session attracted a very large audience. I was nervous. After my presentation it appeared to me that half of the audience jumped to their feet with objecting questions. The idea of enzymes in the membrane was simply foreign to the orthodox view of an inert, passive membrane. At that time the use of isotopes for assessing cellular compartmentation was also new. I defended myself as best I could, but the objections went on and on until finally the Chairman, Dr. Carl Cori, a Nobel Prize winner and a most eminent authority on metabolism and phosphorylation, came to my rescue. He closed the prolonged discussion by summarizing my evidence and proclaiming my interpretation as correct. I have been forever grateful to him. I later demonstrated the presence of other cell-surface enzymes [26,27] and subsequently wrote a monograph on the subject [28].

For the next decade my research life went smoothly. I continued investigations on yeast, concerned with sugar transport and metabolism with collaboration of Harry Berke, Joe Demis, Ivan Bihler [29–33] and returned to K^+ transport, collaborating with Barry Jones and Bill Armstrong [34–39]. I also diversified into transport of PO_4^{2-} with Joan Goodman and Chan Jung [40,41] and Mg^{2+} (and other bivalent cations) with David Jennings and Fred Fuhrmann [42–45], an unexplored area at that time. I even briefly explored muscle transport with Joe Demis [46,47], and intestinal transport with Tom Clarkson [48,49]. Productivity was high; financial support through the AEC was generous; teaching loads were reasonable. Then yet another unpredictable outside event was to have a major impact on my research direction. About 1961 the AEC called us about a potentially serious toxicological problem at Oak Ridge. One of their installations was using tall columns of metallic mercury and the metal was working its way through the seams so that minute droplets were spreading over the floor. The hazards of inhaled mercury vapor were of serious concern. I was delegated to develop a research program to evaluate the important parameters involved in mercury metabolism. We (Tom Clarkson, Bob Sutherland, and I) proceeded to provide the first definitive assessment of the metal's deposition, distribution and excretion patterns; its half times in the body and its steady-state build up in various tissues, all with animals [50–53]. The instigators of our effort, from Oak Ridge and AEC, however, never asked to see or discuss all this marvellous information we accumulated. Their engineers had, in the interim, modified the ventilating system so that the mercury vapor level was reduced below the toxic level. My enforced exposure to mercury did, however, have an unexpected positive influence on my future research and productivity. I became curious about the effects of mercurials at the cellular level. The only relevant information concerned the action of mercurial diuretics on kidney function. Clearly these compounds acted on ion transporting systems, but I considered the kidney to be much too complex an organ to allow easy elucidation of specific membrane effects. Our first effort was carried out by Hermann Passow with yeast cells. He found that mercuric chloride induced an irreversible K^+ leak that resulted in cell death [54]. The results did not, however, encourage us to further studies along these lines with yeast. Our experiences with the mercurials and prior experiences with uranium did, however, encourage us (Passow, Clarkson and myself) to write the first definitive review of the cellular effects of heavy metals, with particular emphasis on the membrane as a primary target for toxicological actions [55]. For Passow and myself this was the start of a long continuing scientific (and personal) relationship.

The early part of my research career on cell membranes had been concerned almost exclusively with transport by yeast cells. The yeast cell has many advantages for such studies. It is a heavily walled, 'tough' cell, tolerant to virtually all experimental conditions. It lives in water, requiring no osmotic protection. As with other freshwater forms, its permeability is extremely low, so that diffusion-driven losses of solutes are minimal [56,57]. Fluxes are largely active (uphill) flows driven by metabolic energy. It is, therefore, an excellent object in which to study active transport processes, but is of little value in assessing passive permeabilities. In looking back it is clear that by this time (mid 1960s) I had already 'mined out' the more obvious aspects of yeast transport that could readily be evaluated with existing technologies and within the framework of existing concepts. I was still publishing well but not really breaking new ground. I did not fully recognize that it was time for a change. Fortunately, fate again intervened in the form of Dr. Larry Young, the Chief of Medicine at Rochester. He was grooming some of his brighter young staff to become Division Heads. He encouraged them to take a year off to do basic research and asked me to supervise Bob Weed, a hematologist. Because it made no sense for him to study yeast cells we initiated work on the effects of mercury and mercurials on red blood cells [58]. The experience was a stepping stone for him. He soon became the Head of the Hematology Division and a leader in his field. It opened new pathways of research for me. Soon yeast was dropped and the red cell became my model system. Furthermore, because the interactions of mercurials are highly specific for sulfhydryl groups, their actions can be attributed to disturbances of functional proteins. The sulfhydryl map of the red cell is simple, involving two compartments: (a) the membrane, and (b) the cytoplasm, in which most of the sulfhydryl is associated with a single defined protein, hemoglobin. There are no internal structures or large numbers of functional proteins to confound the situation. The red cell turned out to be an excellent choice for evaluating the roles of protein sulfhydryl groups in transport, using mercurials. Almost overnight I became an 'expert' on protein sulfhydryl groups and their role in transport. My interest in membrane proteins led ultimately to the discovery of Band 3, the anion transport protein (to be discussed later).

Mercurials influence most protein-mediated membrane functions and they are available in many forms some of which penetrate very slowly and others rapidly [59–63]. We (John van Steveninck, Alcides Rega, Bob Weed, Bob Sutherland and I) exploited the differences in penetrability to assess locations of 'sensitive' sites. The non-penetrating forms could only react with ligands exposed on the outer face of the membrane, whereas those that penetrated slowly could also react with ligands within the membrane or on its cytoplasmic face, with a predictable time-sequence of events. Paradoxically, those mercurials that penetrated most rapidly would quickly reach the cytoplasmic compartment where they would be 'soaked up' by interaction with the massive amounts of hemoglobin, and would thereby have little toxic effect. Using mercurials we could roughly map out the location of functional groups involved in sugar and cation transport, and crudely estimate their numbers.

As noted, another particularly useful property of the mercurials is their exquisite specificity and high affinity for reaction with sulfhydryl ligands, so their actions in the membrane can be attributed to protein sulfhydryl groups. At the time this property was important because we knew so little about membrane proteins, their functions, numbers, or arrangements. The technologies for their study using detergents and acrylamide gel electrophoresis were not to be invented for a number of years. The finding that most membrane functions could be perturbed by mercurials, indicated that membrane proteins were essential components of most permeation pathways [reviewed in 62,63].

I was particularly frustrated by the absence of useful technologies for the study of functional membrane proteins. We (Rudy Juliano, Jope Hoogeveen, Alcides Rega, Bob Weed, Masazumi Takeshita,

Phil Knauf and I) attempted labeling with available covalent probes and separation of membrane proteins by alcohol solubilization [64–67]. The work turned out to be relatively unproductive. A major turning point came about 1971 when an important procedure became available for 'dissolving' and, for the first time, separating the proteins of the red cell membrane, the now familiar SDS-acrylamide-gel electrophoresis [68]. Unfortunately the mercurials were not particularly useful for 'labeling' functional proteins because they were not covalently bound and, therefore, would dissociate during the separation procedures. Furthermore, because sulfhydryl groups are almost ubiquitous in proteins we were soon faced with the knowledge that our 'favorite' probes were not likely to be useful in identifying specific functional proteins. Fortunately, we were 'steered' by circumstances into using other kinds of 'probes', those that react with amino groups. One of the most striking effects of mercurials was to render cells 'leaky' to K^+, attributable to sulfhydryl groups located near the inner face of the membrane [61]. A Ph.D. candidate, Phil Knauf, was examining a similar effect produced by amino-reactive reagents. These compounds were different from the mercurials, however, inasmuch as they also inhibited anion transport (the mercurials did not). In trying to sort out these phenomena it seemed important to find a nonpenetrating amino-reactive probe. Knauf found one in SITS*, a fluorescent substance, chemically a disulfonic stilbene. It had been used to label membranes so they could be followed during purification [69]. This compound turned out to be a highly specific, potent inhibitor of anion transport with no effect on K^+ permeability [70,71]. The former function was associated with ligands located on the outer surface of the membrane, whereas the latter was associated with ligands that were inaccessible to our nonpermeant probe. I've always been an opportunist in research. I very quickly switched from a major interest in K^+ transport to a major interest in anion transport. Another Ph.D. candidate, Joav Cabantchik, who was an excellent organic chemist, undertook to prepare radioactive SITS*, to be used as a label to try to identify the anion-transport protein. Unfortunately, SITS* turn-

* 4-Acetamido 4′-isothiocyano-2,2′-stilbene disulfonic acid.

ed out to be a poor labeling agent because, contrary to the literature, it was only partially covalently bound [72]. Cabantchik then undertook the heroic task of obtaining or synthesizing some 16 derivatives of disulfonic stilbenes, and testing them for inhibitory action. The most potent covalently bound form was DIDS*, which has since become a 'standard' inhibiting and labelling reagent for anion transport systems. We found that DIDS* binds almost exclusively to an abundant membrane protein with the prosaic name, Band 3 (it was the third major band from the top of SDS-acrylamide gels [73]). On this basis, with other supporting data, we proposed that Band 3, an abundant membrane protein of about 100 000 daltons, mediates anion exchange. It was one of the first transporters to be identified. This finding stimulated a great deal of interest and resulted in many invitations to speak at symposia and to write reviews. Several hundred papers on the topic have been published to date. It was one of the more rewarding discoveries made in my laboratory. It was also at this point in my career that my decision was made to move from Rochester to Toronto.

The Hospital for Sick Children – Toronto

Early in 1971, while spending a few weeks in Los Angeles trying to learn how to culture liver cells for a research project long forgotten, I received a long-distance phone call. A strange voice introduced itself as belonging to Mr. Snedden, the Executive Director of the Hospital for Sick Children in Toronto. I had never heard of the hospital. The voice explained that the Hospital was searching for a Director for its Research Institute, and that my name had emerged as a possible candidate. I thought it was all a mistake, explaining that I was not an MD, not a pediatrician, and not knowledgeable about pediatric research. I was doubtful that the Hospital and I should be seriously interested in each other. The voice insisted that there was no mistake and it talked me into visiting the Hospital after I finished my stay in Los Angeles. It turned out that Dr. Irving Fritz, of the University of Toronto, had visited the AEP in Rochester to give a

---

* 4,4′-Diisothiocyanostilbene-2,2-disulfonic acid.

seminar. He was impressed with its organization and with the quality of its staff and its science. He felt that I might be able to do a good job of reorganizing the Research Institute at the Hospital and that the specifics of my research interests and experience were secondary. It was he who had submitted my name to the Search Committee.

Out of curiosity, perhaps mixed with an element of self-flattery, I agreed to visit Toronto. When I arrived I was surprised to find that the Hospital was most impressive, in fact the largest Pediatric Hospital in North America (at the time it had over 800 beds). It had established a Research Institute some years before (1957), but had a long history of research starting in the 1920s with development of the baby food, Pablum. At first I was not very interested. The Institute was controlled by the Clinical Chiefs who, although well meaning, were not particularly experienced in research. The research staff were by and large recruited from those on the clinical staff who displayed any interest in research. Although some of these individuals turned out to be talented researchers, others did not. The research was very clinically oriented. There were some good programs but overall it was not an exciting place at the forefront of biomedical research. In fact, when my wife asked me, I told her that the situation was a 'can of worms' and that the Director of the Institute would be under the control of the Hospital Chiefs. Under those circumstances, I had told the Search Committee that I was not particularly interested and I, therefore, proceeded to be rather blunt about what I felt was wrong and what had to be done to build a first-rate institute. It seemed, however, that my rather critical attitude impressed them, for when I expressed my disinterest they asked me to visit another time as a 'Consultant' to help the Committee outline a plan for the development of the Institute. On my second visit, they offered me substantial space and financial resources to expand the Institute, and more importantly, they offered me, as Director, virtually complete autonomy. I would have control of research positions and budget, and would report directly to the Board of Trustees as well as to the Executive Director. They fully accepted my premises that they had to invest heavily in basic research as a base for the more clinically and disease-oriented research; that they had to recruit, on a world-wide basis, a core of 'full-time' individuals chosen specifically on the basis of their exceptional research talent; and that much more emphasis had to be placed on modern cell and molecular biology. They essentially wrote me a 'blank check'. A new wing had just been completed; several floors would be devoted to new research laboratories. Lots of money were available. If I did not like the administrative and reporting structure I could rewrite the Institute's by-laws. It was an appealing package. It was a bargain I could not refuse: space, money, and power.

I was actually ready to make a move. When Bill Neuman and I had accepted positions as Co-Directors of the Rochester AEP, we had reached a mutual understanding that we would administer it for eight years, sufficient time in our estimation to finish converting it from a war-time project into a first-class academic operation. When the Toronto opportunity surfaced seven of the eight years had already passed. Our reorganization of the Project was almost completed. It was now firmly established as an integral part of the Medical School, and running smoothly. I was 54 years old and ready for one more major effort at building an important research establishment. Furthermore, I had been in Rochester for my whole career so it was time for a new adventure. I set a ten-year timetable for myself (it turned out to be 14 years).

When I discussed my interest in Toronto with Evelyn, she was quite surprised because I had not been particularly enthusiastic after my first visit. She was, however, ready to give it a try. Our youngest son was getting ready to go to college. The other two were already gone, so we would soon be on our own, ready to embark on a new, less complicated life. Furthermore, Toronto was an attractive city. Soon we were busy apartment hunting, and I found myself sitting in a gigantic office with a private bathroom, inherited from my predecessor, wondering what I should do. It was very lonely at first. I had met a few individuals during the search process, but otherwise knew no one. Nor did I know much about the ongoing research activities except in a superficial way. I certainly knew little about Pediatric Research. It was several months before I began to feel at home; before I had sufficient familiarity with the hospital and its staff to start the rebuilding program.

My general plan involved several aspects. First, I wanted to aim at the highest possible quality to be achieved by intensive searches for outstanding individuals, usually younger investigators of high promise. Second, I wanted to concentrate a substantial fraction of resources in a few chosen areas, so that critical masses of effort could be mounted. Third, the areas to be chosen should be based on their relevance to Pediatrics, on their scientific breakthrough potential (as for example, molecular genetics and immunology), and to some degree on opportunities to recruit outstanding individuals or to mobilize resources. Fourth, I wanted a mix, in each chosen area, of basic, disease-oriented and bedside research. Fifth, I wanted the research to be closely interdigitated, wherever feasible, with the clinical operation; the Hospital with its unique patient resources was to become our laboratory and the research information was to be immediately available for clinical application. Sixth, I wanted to expand the training aspects of research, at the graduate, postdoctoral and even the clinical residency levels. These goals may not appear to be particularly unusual for a hospital-based research establishment, but their implementation required major changes in operating philosophy and 'gestalt' at the Hospital. Where the Research Institute had been in a subsidiary position, it was to become much more autonomous and an equal partner (a favored partner in the eyes of some).

During my tenure all of the objectives outlined above have, more or less, been met. By the time I retired from the position of Director, in 1986, the Institute had become the pre-eminent hospital-based research establishment in Canada and one of the two major Pediatric Research establishments in the world (the other being the Children's Hospital Medical Center in Boston). I will not bore the reader with the details, but I feel impelled to at least briefly describe what I consider to be one of my important career accomplishments.

The first years at Toronto were a period of rapid expansion. Search committees were meeting constantly. I seemed to be having dinner with new recruits almost every evening, trying to convince them that their future was with the Institute. Funding and space were not a problem. Later, as the Hospital began to suffer deficits and as inflation began to erode our 'hard' money, financial restraint became the watchword. We, nevertheless, continued to recruit and expand until the end of my term, because we were able to generate more and more external funding. During the last period of my tenure, we ran out of space. Our labs became seriously overcrowded, inhibiting further growth. We still managed some development, using space borrowed from the University in which we 'parked' our latest recruits. As I retired, a new building neared completion, a large fraction of which was devoted to meeting the space deficit in research.

It is clear the above that my term was characterized by considerable expansion. At the present time the Institute has over 700 individuals associated with it. To a large degree research now permeates the professional activities of the Hospital; 35 divisions have a meaningful involvement. The success of the Institute is evident by many criteria such as: generous external funding, recruitment of excellent staff, their professional recognition, their publication record, and public acclaim. One form of recognition that I especially treasure is imitation. There has been a virtual epidemic among Canadian Hospitals in expanding or initiating in-house research establishments. These have often been patterned on our Institute. In fact, I have been a Consultant to Hospitals across Canada and even in the U.S.A. (at least a dozen) and have been appointed to the Boards of several institutes. Although I take much pride and satisfaction in our success, I fully recognize that many, many others must share that credit: trustees, administration, department heads, and members of the Institute. When I retired I was satisfied that I had left a lasting mark on our and on many other Canadian hospitals; that my time as an Administrator was well spent.

Administration is often considered to be a dull activity, a necessary evil with which one impatiently learns to live. This attitude is justified to some degree, especially in heavily bureaucratized organizations. On the other hand, I found pleasure, satisfaction and excitement in orchestrating the growth and development of the research establishments both in Rochester and Toronto. Patience was required because progress in institution building is slow, but in the end the pride of accomplishment has made the effort more than worth while. I feel proud whenever I hear or read a complimentary comment concerning 'my' Department or 'my' Institute.

Research – Toronto

I moved to Toronto just at the time that Cabantchik and I were hot on the trail of the anion-transport protein. In fact the final identification of Band 3 as that protein was actually completed in Toronto. We (Cabantchik had completed his Ph.D. and was now a Postdoctoral Fellow; others, who made major contributions were Phil Knauf, Eze Shami, Michael Balshin, Sol Ship, Sergio Grinstein, Mohabeer Ramjeesingh and Ann DuPre) then proceeded to correlate its structural arrangement in the membrane with its functional properties [74–83]. The key to such studies was the use of 'bimodal' inhibitory probes [reviewed in 76]. These were chemical agents that could bind to Band 3, either reversibly or irreversibly (by covalent binding), depending on the experimental conditions. In the former case, the probe was used to determine, by kinetic analysis, the nature of the inhibition. In the latter case, probe-labeled Band 3 was isolated and the number of binding sites determined. We identified three useful bimodal probes: DIDS, which was reversibly bound, at 0° and irreversibly at 22°C [77]; pyridoxal phosphate which is irreversibly bound after addition of borohydride [79]; and NAP-taurine with which the irreversible reaction is triggered by exposure to light [80]. Within a few years the general features of Band 3 operation were characterized with contributions from our own and other laboratories. In a few years anion transport had moved from concept to reality, from hypothetical 'carriers' to permeation via specific transmembrane proteins, involving particular anion-binding sites associated with specific ligands and conformational rearrangements [76,83]. With Grinstein and later Ramjeesingh we then concentrated our efforts on mapping the arrangement of Band 3 in the membrane (it has multiple crossings of the bilayer), and the locations of functional sites and their mode of operation [84–93]. We kept publishing on Band 3 (except during a pleasant year's sabbatical in Zurich in 1977 with G. Semenza, involving collaboration on intestinal sugar transport [94,95]) but it was clear that there were limitations to our relatively primitive techniques (use of various covalent probes followed by proteolytic and chemical cleavage of Band 3). Finally, about 1982, after ten productive years, I left Band 3 to other more eager investigators who were applying new technologies such as monoclonal antibodies, cloning and sequencing.

In part, I dropped the Band 3 research because I had already initiated another set of projects that turned out to be more exciting. Some ten years earlier a Ph.D. candidate, Linda Wotring (later Linda Roti Roti), had become interested in the capacity of cultured lymphoblasts to tolerate exposure to hypo-osmotic media. The cells did not appear to swell in diluted solutions. At that time, the measurements were made by packing cells in a calibrated tube by centrifugation, equivalent to hematocrit determinations for red-cell volumes. This is, however, a slow procedure. The Coulter Counter had just become available, allowing more rapid estimates of cell volume. Roti Roti found that immediately after exposure to hypotonicity the lymphocytes did indeed undergo the expected osmotic swelling, but that soon thereafter they would spontaneously shrink back toward normal (isotonic) volume [96]. She demonstrated that the shrinking phase was associated with a large increase in K^+-permeability and a consequent loss of KCl and osmotically obliged water. After exposure to hyperosmotic media the expected initial shrinking was followed by a slow reswelling toward normal size. A similar phenomenon had just been demonstrated in bird erythrocytes [97] and it has been subsequently demonstrated that most cell types respond similarly [see reviews 98–100]. There has been much speculation that these volume readjustments following exposure to anisotonic media represent activities of cell-volume regulating systems. This assumption is inherent in the naming of the phenomena, RVD (regulatory volume decrease) and RVI (regulatory volume increase).

Some ten years after the publication of our initial studies I returned to the phenomena of RVD and RVI, stimulated by interactions with Dr. I. Gelfand, Head of the Hospital's Immunology Division, who wanted to know about electrolyte metabolism and volume changes that might be related to lymphocyte activation. I enlisted the collaboration of Sergio Grinstein (now a member of the Institute's staff) and he soon became the spearhead of the research effort. We demonstrated that in RVD, the increased K^+ permeability appeared to be Ca^{2+}-activated [101,102], similar to Ca^{2+}-

dependent K^+ permeability present in many or perhaps all cells. We also demonstrated that cell swelling activated a Cl^- permeability that was even larger than the increase in K^+ permeability, so that depolarization occurred [103,104]. The volume-activated K^+ and Cl^- permeabilities were extensively characterized when Balasz Sarkadi from Budapest spent a year in our laboratory [105–109].

Although the phenomenon of RVD was interesting, it turned out that the RVI was even more so. We demonstrated that the uptake of NaCl that caused the reswelling, resulted from activation of Na^+/H^+ exchange [110]. The parallel operation of this exchanger with Cl^-/HCO_3^- exchange resulted in a net gain of NaCl. In other cell types, however, the NaCl uptake resulted from operation of a NaCl co-transport system [98,99]. It is interesting that the two mechanisms used in the kidney and in various glands for transepithelial movement of salt and osmotically obliged water, are also used by many cells for uptake and control of salt and water content.

The activation of Na^+/H^+ exchange by osmotic challenge indicated a linkage between cell volume regulation and cell pH regulation. We demonstrated that in lymphocytes the exchanger is a major factor in the cell's capacity to adjust to acid loads [111]. The link between volume and pH regulation is bicarbonate. In its absence activation of the Na^{2+}/H^{2+} exchanger results in alkalinization but in little volume change. In the presence of bicarbonate, the Cl^-/HCO_3^- exchange buffers out much of the pH change, but the net result is an uptake of NaCl and osmotic swelling [112].

We devoted considerable effort to characterizing the Na^+/H^+ exchange system [113–116] and its regulation [100,117–123]. As in other cells [124], the lymphocyte exchanger is controlled by a 'modifier' site on the cytoplasmic side of the membrane that is sensitive to cytoplasmic pH. As a result of its action the exchanger is quiescent at 'normal' cytoplasmic pH (about 7.1), but becomes more and more active as the pH is reduced. It, therefore, acts as a pH stat that maintains pH at its 'set-point'. When we examined the activation by hypertonic shrinking, we found that it was not due to acidification, but rather that the pH became more alkaline than normal. The mechanism turned out to involve a change in the 'set-point' of the modifier to a more alkaline value. This finding led us unexpectedly into a whole new field of transport regulation by protein kinases and into comparisons of hypertonic activation of the exchanger with its activation by growth factors, hormones and other interesting agents [100]. Why hypertonicity should activate kinases and why so many cell activators should stimulate the Na^+/H^+ exchange systems are fascinating but as yet unanswered questions. Dr. Grinstein has continued productive activities in this interesting field, but I have recently moved into studies of Cl channels and volume regulation in epithelial cells.

Perspectives on research success

I have already mentioned that at the time that I started publishing, only one other laboratory was concerned with yeast transport, that of E.J. Conway, an eminent Irish biochemist. I, therefore, had a semi-monopoly on the yeast transport field. This was a great advantage for me. Our yeast studies were productive, so that whenever a meeting was to be held on transport, the organizers felt impelled to include me on the program. I was the yeast person. Thus I travelled a great deal and became well acquainted and friendly with most of the important figures in the transport field. There is a great advantage to holding a monopoly position in some field of research.

The total number of membrane researchers was small in those early times, and it was possible to know almost everyone fairly well. I once examined the growth of membrane research by estimating the number of papers published each years. From about 50 papers in 1948 the number had reached over 400 by 1968, representing an 8% compounded growth. After 1970 the growth rate increased substantially, as a result of major technical breakthroughs that allowed identification and isolation of membrane proteins [125]. By 1987 over 5000 membrane papers were published. In the 1950s and 1960s the membrane field was relatively homogeneous, with most interest focussed on transport parameters, particularly fluxes and their electrical manifestations, with some lipid structure and some morphology (electron microscopy). By the mid 1970s and certainly by the 1980s, the number of investigators had grown substantially and the field had split into many special areas. Growth and diversification has resulted in fragmentation. It is no longer possible to meet and know everyone in the field. It has become difficult to even keep up with major developments, despite the aid of computer searches.

The quite sudden change in the amount of membrane research and its directions that occurred in the early 1970s, can be attributed to a series of technical innovations [125]. Perhaps the most important were those that allowed solubilization of membrane proteins using detergents, and their separation by acrylamide gel electrophoresis. Before that time it was virtually impossible to attribute transport to particular proteins, or to even think of isolating them. An important exception was the Na/K ATPase whose enzyme activity allowed its identification and its partial purification by more standard biochemical procedures [4]. As a result, interest in membrane proteins has become one of the main foci of membrane research. New groups of investigators such as biochemists, cell biologists, immunologists, endocrinologists and others have been attracted with interests not only in transport proteins, but also in proteins that act as receptors, structural elements, antigens, or in recognition phenomena etc. In the past, technologies have also had an enormous impact. In the late 1940s, for example, the application of isotopes and of sophisticated electrical measurements led to substantial advances in understanding of transport phenomena. From these cases it is clear that advances in technology may be just as important as the formulation of new concepts, in the development of research. In fact, technical breakthroughs usually precede conceptual advances. They allow new sets of information to be generated that in turn allow development and testing of new concepts. Those who are opportunistic in exploiting new technologies can end up leading the research pack.

As I look back on my own career it is clear that some of my more important 'discoveries' resulted from luck, accident, or other forms of serendipity [126]. Unexpected results sometimes led me into exciting unanticipated directions, whereas the logical approach, the asking of questions, and formulation and testing of hypotheses led me into slower, incremental advances down more expected pathways. I have already mentioned some examples. The use of metals such as uranium and mercury as probes of membrane function did not arise out of my intellect but resulted from my involvement in practical toxicological studies. The accidental use of detergent containing hexametaphosphate led me into the concept of metal binding ligands on the surface of the yeast cell. I renewed my interest in volume regulation in lymphocytes because I was being 'bugged' by the Head of our Immunology Division to look at electrolyte metabolism in that cell type. In turn the studies on volume regulation led to a series of unexpected but interesting investigations on regulation of cell pH and potentially into factors regulating growth. Perhaps the most important factor in my research progress has been my ability (if it is an ability), or my luck, in attracting creative students, fellows and colleagues. Many of the important research efforts for which I received some credit were generated as a result of my interactions with such individuals. Although it is generally recognized that good companions, luck, serendipity, and unexpected results, all nonlogical factors, play a very important role in research, the granting agencies show no recognition of these factors. If I were to apply for a grant to 'fool around' with lymphocytes or to 'brainstorm' with colleagues, until some lucky result or unexpected idea sent me down a research path that would, after the fact, be labeled original and exciting, I would get no support. It pays to be lucky but one must also be clever in recognizing opportunities as they present themselves and in exploiting them. For success, be clever, be opportunistic, stay close to creative companions, and when necessary, work hard. It also helps if one can write a grant proposal that sounds exciting, innovative and original to the referees.

Essential companions

Everyone knows that students and collaborators can contribute extensively to research productivity. For those, like myself, with commitments to administrative tasks, companions are essential. As my administrative role increased in dimension and my available time shrank and became more unpredictable and fragmented, I became more and more dependent on the cooperation of colleagues;

students, postdoctoral Fellows, and friendly collaborators. I became more and more removed from the 'bench', so that my involvement was largely at the level of planning of protocols, analysis of results, brainstorming for ideas, and writing or cowriting of papers and grants. Fortunately, I was blessed with a long series of outstanding Fellows and collaborators, without whom I would have been unable to carry on a meaningful research activity. Many of them have become well recognized scientists in their own right. I am grateful to them, proud of them and consider them friends. It was one of my great pleasures to be associated with them and to watch them succeed. I have mentioned a few in my narrative in discussing particular research activities. Their names are listed with mine in our joint papers; all of them deserve full credit. I list their names here as an inadequate 'thank you':

Arnon Aharoni; William McD. Armstrong; Michael Balshin; Harry Berke; Maths Berlin; Ivan Bihler; G.W.F.H. Borst-Pauwels; William Breuer; Joav Cabantchik; Thomas W. Clarkson; Joseph Demis; Mark Deziel; Richard A. Dilley; Jerome Donlon; Alexander Dounce; Ann DuPre; Gunter F. Fuhrmann; Jorgen Funder; Irwin Gelfand; Joan Goodman; Sergio Grinstein; J. Thomas Hoogeveen; Leon Hurwitz; David Jennings; W.B.G. Jones; Rudolph Juliano; Chan Jung; Amira Klip; Philip Knauf; Fredericka Dodyk Kundig; Paul L. LaCelle; Avinoam Livne; Paul Morrow; Shizuo Odashima; Thomas J. Ostwald; Hermann Passow; Antonio Pena; Mohabir Ramjeesingh; Alcides Rega; Esther Reichstein; Connor Reilly; Tim Rink; Linda Roti Roti; Balasz Sarkadi; Thomas Scharff; Giorgio Semenza; Yehezkel Shami; Saul Ship; John van Steveninck; Robert Sutherland; Masazumi Takeshita; Robert Weed; Christoff Weiss; J.F. White.

Family matters

I married while still in graduate school. That might not seem like a revolutionary situation today, but in 1940 it was unusual. The pattern, at that time, was for the male to establish himself in a career and then to offer himself to his chosen at an advanced age. Marriages for professionals were, therefore, usually undertaken when one was about thirty. I have already pointed out how my pending marriage to a musician determined the location of my graduate studies. As our family grew (ultimately to three childeren and to a total of 14, counting spouses and grandchildren) I tried to save a reasonable amount of time for them. Unlike many colleagues, I never, short of an emergency deadline, took work home; nor did I spend weekends and evenings at the laboratory. Evelyn's and my professional lives were distinct; music and medical research have little in common. Nevertheless we were always strongly supportive of each other's activities. When I started to receive invitations to travel to meetings, Evelyn usually arranged to travel with me, so she soon became friends with my scientific companions. She was not very familiar with the intricacies of science, but she was quite familiar with scientists. My life as a scientist must have been pleasant enough because my children chose the same path, more or less. My two sons have become molecular biologists, and my daughter was trained in psychology, a sort of science. She married a psychiatrist, but he, at least, had the good sense to become an academic, so I count all my children as following in my footsteps, which I take as a great compliment and a signal that my career was consistent with a happy family life.

REFERENCES

1 A. Rothstein, J. Cell Comp. Physiol., 28 (1946) 221-230.
2 A. Rothstein and H. Enns, J. Cell. Comp. Physiol., 28 (1946) 231-252.
3 E.J. Conway (1955) Int. Rev. Cytol., 4, 377-410.
4 J.C. Skou (1965) Physiol. Rev., 45, 596-618.
5 A. Rothstein, in G.E.W. Wolstenholme and C.M. O'Connor (Eds.), Regulation of the Inorganic Ion Content of Cells, CIBA Found. Study Group No. 5, Churchill, London, 1960, pp. 53-68.
6 E. Adolph et al., Physiology of Man in the Desert, Interscience, New York, 1947.
7 C. Voegtlin and H.C. Hodge, Pharmacology and Toxicology of Uranium Compounds, National Nuclear Energy Series, Div. VI, Vol. 1, 1949.
8 C. Voegtlin and H.C. Hodge, Pharmacology and Toxicology of Uranium Compounds, National Nuclear Energy Series, Div. VI, Vol. 2, 1953.
9 A. Rothstein and C. Larrabee (1948) J. Cell. Comp. Physiol., 32, 247-260.
10 A. Rothstein, A. Frenkel and C. Larrabee (1948) J. Cell. Comp. Physiol., 32, 261-274.
11 A. Rothstein, R. Meier and L. Hurwitz (1951) J. Cell. Comp. Physiol., 37, 57-81.
12 A. Rothstein, R. Meier and C. Larrabee (1951) J. Cell. Comp. Physiol., 38, 245-270.
13 L. Hurwitz and A. Rothstein (1951) J. Cell. Comp. Physiol., 38, 437-450.
14 H. Passow, A. Rothstein and T.W. Clarkson (1961) Pharmacol. Rev., 13, 185-224.
15 D. Branton and R.B. Park, Papers on Biological Membrane Structures, Little, Brown, Boston, 1984.
16 S.J. Singer and G.L. Nicholson (1972) Science, 175, 720-731.
17 A. Davson and J.F. Danielli, The Permeability of Natural Membranes, Cambridge University Press, Cambridge, U.K., 1943.
18 E. Gortner and F. Grendel (1925) J. Exp. Med., 41, 439-443.
19 A. Rothstein (1954) Symp. Soc. Exp. Biol., 8, 165-201.
20 W. Wilbrandt and T. Rosenberg (1961) Pharmacol. Rev., 13, 109-172.
21 J. van Steveninck and A. Rothstein (1965) J. Gen. Physiol., 49, 235-246.
22 A. Rothstein and J. van Steveninck (1965) Ann. N.Y. Acad. Sci., 237, 606-623.
23 S. Roseman (1969) J. Gen. Physiol., 54, 138-152.
24 A. Dounce, A. Rothstein, G. Beyer, R. Meier and R. Freer (1948) J. Biol. Chem., 174, 361-370.
25 A. Rothstein and R. Meier (1948) J. Cell. Comp. Physiol., 32, 77-96.
26 A. Rothstein, R. Meier and R. Scharff (1953) Am. J. Physiol., 173, 41-46.
27 D. Demis, A. Rothstein and R. Meier (1954) Arch. Biochem. Biophys., 48, 55-62.
28 A. Rothstein, Enzymology of the Cell Surface. Protoplasmatologia, Band II, E.4., Springer-Verlag, Vienna, 1954.
29 A. Rothstein and H. Berke (1952) Arch. Biochem. Biophys., 36, 195-201.
30 A. Rothstein and H. Berke (1952) Proc. Soc. Exp. Biol. Med., 81, 559-563.

31 A. Rothstein and C. Demis (1953) Arch. Biochem. Biophys., 44, 18–29.
32 H. Berke and A. Rothstein (1957) Arch. Biochem. Biophys., 72, 380–394.
33 I. Bihler, A. Rothstein and L. Bihler (1961) Biochem. Pharmacol., 8, 289–299.
34 A. Rothstein in A.M. Shanes (Ed.), Electrolytes in Biological Systems, American Physiological Society, Washington, DC, 1955, pp.65–100.
35 A. Rothstein and M. Bruce (1958) J. Cell. Comp. Physiol., 51, 145–159.
36 A. Rothstein and M. Bruce (1958) J. Cell. Comp. Physiol., 51, 439–455.
37 W.M. Armstrong and A. Rothstein (1964) J. Gen. Physiol., 48, 61–71.
38 W.B.G. Jones, A. Rothstein, F. Sherman and J.N. Stannard (1965) Biochim. Biophys. Acta, 104, 310–312.
39 W.M. Armstrong and A. Rothstein (1967) J. Gen. Physiol., 50, 967–988.
40 J. Goodman and A. Rothstein (1957) J. Gen. Physiol., 40, 915–923.
41 C. Jung and A. Rothstein (1965) Biochem. Pharmacol., 14, 1093–1112.
42 D. Jennings, D. Hooper and A. Rothstein (1958) J. Gen. Physiol., 41, 1019–1026.
43 A. Rothstein, A. Hayes, D. Jennings and D. Hopper (1958) J. Gen. Physiol., 41, 585–594.
44 F. Fuhrmann and A. Rothstein (1968) Biochim. Biophys. Acta, 163, 331–338.
45 F. Fuhrmann and A. Rothstein (1968) Biochim. Biophys. Acta, 163, 325–330.
46 D. Demis and A. Rothstein (1954) Am. J. Physiol., 178, 82–84.
47 D. Demis and A. Rothstein (1955) Am. J. Physiol., 180, 566–574.
48 T.W. Clarkson and A. Rothstein (1960) Am. J. Physiol., 199, 898-906.
49 T.W. Clarkson, A. Rothstein and Cross, A. (1961) Am. J. Physiol., 200, 781–788.
50 A.D. Hayes and A. Rothstein (1962) J. Pharmacol. Exp. Ther., 138, 1-10.
51 T.W. Clarkson and A. Rothstein (1964) Health Phys., 10, 1115-1121.
52 A. Rothstein and A. Hayes (1964) Health Phys., 10, 1099–1113.
53 T.W. Clarkson, A. Rothstein and R. Sutherland (1965) Br. J. Pharmacol. Chemother., 24, 1–13.
54 H. Passow, A. Rothstein and B. Loewenstein (1959) J. Gen. Physiol., 43, 97–107.
55 H. Passow, A. Rothstein and T.W. Clarkson (1961) Pharmacol. Rev., 13, 185–224.
56 A. Rothstein in J.F Hoffman (Ed.), The Cellular Functions of Membrane Transport, Prentice-Hall, Englewood Cliffs, NJ, 1963, pp. 23–39.
57 A. Rothstein in L.B. Guze (Ed.), Microbial Protoplasts, Spheroplasts and L-Forms, Williams and Wilkins, Baltimore, 1965, pp. 174–185.
58 R. Weed, J. Eber and A. Rothstein (1962) J. Gen. Physiol., 45, 395–410.
59 J. van Steveninck, R.I. Weed and A. Rothstein (1965) J. Gen. Physiol., 48.
60 A.F. Rega, A. Rothstein and R.I. Weed (1967) J. Cell. Physiol., 70, 45–52.
61 R.M. Sutherland, A. Rothstein and R.I. Weed (1967) J. Cell. Physiol., 69, 185–198.
62 A. Rothstein in F. Bronner and A. Kleinzeller (Eds.), Current Topics in Membranes and Transport, Vol. 1, Academic Press, New York, 1970, pp. 1–76.
63 A. Rothstein in D.F.H. Wallach (Ed.), The Function of Red Blood Cells: Erythrocyte Pathobiology, Alan Liss, New York, 1981, pp. 105–131.
64 J.Th. Hoogeveen, R. Juliano, J. Coleman and A. Rothstein (1970) J. Membr. Biol., 3, 156–172.

65 A. Rothstein, M. Takeshita and P.A. Knauf in F. Kreuzer (Ed.), Passive Permeability of the Cell Membrane, Plenum, New York, 1971, pp. 393–413.
66 A.F. Rega, R.I. Weed, C.F. Reed, G.G. Berg and A. Rothstein (1967) Biochim. Biophys. Acta, 147, 297–312.
67 R.L. Juliano and A. Rothstein (1971) Biochim. Biophys. Acta, 249, 227–235.
68 G. Fairbanks, T.L. Steck and D.F.H. Wallach (1971) Biochemistry, 10, 2606–2617.
69 A.H. Maddy (1964) Biochim. Biophys. Acta, 88, 390–399.
70 P.A. Knauf and A. Rothstein (1971) J. Gen. Physiol., 58, 190–210.
71 P.A. Knauf and A. Rothstein (1971) J. Gen. Physiol., 58, 211–223.
72 Z.I. Cabantchik and A. Rothstein (1972) J. Membr. Biol., 10, 311–330.
73 Z.I. Cabantchik and A. Rothstein (1974) J. Membr. Biol., 15, 207–226.
74 Z.I. Cabantchik and A. Rothstein (1974) J. Membr. Biol., 15, 227–248.
75 A. Rothstein, Z.I. Cabantchik, M. Balshin and R. Juliano (1975) Biochem. Biophys. Res. Commun., 64, 144–150.
76 Z.I. Cabantchik, P.A. Knauf and A. Rothstein (1978) Biochim. Biophys. Acta, 515, 239–302.
77 S. Ship, Y. Shami, W. Breuer and A. Rothstein (1977) J. Membr. Biol., 33, 311–323.
78 Y. Shami, A. Rothstein and P.A. Knauf (1978) Biochim. Biophys. Acta, 508, 357–363.
79 Z.I. Cabantchik, W. Breuer, M. Balshin and A. Rothstein (1975) J. Biol. Chem., 250, 5130–5136.
80 Z.I. Cabantchik, P.A. Knauf, T. Ostwald, H. Markus, L. Davidson, W. Breuer and A. Rothstein (1976) Biochim. Biophys. Acta, 455, 526–537.
81 P.A. Knauf, S. Ship, W. Breuer, L. McCulloch and A. Rothstein (1978) J. Gen. Physiol., 72, 607–630.
82 P.A. Knauf, W. Breuer, L. McCulloch and A. Rothstein (1978) J. Gen. Physiol., 71, 631–649.
83 A. Rothstein, Z.I. Cabantchik and P. Knauf (1976) Fed. Proc., 35, 3–10.
84 M. Ramjeesingh, S. Grinstein and A. Rothstein (1980) J. Membr. Biol., 57, 95–102.
85 S. Grinstein, L. McCulloch and A. Rothstein (1979) J. Gen. Physiol., 73, 493–514.
86 M. Ramjeesingh, A. Gaarn and A. Rothstein (1980) Biochim. Biophys. Acta, 599, 127–139.
87 M. Ramjeesingh and A. Rothstein (1981) Biochim. Biophys. Acta, 641, 173–182.
88 A. DuPre and A. Rothstein (1981) Biochim. Biophys. Acta, 646, 471–478.
89 M. Ramjeesingh, A. Gaarn and A. Rothstein (1981) J. Bioenerg. Biomembr., 13, 411–423.
90 M. Ramjeesingh and A. Rothstein (1982) Membr. Biochem., 4, 259–270.
91 A. Rothstein and M. Ramjeesingh (1982) Phil. Trans. Roy. Soc. Lond. B, 299, 497–507.
92 M. Ramjeesingh, A. Gaarn and A. Rothstein (1983) Biochim. Biophys. Acta, 729, 150–160.

93 M. Ramjeesingh, A. Gaarn and A. Rothstein (1984) Biochim. Biophys. Acta, 769, 381-389.

94 C. Tannenbaum, G. Toggenburger, M. Kessler, A. Rothstein and G. Semenza (1977) J. Supramol. Struct., 6, 519-533.

95 G. Toggenburger, M. Kessler, A. Rothstein, G. Semenza and C. Tannenbaum (1978) J. Membr. Biol., 40, 169-190.

96 L. Roti Roti Wotring and A. Rothstein (1973) Exp. Cell Res., 79, 295-310.

97 F.M. Kregenow (1971) J. Gen. Physiol., 58, 372-395.

98 E.K. Hoffmann (1986) Biochim. Biophys. Acta, 864, 1-31.

99 F.M. Kregenow (1981) Annu. Rev. Physiol., 43, 493-505.

100 S. Grinstein and A. Rothstein (1986) J. Membr. Biol., 90, 1-12.

101 S. Grinstein, A. DuPre and A. Rothstein (1982) J. Gen. Physiol., 79, 849-868.

102 S. Grinstein, B. Cohen, B. Sarkadi and A. Rothstein (1983) J. Cell. Physiol., 116, 352-362.

103 S. Grinstein, A. DuPre, C.A. Clarke and A. Rothstein (1982) J. Gen. Physiol., 80, 801-823.

104 S. Grinstein, C.A. Clarke, A. Rothstein and E. Gelfand (1983) Am. J. Physiol., 245, C160-C163.

105 B. Sarkadi, E. Mack and A. Rothstein (1984) J. Gen. Physiol., 83, 497-512.

106 B. Sarkadi, E. Mack and A. Rothstein (1984) J. Gen. Physiol., 83, 513-527.

107 S. Grinstein, A. Rothstein, B. Sarkadi and E.W. Gelfand (1984) Am. J. Physiol., 246, C204-C215.

108 B. Sarkadi, L. Attisano, S. Grinstein, M. Buchwald and A. Rothstein (1984) Biochim. Biophys. Acta, 774, 159-168.

109 B. Sarkadi, R. Cheung, E. Mack, S. Grinstein, E.W. Gelfand and A. Rothstein (1985) Am. J. Physiol., 248, C480-C487.

110 S. Grinstein, C.A. Clarke and A. Rothstein (1983) J. Gen. Physiol., 82, 619-638.

111 S. Grinstein, S. Cohen and A. Rothstein (1984) J. Gen. Physiol., 83, 341-369.

112 S. Grinstein, S. Cohen, J.D. Goetz and A. Rothstein (1985) Fed. Proc., 44, 2508-2512.

113 S. Grinstein, J.D. Goetz and A. Rothstein (1984) J. Gen. Physiol., 84, 565-584.

114 S. Grinstein, J.D. Goetz and A. Rothstein (1984) J. Gen. Physiol., 84, 585-600.

115 S. Grinstein, J.D. Goetz, W. Furuya, E.W. Gelfand and A. Rothstein (1984) Am. J. Physiol., 16, C293-C298.

116 S. Grinstein, S. Cohen and A. Rothstein (1985) Biochim. Biophys. Acta, 812, 213-222.

117 S. Grinstein, J.D. Goetz, S. Cohen, W. Furuya, A. Rothstein and E.W. Gelfand (1985) Mol. Physiol., 8, 185-198.

118 S. Grinstein, S. Cohen, J.D. Goetz and A. Rothstein (1985) Cell. Biol., 101, 269-276.

119 S. Grinstein, S. Cohen, J.D. Goetz, E.W. Gelfand and A. Rothstein (1985) Proc. Natl. Acad. Sci. USA, 82, 1429-1433.

120 S. Grinstein, S. Cohen and A. Rothstein (1985) J. Gen. Physiol., 85, 765-787.

121 S. Grinstein, S. Cohen, J.D. Goetz, A. Mellors and E.W. Gelfand (1986) Curr. Top. Cell Reg. Transp., 26, 115-136.

122 S. Grinstein, J.D. Goetz, S. Cohen, A. Rothstein and E.W. Gelfand (1986) Ann. N.Y. Acad. Sci., 456, 207-209.
123 S. Grinstein, S. Cohen, J.D. Goetz, A. Mellors and E.W. Gelfand (1986) Curr. Top. Cell Reg. Transp., 26, 115-136.
124 Aronson, P.S. (1985) Annu. Rev. Physiol., 47, 545-560.
125 A. Rothstein (1984) Can. J. Biochem. Cell Biol., 62, 1111-1120.
126 A. Rothstein (1986) Can. J. Biochem. Cell Biol., 64, 1055-1065.

G. Semenza and R. Jaenicke (Eds.) Selected Topics in the History of Biochemistry: Personal Recollections, III. (Comprehensive Biochemistry Vol. 37)

Chapter 8

Never a dull moment Peripatetics through the gardens of science and life

HENRYK (HEINI) EISENBERG

Polymer Department, The Weizmann Institute of Science, Rehovot 76100 (Israel)

Childhood and family (1921–1939)

It is said that there are moments in a human life when the panorama of your whole existence, and that of your family as well, passes before your astounded eyes, in an act of sublime remembrance. I felt this way late in June 1984, tucked away in my narrow Austrian Airlines seat, flying from Vienna to Moscow. Though carrying an Israeli passport, I had been given a visa to attend the 16th meeting of FEBS, the Federation of European Biochemical Societies, to take place in Moscow. My last visit to the capital of the U.S.S.R. was in the Fall of 1926, 58 years earlier. I came, by train that time, with my mother, Inna, to spend six months, including the very cold winter of 1926/1927, to visit her own mother, sisters and brothers and their families, whom she had not seen in a long while. Until that time we were living in Chemnitz, now Karl-Marx-Stadt, where my father Issay, Simha in official documents, was active in the textile industry, for which Chemnitz was famous. I was born in Berlin on March 7, 1921. My father had recently received an offer to establish a big textile factory in Cernăuţi, Roumania, called Czernowitz, under Austrian government before World War I, and Chernovtsy,

Plate 10. Henryk (Heini) Eisenberg. (Courtesy Photo Ben Zvi, Rehovot.)

U.S.S.R. after World War II. Cernăuţi was the capital of the beautiful Bucovina, Land of Beeches or Buchenland, and was a pleasant provincial university city, a cultural, religious and administrative center on the river Pruth, still boasting, between the two wars, most of the characteristics now long gone, of the k. und k. Habsburg empire; it featured a large Jewish population, roughly half of a cosmopolitan mix of about 120 000, including Ruthenians, Swabians, Roumanians, Gipsies, Hungarians, Poles and many others. The plan in 1926 was for my father to move to Cernăuţi to start his business while my mother and I were travelling, and to receive us there upon completion of our Moscow trip.

Now, in 1984, I would be seeing my aunt Liza again, who in 1916, in the midst of war and revolution, introduced my father to her elder sister, Inna, my mother to be. I would be visiting the grave of my aunt Tanya, my mother's eldest sister, buried with her husband, the sculptor Ivan Shadr, in Novodievece Cemetery. I would recall the past and discuss the present with my cousin Luba, daughter of Liza, whom I well remembered from my 1926 trip.

Much of my family's business turned around textiles. My grandfather Leon-Yehuda Lev-Eisenberg, owned a textile factory in Warsaw, and my father, who was his eldest son, travelled widely in the East, between Riga and Moscow, to sell his wares. This is how he first met my beautiful mother, though I have no explanation how the paths of a young Jewish textile merchant and that of the daughter of a little-known Russian gentleman playwright crossed, in a city essentially closed to Jews. My grandmother, Roza Frischman by birth, came from the important textile center in Poland, Lodz, and was a first cousin to David Frischman, a major writer of classical proportions, responsible to a large extent for the renaissance of the Hebrew language. One of Roza's sisters, Rahel, was an unusual person, who quite early in her life moved to Holland, married a well-known professor of psychology, divorced him quickly after giving birth to their two daughters, and then went on to establish a large factory in Amsterdam, for, believe it or not, ladies dresses. In the 1930s, Rahel and her second husband moved to Tel Benyamin, near Tel Aviv in Palestine and were known then as the 'rich Dutch'.

Meir Dizengoff, the Lord Mayor of Tel Aviv, was extremely

excited that a cousin of the great hebraist had settled in Tel Aviv and received her in the street carrying David Frischman's name. Yet Rahel did not know and never learned a word of the reborn Hebrew language. When I moved to Palestine in the late Fall of 1939, to study chemistry at the Hebrew University in Jerusalem, my father told me, you must go and look up my Aunt Rahel in Tel Aviv. This indeed I did at the first suitable occasion and when I rang the bell she opened the door. I said, 'I am Heini, Issay's son.' She said, 'Yes, do you play bridge?' 'Yes,' I said. She said, 'Good, we need a fourth hand.' This is how I, a mere 18 years old at that time, came to know Tel Aviv's banking and social haute volée in those early days.

My father had two brothers and eight sisters. His younger brother Zeev Barzilai was a journalist, writer and pioneer, who settled in Palestine after World War I, participated at the founding meeting of the Histadrut Labor Confederation in Haifa in 1920, and became the first secretary of the newly founded Jaffo Labor Council. He attended the Zionist Congress in Carlsbad in 1921, and wrote extensively on problems of the day in the early Zionist labor press in Jaffo and Tel Aviv. He was a serious budding leader and would have no doubt strongly affected the turn of future events in Palestine had he not drowned in June 1922, weakened by malaria, in a freak boating accident, in the Yarkon river, near the Seven Mills. The worker's library in Tel Aviv carries his name.

I have exhausted the description of unusual personalities from my father's home. One of his sisters, Irma Green her stage name, was a movie actress, beautiful, but only little known. Most of the family living in Poland was exterminated in the Nazi holocaust, and two sisters only fled east to Russia during the World War II and managed to reach Israel and establish themselves later there. My father was a devoted Zionist and a wonderful man. In Cernăuţi, and later in Bucharest, in the grim days of the war and of Nazi, Roumanian fascist and Soviet persecution of the Jews, he helped untold numbers of people to hide, to survive and move on to freedom. He set up schools and homes to teach aspiring surviving youngsters the bare essentials required for survival. He never reached the land of promise. In 1948, when he was supposed to come to the newly founded State, the realization of his dreams, he was felled by cancer of the liver, in the shortest of time. His grave, with a beautifully sincere inscription, is still in the old Jewish cemetery in Bucharest, unless it has been removed to make room for yet another one of those horrible projects of the present ruler. My mother came to Israel with my sister Erika in 1948, but did not live much longer. She fell victim to the consequences of a vicious breast cancer.

My mother, Inna Vladimirovna Gurieva, fell in love with my father in Moscow in the midst of the momentous events of a disastrous war and the emergence of the new revolutionary communist forms. They came from two totally different worlds and believed in a future bound to western enlightenment. They did not foresee the tragedy of World War II. When I went to study in Jerusalem in September of 1939 I took with me a bunch of family pictures, which made my mother weep, as this weakened her expectation that I would come home the next year for summer vacations. Though my father came from a strict Jewish background and was a devoted Zionist, he was very liberal in sincere justification of his marriage to my mother. I myself have always put the happiness and welfare of man before limiting religious or nationalistic dogmas and concepts of any kind, though accepting the deep manifold cultural connections to one's own past. It was important for my mother to remind me that I was Russian, too, and should know about my past and that Pushkin, for instance, should not be a total stranger to me. At the same time she strongly believed in the culture of France, and I owe her my strong connection with French literature and language, following 12 years of intensive private studies. I was indeed inscribed in the 'Lycée Saint Louis' in Paris for a year of 'Mathématiques Spéciales' in 1939, but history willed otherwise.

I knew my mother's elder sister, Tanja, well, a woman of unusual culture and knowledge of foreign languages, married to the sculptor Ivan Shadr. Shadr was born in 1887 in Shadrinsk, on the White Sea, in the far North. He had spent a year with Auguste Rodin in 1910 and then went to Rome. His creativity was at a peak at the time of the Soviet revolution and although I believe he did not create a new modern style under difficult circumstances, he did not follow the stereotype style of Soviet totalitarian art, but made a genuine contribution to the expression of man, his efforts and work, and the beauty of the human form. From his sculptures of peasants and workers the first Soviet banknotes and stamps were fashioned, Lenin got his due but also Pushkin and Gorky, the beauty of his wife Tanja, and many others including the tombstone of Stalin's wife, Alleluyeva. For the realization of Lenin's death mask he was accorded a study trip West, and that is when I first met him in Germany, on his visit there. I was only five years old, but I remember his grave voice in addressing the barber in Chemnitz 'Rasierén Sie mich, bitte . . .'. A year later I was allowed to play around in his studio in the Soviet State Mint in Moscow. He made a sculpture of me which unfortunately was lost, because the plaster model was not cast in bronze in due time. Shadr died a young man in 1941 shortly after Tanja's visit to my parents in then Soviet-occupied Cernăuţi. On my visit to Moscow in 1984 I was happy to see much of his work in galleries and public squares, and a commemorative plaque on the side of his old home in the center of the city.

When my father and mother left Moscow at the end of the World War I they first went to Warsaw, to meet parents, brothers and sisters and then proceeded to Germany to establish a home. The war was over and a bright future was expected. I was given a German name and we felt very German. How lucky we were to leave Germany in 1926 before the black specter of Naziism descended on it. Czernowitz became my home for the next 12 years, all during elementary and high school. Czernowitz sits on a hill surrounded by other hills. From the main railroad station downtown, an area in which many industrial enterprises, such as my father's 'Hercules' textile works, were located, a long road of a few kilometers steadily climbing upward led to the 'Ringplatz', the center of the town (Plate 11), displaying the 'Rathaus', the Hotel 'Schwarzer Adler', the monument for the victims of World War I, and the entry into the fasionable 'Herrengasse'. The straight road continued for many more kilometers, flattened out, past the orthodox Cathedral, the military barracks, the big public park, and finally reached the small railroad station at the southern end of town. My father, who travelled a lot to visit the 40 odd stores selling the wares of the factory, did not like to waste time by arriving early at the station. He often missed the train by a minute or two and then drove madly through town in a 'Fiaker'

Plate 11. 'Ringplatz' of Cernăuţi. (Painting by Berthold Klinghofer.)

drawn by two strong horses, to catch the train before its departure from the other station in town. The city boasted the 'Habsburgshöhe' and the beautiful 'Schillerpark' on steeply descending hills, an imposing Jewish temple, the extensive archbishop's residence, the town theater on the market place and the university buildings. Life was on a high cultural level and close contacts were maintained with Prague, Vienna, Paris and other major European centers of culture. In this microcosm of Austro-Hungarian 'Überbleibsel', F.C. Maccabi met F.C. Jahn on the soccer fields, *Drei von der Tankstelle* and other German movies were the hits in the cinema, and the impact of Roumanian legal dominance was felt only in police stations or in some of the official high schools. At one time a law was passed that German was forbidden and one was sent to jail for not speaking Roumanian in the evening show-off promenade on the 'Herrengasse'. I must say I have forgotten its Roumanian name. Gregor von Rezzori has immortalized the city of my youth in the fictional *Ein Hermelin in Tschernopol* [1]. In reality it has enriched our world with, among others, Josef Schmidt, the diminutive opera singer, emerging from his cantorial background, Itzik Manger, the Yiddish poet of the Purim 'megillah', the powerful sculptor Bernard Reder, drawing from the experience of the ancient graveyards, Paul Celan, who became the foremost postwar poetic figure and enigmatic conscience of a nation who had hit him and his family with utmost brutality in the days of the horrors of World War II [2].

I owe much of my cultural background to the public high school in the center of town, the Liceul Aron Pumnul, which I attended for eight years. Many years ago it had been the school of the great Roumanian poet Mihai Eminescu and successfully continued a tradition steeped in European learning of the highest calibre. Both teachers and students ranged over many nationalities and a successful matriculation certificate meant qualification for advanced studies. Unlike high school today, in the United States for instance, a high school education represented a complete education with no ends left dangling. Languages, science, law, it was all included. With Kallos we studied the French classics and romantics, while Balan exposed us to the Roumanian language and literature, deeply steeped in its Latin past; Italian ranged from Dante to D'Annunzio, and in physics Nedelcu used Grimsehl as a text. At our first meeting Silberbusch, professor of mathematics, grabbed me out of my bench and dragged me all around to convey the concept of a function, which has not left me since. Dragan was a superior teacher of the natural sciences; in summer and winter he wore only a simple net shirt and was said to start his winter mornings by breaking a hole in the ice of the frozen river Pruth, for his daily dip. Sireteanu, with whom we read Ovid in Latin, was particularly strict in examinations and reduced the three mandatory examination questions (if the unfortunate candidate did not know the answer) to: when was Caesar born, how many years before Christ was Caesar born, and the third version of this form of class-room torture escapes me. In geography the great continental rift was introduced dramatically by Pochmarsky, as a lesion akin to the deep facial mark about which duelling students were proud. An unusual teacher was Toma, responsible for the studies of law. He was very bright and clever, yet he was an avowed antisemite, in favor of Codreanu's Iron Guard nationalistic group. As a matter of fact there was a group of fellow students in our class who, at one time around 1938, wounded or killed a government official in an act of terror, and were arrested by the police. I do not remember how their trial ended. With Toma we, meaning the Jewish students in the class, had long political conversations, and when one of us said he was emigrating to Palestine, Toma asked:

> 'Well, tell me, what are you going to do there? Who are you going to deceive? Are you going to deceive each other?'

Unfortunately, this question has stayed with me ever since and has been assuming unpleasant proportions in recent days.

Cernăuţi as we knew it is no more. The Soviet Union annexed it after World War II in their massive physical expansion into the West. Practically none of the original population is still there. No more is one likely to hear on a street corner the strange linguistic warning such as:

> 'Albu gehn Sie, albu stehn Sie, aber machen Sie keine Kupkele nicht . . .'

My parents left Cernăuţi fleeing to Bucharest in 1941, when the Soviets were driven out by the Germans and their Roumanian allies. A few survivors among my classmates are spread throughout the world and we occasionally meet and exchange reminiscences. We used to swim in the Pruth in summer and ski in winter down the slopes of the Ţeţina, reached by a long drive on horse-ridden carts. We used to try to sneak into the operahouse to hear Aida or Carmen without paying for the tickets. We courted the pretty girls from the separate girls' high school while skating to the sound of a waltz by Strauss or Waldteufel. In September 1939 I returned the Rathsprecher Maccabi junior tennis cup I had won that summer to the club, though it was never to be contested again, said good bye to my family and friends and took the ss 'Transsilvania', to carry me from Constanţa to Haifa, to start a new life.

Years later, in 1957, I met in Vienna the best student of our class, who did not make it into academe. At the end of the war, stranded in Vienna, he owned the cabaret 'Casanova', a strip joint and location of the famous movie, *The Third Man*. History has its own twists and his son is now well on his way to become an outstanding neurobiologist after obtaining his doctorate in Jerusalem.

Palestine and World War II (1939–1946)

Arab violence against Jews in Palestine was suspended in 1939 with the outbreak of the war. When I thus descended the gangplank in Haifa in September the country was quiet on the internecine strife level. Still, when I unsuspectingly one day returned on foot from the Hebrew University campus on Mount Scopus through Arab-inhabited town quarters into the Jewish new city, my friends were struck with consternation. Yet peaceful coexistence continued for the next few years and my first and second year at the University created the impression of a world at rest, though momentous events in Europe were already following their fateful course.

My first exposure to science on a university level were the impressive introductory lectures to physics, by Samuel Samburski, who went on to become an outstanding contributor to the history of ancient Greek science, and mathematics seminars with Avraham Halevy Frenkel, Benjamin Amira and Jakob Levitsky. Inorganic and analytical chemistry in Max Bobtelsky's laboratory was a nerve-racking bore, and stories went round about an enterprising student trying to discover the anion bottle and the cation bottle, hidden by a laboratory assistant, with the respective unknown concentrations marked on the labels. At the laboratory benches and even in the hydrogen sulphide 'Stinkraum' friendships were forged, sometimes long lasting. Ours was a varied student group coming from Poland, Roumania, Hungary, Palestine, and we even had an Arab student colleague, scion of the well known Nashashibi family. The whole population of the University at that time was only a few hundred, and everybody knew everybody else. There were rumours that a famous young Italian physicist, Giulio Racah, had arrived and was busy studying Hebrew. I only met him many years later when I visited his house with Aharon Katchalsky and asked him where his books were. He answered that he had no books, when he needed a relation, he derived it. I do not remember if he was serious or not. I also did meet in those days the Farkas brothers, Adalbert and Ladislaus, known for their work on heavy water, who brought a high level of physical chemistry into the University, and were responsible for laying the foundations for exploring the mineral riches of the salt-saturated Dead Sea.

Many memories come back from these early and innocent days. Jerusalem had a large number of bookstores owned by 'Yekke' proprietors and I remember being offered under the counter Henry Miller's *Tropic of Capricorn*, printed in France. For chemists it was a good joke that the physico-organic text on *Free Radicals* was picked up by the censorship. We spent tea time in Cafe 'Atara' ogling pretty girls and we went dancing in the evenings in Café 'Rehavia', or, if we felt very posh, in the 'King David Regence Bar'. Jerusalem was not a very exciting city. It is said that the Hotel Eden owner, Mr. Lifshitz, was once asked by a tourist: 'Is there a night life in Jerusalem?' Whereupon the answer was: 'Yes, there was one, but she went to Tel Aviv.' On Saturdays we went to eat good Arab food in Arab restaurants around the main post office or on Mamilla road.

Trees had not yet been introduced in those days by Teddy Kollek into Jerusalem, and if one wanted to spend a shady Saturday afternoon of relaxation one went to the only then existing garden suburb Beth Hakerem. On an occasional trip to Tel Aviv one went to see Hanna Rovina in *Mirele Efros*, in the Habimah Theater, situated in the cellar of the 'Mograby' cinema on Mograby square, one dipped in the sea and had coffee and cake at Café 'Pilz'. Asking for directions one was smilingly directed to 'Rechov Allenby Strasse'; and I remember the story of the guy hearing a strange noise in the country side, close to a building site. When getting closer he saw a line of well behaved workers, passing bricks along the line, accompanied by a continuous: 'Bitte schön, Herr Doktor, danke schön, Herr Doktor . . .' Those were the days.

As time went on, an increasing number of British, Australian and New Zealand troops appeared in the streets and in public places. Italian planes bombed clearly civilian areas in Haifa and in Tel Aviv. An event stands out in my memory. It was announced that on a certain day, which I cannot place precisely, at 8 o'clock in the evening lights would go out in Jerusalem and that blackout would stay in effect to the end of the war. As it was already dark at that hour we went up on Mount Scopus to watch and appreciate the symbolic implication of this act. Lights had gone out over Europe and were to reappear only much later. We heard the famous words of Winston Churchill: 'We shall fight on the beaches . . .', broadcast on short waves by the BBC, and we realized that serious dangers and events were closing in on us. The German and Italian armies were aiming at the conquest of North Africa and Egypt, and that put the whole Middle East, including Palestine and its Jewish population in dire danger of occupation and destruction. Many Australians for instance, on short leave in Palestine, were returning to Tobruk, a lone outpost in the Western desert, not far beyond the port of Alexandria and the Egyptian border. The Jewish community in Palestine felt the need to play a part in the defence of Palestine and the Middle East and about 20 000 Jews volunteered for service in the Palestinian units of the British Eighth Army. Neither Palestinian Arabs nor Egyptians followed a similar course.

Following completion of my second year of study at the University I went to work for a few months as a night porter and assistant barman at the Zion Hotel in Haifa, owned by a friend of my father, and then joined the Royal Engineers of the British army in February of 1942. At that time General Rommel had reached El Alamein, barely 100 km west of Alexandria and the situation was grim. The Jewish Agency was trying to influence the British to form a Jewish fighting brigade, but this was delayed for political reasons, and the Jewish Brigade came into action only much later, and very briefly, before the end of the war.

Upon recruitment in Sarafand, near Tel Aviv, I passed a draughtsman's exam, on a hunch, and was qualified draughtsman A3, which carried with it a huge salary in the context of those days. One day we were moved westwards, arrived in pitch darkness and woke up in our tents the next morning in a wide desert in which a big boat was slowly moving, on the Suez canal, of course. We passed our basic training near Ismailia building bridges and taking them apart and, as a result of my connection with chemistry, I became responsible for poison gas defence drill practice. Upon 'graduation' I was moved near Cairo, in the Tura caves between Maadi and Helwan, to a unit in formation, the 524 Palestinian Field Survey Company of the Royal Engineers.

The '524' was a very unusual unit and consisted of many unusual people. Its purpose was to survey, draw maps from surveys or aerial photographs, and reproduce them in print. I belonged to the reproduction section which photographed on big glass wet plates, prepared grainy zinc plates from blanks onto which the photographs were transferred, a process in which I specialized, and then printed the plates by an offset process. Our unit was supposed to be attached to the Eighth Army, General Montgomery's advanced headquarters - a process which, again for political reason, was never completed. All heavy photography and printing equipment was loaded onto huge lorries so it could be moved quickly to new locations, and unfolded upon arrival. The people in our unit were an unusual lot of graphic artists and designers, photographers, architects, engineers and geographers who provided intellectual stimulation and later went on to occupy important positions in the State of Israel. I remember, for instance, studying Hardy's *Pure Mathematics* with a junior officer, Elisha Netaniahu, who later became professor of mathematics at the Technion in Haifa.

We became friendly with our British officers and NCOs, later replaced by our own people, including the commanding officer and sergeant major. In that fateful summer we worked very hard overprinting maps which had to be changed following aerial bombardments and photography by the Allied Air Force prior to the fateful battle of El Alamein in October of 1942, when the German threat to the Middle East was once and for all removed. Much of what happened in Cairo, in the Western desert and in the Middle Eastern 'Hinterland' has been dramatically described by Olivia Manning in her *Levant Trilogy*, and I will not compete with her literary genius, nor with the TV rendition of her books [3]. At any rate, once the battle moved away for good, we were free more often to take the train from El Masara to the Bab-el-Luk station in Cairo, to spend our off-duty hours in this huge metropolis.

Much of native Cairo was considered dangerous and out of bounds to the British soldiers, such as the mysterious Muski bazaars, for instance. Memories go back to the 'Groppi' coffee house with exquisite Italian cakes on Kasr-el Nil Street and to the 'Music for All' concerts given by visiting artists including the pianist Pnina Salzmann and the Fenyves-Vince Chamber Music Quartet from the Palestine Orchestra. The Palestine Orchestra itself under Malcolm Sargent performed in the large King Farouk University auditorium, with British soldier trumpeters placed inside the chandeliers as special effects at a gala performance. In the Jewish soldiers club one day the course of my life changed as I met and fell in love with my wife to be, Nutzi. She was a member of the ATS, the Palestinian Auxiliary Territorial Services and her camp was in Abbassia, near Heliopolis, on the other side of town. We often met in Cairo and also at the homes of local Jewish families. We were married in uniform on September 19, 1943 by a British military rabbi in the large Sephardi Synagogue in Cairo (Plate 12) and our wedding dinner was at the 'Auberge des Pyramides', a favorite hide-out of King Farouk. Our honeymoon was started at the Mena House Hotel, opposite the Gizeh pyramids and continued on the Alexandria 'Corniche'. We often went back to the Mena House on weekends, except once when the hotel was occupied by the Churchill-Roosevelt-Chiang Kai-shek conference. Correspondence with my parents was by infrequent Red Cross mail only and at one time they thought I had fallen prey to a belly dancer in Egypt, though they later were relieved to find out that Nutzi came from Galaţi in Roumania and had very nice relatives in Bucharest.

The front in the war had now moved from the Middle East to Europe and, in the summer of 1944, less than a year after our marriage, I received sailing orders to Italy. We landed in Bari, continued to Naples, then North to Siena where, near Poggibonsi, the final quarters of our unit were established. The front line at that time was between Florence and Bologna. Later, the German front collapsed and on May 8, 1945 V Day was celebrated in the streets and squares of Siena. We participated at the first postwar Palio, the traditional Sienese horse race on the 'Piazza del Campo', dedicated to the Allied troops. I travelled widely over Italy and learned to love its people, its art and its opera. I remember in particular a *Rigoletto* performance at noon in Naples, because there was no electricity, applauded and booed by an enthusiastic audience, and a song recital by Maria Caniglia in a burnt-out church. There was only very little food at that time, and most of it coming from Army sources, undergoing a metamorphosis into magnificent Italian-style dishes. On a more serious level we made contact with and accommodated in our camps Jewish refugees, survivors of the Nazi holocaust, seeking

Plate 12. Wedding of Nutzi and Heini.

ways to move into Palestine, still closed by the British administration. Here I refer to the Italian Jewish writer Primo Levi for poignant descriptions of refugee migration and suffering [4].

In another unique experience toward the end of my stay in Italy, I managed to undertake an unauthorized visit, in army uniform, to my parents and sister in Bucharest, for one week, and to get to know Nutzi's family. We, myself and two colleagues, travelled from Vienna to Bucharest via an almost totally destroyed Budapest, in an overcrowded Soviet military train. I did not know that this would be the last time I was to see my father.

Not long thereafter I received sailing orders to Palestine and was released from army service in the summer of 1946 in a British army camp in Rehovot, not far from where we have now lived for many years. Nutzi had been released earlier and was waiting for me in our modest one-room apartment in Jerusalem. I could now go back to continue my studies and these memories can switch from a description of the impact of the fateful events of World War II on the course of our lives, to matters of science. Little did we know that momentous events linked to the establishment of the State of Israel would take hold of us soon again and would relentlessly grip ourselves and our children, following us without respite to the present day.

Hebrew University and War of Independence (1946–1948)

In the summer of 1946 I was thus back at the University and ready to catch up five years of study lost. There was no time to lose. We indulged in the euphoria of the victory over the forces of evil but we realized that there lay struggles ahead. Slowly we became aware of the brutal damage wreaked by the Nazis and their allies on mankind and on the Jewish people in particular, and we realized that it was essential that our endeavors towards the re-establishment of a Jewish homeland in the near future should gain increased momentum. The renewed influx of immigrant survivors was blocked by the mandatory authorities and the conflict between Jews and the British authorities was renewed on both the official Hagana level, and that of the more extreme splinter organizations. One day, while I was bus-riding up Mount Scopus, the air filled with the sound of the exploding King David Hotel wing, occupied by British governmental offices. Ben Oyserman, press photographer and pioneer movie producer, then married to Aunt Rahel's daughter Nini, was almost immediately on the spot to film the event. Years later Ben became my brother-in-law by marrying my sister Erika. He died in the 1967 war in Gaza by an exploding Arab bomb. Aunt Rahel, the eternal optimist and totally divorced from reality, one evening came to Jerusalem and suggested we all go somewhere. As it was the time of the full moon it was decided that Ben would drive us down for dinner at the Kallia Hotel, on the northern tip of the Dead Sea, a fantastic trip in a moonlit night. The exact date of this event can be fixed, as you shall soon see. As we drove down - this is a 1200-meter drop from plus 800 to minus 400 (below sea level) - darkness surrounded us, but the moon did not appear. The surrounding hillsides were lit by a succession of bonfires. A British army patrol finally stopped us and advised us to go back. That night there was a full moon eclipse. The bonfires were lit by Arabs to celebrate the return of the Mufti to Jerusalem. Another event that comes to mind is when we woke up one Saturday morning to hear on the radio that the whole Jewish Agency official leadership had been arrested by the authorities and placed in a concentration camp.

At the University I completed labs and passed examinations. Maybe I was a little too much in a hurry and tried to improvise. Thus I remember that in a biochemistry exam I wrote down that butyric acid is the main component of butter. In physical chemistry I somehow managed to convince Adalbert Farkas, the younger of the two brothers, that I knew what entropy was. In organic chemistry my mentor was Moshe Weizmann, a brother of Chaim Weizmann, who could be found daily in Café Vienna in Zion Square, presiding over political discussions. He used to say that in their family an error had occurred, because he should have become a politician, and Chaim should have stayed a chemist. In the laboratories and in the classrooms I made new friedships with students in the natural sciences, five years younger than myself, in addition to the few returnees from the war, who were continuing their education.

When it became time to choose a tutor for my master thesis my friend Pnina Spitnik advised me strongly to talk to Aharon Katchalsky, a remarkable young assistant, working along with his younger brother Ephraim, on problems of biological interest. I went to talk with Aharon and was tremendously impressed by him. Science and culture as a rule are the outcome of a long tradition in which knowledge and style are transmitted from generation to generation, sometimes over centuries. We have become used to the fact that the great centers of learning in Europe took hundreds of years to develop and even not so long ago New World U.S. scientists went to England and Germany for their postdoctoral training. Yet here was a University, founded only a few decades ago, in 1925, in which some good and many not so good refugee teachers taught, in difficult times and under limited circumstances, a motley mixture of students coming from a variety of backward, and maybe not so backward countries. Aharon and his brother had come to Jerusalem as very young children from a small town in Poland and their parents owned a small fancy goods store in Yellin street in Jerusalem. In no way could those circumstances explain how this young man, almost self-taught in science and culture, reached a level of inquisitive thought, judgment and originality which in a few years would propel him onto the center stage of modern research. Aharon had studied botany and zoology with Alexander Eig and I believe he even published a catalog of Palestinian butterflies. He became deeply interested in solving the major problems of biology, and without a proper training in chemistry or physics, hardly available in Jerusalem in the days before or during World War II, he had come to the interesting idea that active biological materials, proteins, enzymes, sugars and nucleic acids, are macromolecules or polymers, often carrying electrical charges. It should thus be possible, in an initial step, to study synthetic polymers, whose charge could be changed at will, believing that these polyelectrolytes, independently defined by Ray Fuoss at Yale University [5], could teach us some basic facts concerning the properties of biological structures. Aharon had just come back from Basel, where he had spent a few months in the laboratory of Werner Kuhn, well known for his pioneering theoretical studies for the calculation of the dimensions of random coil polymers by extending the concept of Brownian motion, and applying these ideas to experimental methodology, such as viscosity, birefringence of flow, sedimentation in the ultracentrifuge and so forth. Kuhn could thus predict that the intrinsic viscosity of polymers should vary ideally with the square root of the molecular weight, as it was called in those days, and not with the molecular weight, as Hermann Staudinger had erroneously believed, assuming a rodlike structure. Staudinger's unique contribution though lies in establishing the concept of macromolecular structure, as contrasted to badly defined colloidal aggregation, and he may be forgiven for a little mathematical imprecision. Thus, synthetic macromolecules in solution usually coil in random fashion, and their size can be estimated by a number of physical studies. When electrical charges are placed on polymer chains, such as for instance when a carboxylic acid or amino acid group is ionized by addition of base or acid, respectively, a logical conjecture is that the electrostatic repulsion energy will lead to expansion of the polymer chains, the restoring force being provided by the rubber elastic-type force, countering the expansion of the chains. The electrostatic expansion force can be modulated by adjusting the ionic strength, an increase in ionic strength leading to much reduced charge repulsion and expansion. Strong electrostatic acidic groups, such as sulphate, or phosphate, for instance, are always ionized and therefore not subject to ionization-deionization by change of pH. At high electrolyte concentration the possibility arises of specific ion effects. Aharon had worked out an elementary theory in Basel with O. Künzle and Kuhn, which was presented at a meeting in Amsterdam, and appeared in 1948 [6], alongside a similar approach by Hermans and Overbeek [7].

Obviously, when I first talked to Aharon, he was extremely keen to obtain experimental confirmation of these ideas, and he expounded to me what he felt was worth doing. I still have in my files the notes he jotted down, suggesting a possible course of experiments (Plate 13). I believe they represent about ten man-years of work. Thus, expansion of macromolecular chains is accompanied by an increasing viscosity; the viscosity will be a function of the degree of ionization (pH), ion concentration and molecular weight, already a sizeable program. Surface tension is another interesting property.

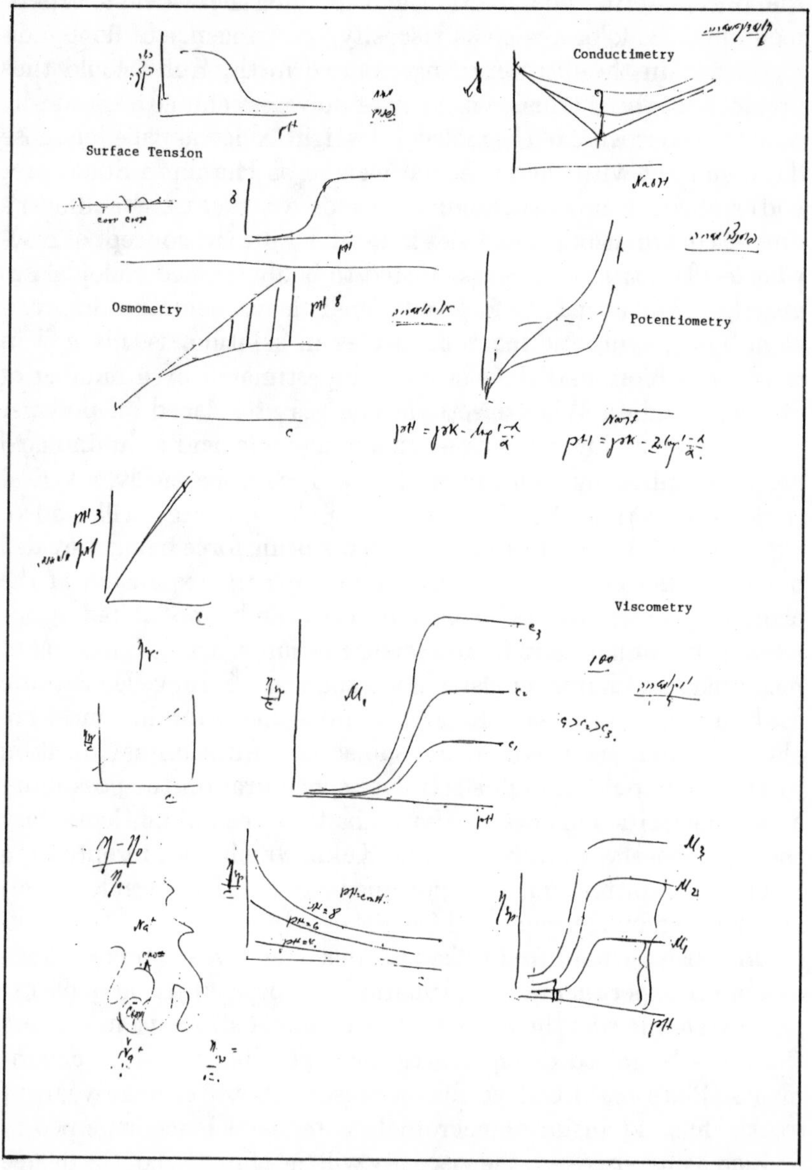

Plate 13. Research program suggested to the author by Aharon Katchalsky in June 1947.

A non-ionized macromolecular acid will be adsorbed on to the surface. Once charges are provided it will be pulled away from the surface. Aharon had also considered electrolyte conductance which obviously depends on charge. To measure conductance Aharon had commissioned Elek Suchoczever, a mechanical genius employed in the university workshop, to build a classical Grinnell-Jones conductance bridge, from a description in the *J. Am. Chem. Soc.*, of 1928 [8]; a rather excellent achievement in that time and day. About Elek I will have to say more a little later. Actually, conductance of polyelectrolytes is a very complex phenomenon and when I much later asked Lars Onsager at a Faraday Society meeting in Oxford in 1957 whether he had thought about calculating the conductance of polyelectrolyte solutions, he said that to do so new mathematics had to be invented. It turned out that the other quantities were not so simple to calculate and even today complete success has not been achieved.

In those days Aharon proudly showed me a slide rule ('do you know what that is, if you now are under 50 years of age, say?') which had an additional very useful reciprocal scale, and of course today everything is calculated by computers and super-sophisticated mathematics. I am writing these memoirs in a very conversational tone as (*a*) I want people 'of all denominations' including my wife, children and grandchildren to read and enjoy them and (*b*) I have recently written an extensive purely scientific historical account on *Thermodynamics and the Structure of Biological Macromolecules*, which can be consulted for technical details [9].

The first method which Aharon and Pnina had actually used was potentiometric titration. It is quite clear that as protons are removed from a polycarboxylic acid, an electrostatic field arises, which increases with increasing dissociation (pH). It becomes increasingly difficult to remove protons, the acid gets weaker and weaker, and the potentiometric titration of the polyacid is different from that of simple acid: that is the theory in a nutshell. As a matter of fact, when I first started working with Aharon, Pnina had gone to work at Brooklyn Poly, Herman Mark's Mecca of polymer science; Ephraim also spent a year there, and my first practical work was to correct a proof Aharon gave me of his first polyelectrolyte paper with

Pnina on potentiometric titration of polymeric acids, which appeared in *J. Polymer Sci.* [10].

A word should be said about the material used in our early studies. This was polymethacrylic acid and people often asked me, 'why did you use polymethacrylic and not polyacrylic acid, for instance?' This reminds me of a story Arthur Kornberg told at a seminar recently. A sales lady comes into his lab, and starts asking: 'what are you working on here?' Answer: 'We are working on DNA replication'. 'Very good', she says, 'we have a kit for that!' Well, in those days there were no kits. We worked on polymethacrylic acid because in those days there was much perspex, polymethylmethacrylate scrap from RAF war-time bombers' protective bubbles. It was possible to depolymerize the resin, hydrolyze these esters easily to the free acid, to polymerize the acid to various molecular weight values, and that is how polymethacrylic acid was obtained. Werner Kern, a pupil of Staudinger, had been working with polyacrylic acid. While many results relating to polyacrylic or polymethacrylic acid are identical, or similar, a few years later, Alex Silberberg, a student of Werner Kuhn working with Aharon, found a very interesting phenomenon. Namely, when solutions of polymethacrylic (not polyacrylic) acid were stirred rapidly, clear solutions gelled into solid-like material (you could turn the test tube over and the stirring rod would not fall out) and the material would regain its liquid form after a while. You see already a very interesting observation, demonstrating an important structural principle relating to hydrophobic interactions. It was given the name negative thixotropy by the authors. When Aharon described this phenomenon at a lecture at the IUPAC Congress in Milan in 1954, an Italian professor got up very excited and started shouting: 'Il miracolo di San Gennaro, . . . il miracolo . . .'. Apparently the blood of the Saint in the church in Naples is solid all year and becomes fluid, in good years, on the day of the Saint.

After my first conversation with Aharon I was all set to go. I prepared a decent amount of polymethacrylic acid fractions and got started on my master thesis. I realized that it would not be easy to determine the molecular weight, or the degree of polymerization of these materials as the simple relationships derived by Paul Flory for nonionic polymers did not apply here in straightforward fashion and a proper calibration was not available. It occurred to us then that if it could be shown that the properties of polymethylacrylate and polymethylmethacrylate are unchanged following acid hydrolysis in acetic acid and re-esterification by diazomethane, then the properties of the esters in organic solvents could be used to establish a calibration scale for the free acids. This procedure turned out to be successful.

While I was thus making my first acquaintance with original experimental investigation, political events started precipitating. Following a series of United Nations investigations, some open to the public at which I watched the charismatic appearance of Chaim Weizmann in the Jerusalem YMCA, for instance, we stayed up all night for the fateful vote of November 29, 1947 in the UN General Assembly, approving the partition plan leading to the establishment of the Jewish State. Joy knew no bounds, bus and taxi rides were free, and everybody was embracing everybody else. Yet soon enough reality was restored when in a first incident a Jewish bus driving on the Jerusalem–Tel Aviv highway through Arab Ramleh was violently attacked and five people were killed. In Jerusalem Ben Yehuda street, the offices and printing plant of the English-language newspaper, *The Palestine Post* and the Jewish Agency buildings were bombed, and the armor-protected No. 9 bus on its way to Mount Scopus was attacked in Sheich Jarrah quarter, under the eyes of the British, with heavy loss of life of staff, teachers and students, including our own instructors and personal colleagues and friends. We then realized that the UN vote was a formal declaration only and that its implementation depended on an intense struggle in the months to come.

Aharon was a leading force in organizing students in chemistry, physics and biology into an army unit to be called 'Hemed', the army science unit. Ora Herzog, wife of the current President of Israel Haim Herzog, was the effective and charming secretary of this unit. We were preparing for the defence of Jerusalem and did a lot of improvising to create defence tools for the coming emergency. We spent much time guarding the Mount Scopus enclave campus, including the Hadassah Hospital, in which many Arabs were treated, and which was to be separated from Jewish Jerusalem until 1967. I was immensely impressed, one night, as I told two elderly workmen of Russian origin about my family and uncle Shadr in Moscow when, on the spot, they produced an issue of the Soviet newspaper *Ogoniok*, including an article on Ivan Shadr, sculptor of the Soviet epoch.

We went to Kibbutz Beth Haarava near Kallia on the Jordan river for military training and were shocked on our return to Jerusalem to hear about the death by Arab hands of the 35 people, mostly students, on their way to protect Kfar Etsyon, near Jerusalem.

On May 15, 1948 the declaration of the State of Israel took place in the Tel Aviv Museum. Seven Arab armies converged on the tiny state, and Jerusalem went into siege. Bombs started falling on houses in the Jewish area, including our house in Kiryat Shmuel, overlooking the Arab quarter of Katamon. Food and water in Jerusalem were low at that time, but the spirits were high. We did everything we could to create tools for protection of the city, cut off, except for an occasional armored convoy, from supplies in Tel Aviv and the rest of the country. Paint shops were raided for aluminium powder and solvents. A supply of potassium chlorate was found, a very dangerous material with which my friend and colleague Israel Miller tried to make explosives, and even set up to synthesize more by oxidation of potash, KCl, for the manufacture of the French explosive 'Cheddite'. It is said that every factory manufacturing it is bound to blow up sooner or later. We were lucky that, in spite of the falling mortar shells nothing happened, though potassium perchlorate would have been a safer choice. I myself climbed on top of the Schneller compound British army camp, the day after it was vacated, to dismantle some warning-trip flares, which we aimed to reproduce. All this would not have been possible without the genius and good humor of Elek, to whose house we went continuously for advise, help with the solution of mechanical problems, and life wisdom. Many of Elek's solutions were taken from a collection of Soviet practical engineering texts which had been created to teach a huge ignorant population many aspects of modern technology. Elek himself copied, from odds and ends, the famous 'Davidka' mortar, which produced more noise than damage, but it seems that the Arabs were very sensitive and ran when they heard a big bang. A huge laboratory-type flame-thrower was also built to help the 'Palmach' unit to which belonged Elek's son Amos, later President of the Technion, in their fights around Ramat Rahel, on the Bethlehem road. Again, more a spectacle of sound and light than a modern weapon, but psychologically very effective at that time. The Jordan Arab Legion had conquered all of the Old City inside the walls, taking prisoner all of the old Jewish religious population living in the Jewish Quarter, and then proceeded to destroy most of the quarter, including century-old synagogues. The Israeli High Command planned to reconquer the old city but did not succeed in those days, with the huge tasks facing them and only very limited means at their disposal. The young 'Hemed' physicists led by Prof. Racah had heard about the effectiveness of hollow charges to provide deep penetration. They therefore proceeded to make a theoretical calculation to design the most effective hollow charge possible. Unfortunately, when placed along the Old City Turkish wall it just bounced off, and did not produce the deep hole required. It seems that, in those early days, improvisation was playing the biggest role.

The short war went its course; Jerusalem, and many other valuable points in the young State were redeemed or protected. In the second armistice, in August 1948, we left Jerusalem and proceeded to Rehovot. For some years now the extension of the small Sieff Institute, dedicated and devoted to organic chemistry and microbiology, founded by Chaim Weizmann in 1933 to accommodate German Jewish refugee scientists, had been planned by friends of Weizmann, to become a multifaceted modern research establishment, the Weizmann Institute of Science. Both Aharon and Ephraim had been hired to assist American polymer scientists and biophysicists in their lavishly planned departments. Most of these scientists, except the well-known applied mathematician Chaim Pekeris, never came, and both Aharon and Ephraim took over the Polymer and Biophysics Departments, staffing them with their students, transferred from Jerusalem. The super-dynamo director of the new Institute was Ernst Bergmann, a pupil of Wilhelm Schlenk, and coworker of Chaim Weizmann for many years, director of the

Sieff Institute since its inception. Bergmann had an encyclopedic knowledge of chemistry and we spoke about him as 'Dr. Beilstein'. He took an interest in everybody's work and, if you had a discussion with him, the next morning at 8 a.m. you would find a full set of literature references and advice on your bench. Whenever a lecture would take place, in any field, he would go round through the building and push everybody into the lecture room. I heard in those exciting days many lectures in organic and physical biochemistry, in biology, mathematics and physics, and I must say that much of my broad interest in science goes back to the stimulation provided by Ernst Bergmann and by Aharon. The activities of 'Hemed' continued for a while, but then moved into more regular channels as the country became established. Having done military service for many years, and being married, I was released from service early in 1949, and could now once more, I hoped, return to full-time studies and scientific research. Somehow we adjusted and raised a family and got used to a life displaying repeated high peaks of violence in 1956, 1967, 1973, 1982, elevated baselines of unrest in between and the continuously receding but never submerging hope that one day there would be peace in these tormented areas and the many human sacrifices had not been in vain.

Weizmann Institute: Polyelectrolytes (1948–1952)

Activities in the Polymer Research Department headed by Aharon moved at a fast pace. A number of young students, liberated from army service, now joined our group and took up activities along the course which Aharon had outlined to me in Jerusalem. Aharon was brimming with ideas and discussions, starting early in the morning around the blackboard in the department's lecture room and going on late into the night at Aharon's house, where it was sometimes freezing cold in the wet winter. For heating we used in those days small kerosene stoves whose smelling fumes negated the benefits of mild heating. If Aharon had an idea, he would discuss it with more than one person, who would go on developing the topic, not realizing that somebody else was doing almost the same thing. Aharon of course did not precisely realize whom he had talked to and this occasionally led to unpleasant occurrences. Aharon was a very gifted lecturer and was very early on invited to present plenary lectures at big polymer meetings. We carefully prepared slides of our work for presentation which he did magnificently, the first one in Torino in 1950. Unfortunately for us, extensive presentation of our data in the printed lectures increased the difficulties of publication in regular form. It seems we had not yet learned all the rules of the game, but this did not distract us from proceeding on a vigorous course.

We were now reasonably convinced that polyelectrolyte coils in solution expand and contract depending on charge and ionic strength. It became possible to visualize this by crosslinking the polymer chains in a process reminiscent of rubber vulcanization. Softer or harder gels were obtained by varying the degree of crosslinking, and the gels behaved consistently with changes of pH and ionic strength. These studies became the subject of my Ph.D. thesis. Aharon wrote Kuhn about this effect and his letter crossed a letter by Kuhn with the same observation. They reported on it side by side in *Experientia* [11,12] and we also submitted a joint paper to *Nature* [13]. By way of curiosity, I also came across a paper by Jacob Riseman and John Kirkwood proposing a model of muscle contraction following the action of phosphate charges on myosin chains [14]. I proceeded to gently phosphorylate crosslinked polyvinylalcohol chains and we observed a very nice swelling effect [15].

In November of 1949 the Institute was officially opened in the presence of Chaim Weizmann, and a number of famous scientists arrived who had contributed to the design of the Institute.

Herman Mark, the ‘Geheimrat’, as he is affectionately called, repeatedly came after that. Mark, one of the pioneers in the development of polymer sciences, was now Director of the Polymer Institute at Brooklyn Poly, and closely associated with big industry and academe. His ingenuity was boundless and he is said to have left Nazi Vienna with his possessions in the form of platinum fashioned into clothes-hangers [16]. From the ‘Geheimrat’ we learned the latest in new fibers and plastics, he brought samples with him which then served us as kitchen and bathroom fixtures, but he also had the knack of writing a very complicated hydrodynamic equation by John Kirkwood on the blackboard and then taking ten minutes to explain what it really meant. All with fantastic Viennese charm. He brought the silicon compound bouncing putty to us and showed us how, depending on the time scale a material can behave both as a liquid and as an elastic solid. An experiment that was really stunning was the demonstration that crystallization of a water-soluble polymer can make it insoluble, a reminder how very subtle changes in our bodies can have very large effects. To a fiber of polyvinyl alcohol, a water-soluble polymer, was tied a 50-g brass weight and the fiber was introduced into a glass cylinder filled with almost boiling water. All was well, but the moment the weight touched the bottom of the cylinder the fiber dissolved and disappeared instantly. It seems that the pull of the small weight was sufficient to generate fiber crystallization forces preventing dissolution: a molecular phenomenon observed on a macroscopic scale. Many years later, when I was still doing my army reserve duties teaching and entertaining soldiers in the frontlines, I used a kit based on the Mark lectures. I lectured from strongholds on the Suez Canal to advanced units in the Golan Heights, where one night I slept in a bunker close to my son Shai, doing his army service. The 'Geheimrat' was extremely well organized. Whenever he had left, after one of his frequent visits, a telegram arrived thanking us for our hospitality. Once we took him to the airport but he had to come back as, for some reasons, his flight was delayed by a few days. The telegram thanking us arrived anyway. The superb ability of Mark to explain was really unsurpassed. In 1954, at the Milan/Torino Polymer Congress, one of the professors from his Institute, Robert Mesrobian, was scheduled to give a plenary lecture, but did not arrive in time because of a flight delay. Mark got up, and said, let me explain what Mesrobian is doing. He then proceeded, without slides or other help, to present in ten minutes a superb and complete exposition of the work of the absent lecturer. When he had finished and was applauded, a breathless Mesrobian rushed into the lecture room straight from the airport, and insisted on giving his talk. He had carefully prepared it in Italian and was not going to give up. He spoke for almost an hour and nobody understood a word. Mesrobian entered our life again a few years later. Herbert Morawetz, also from Brooklyn Poly, was spending a year in our lab, just prior to the 1957 IUPAC Polymer Congress, organized by us in Rehovot. He shocked my wife Nutzi considerably one day, because it was so much against his nature and we were still living with different concepts, by saying that, as she understood it, Miss Robian was coming to stay with him. Nutzi knew that he was happily married to a well-known mathematician, Cathleen. Herbert has recently summarized the history of polymer science [17].

The lectures at the 1949 Weizmann Institute opening symposium were given in the 'moshava' colony, as we are still used to calling the city (now) of Rehovot, in the 'Beth Haam', the only hall then serving as cinema, theater and many other purposes. The Israel Philharmonic came to play on this occasion. There was no proper hall on campus. All smaller lectures were given in our library, which held at most 35 listeners. One day Leopold Ružička, the famous organic chemist from the ETH in Zurich came by and asked: 'Ist der Fieser drinnen?' When I said 'Ja', he said 'Dann bleib ich draussen'. All in good spirit. Ružička gave brilliant lectures and often called his collaborator, 'mein kleiner Jude', which was Oskar Jeger, a well known chemist himself. Both Louis and Mary Fieser, the well-known Harvard organic chemists and textbook authors had come, as they had been instrumental in the design of the new building which we occupied. Mary Fieser was very strict on the color scheme of the new laboratory and the lab benches in the organic labs were finished with a beautiful blue plastic paint, which dissolved upon the slightest touch of acetone or other organic solvents. When Mary was asked how we should handle this situation, she said a good chemist never spills a drop of solvent.

Work in the lab was proceeding satisfactorily but we soon realized no good theory was available, and is not available to the present day, for the determination of polyelectrolyte dimensions. The theories based on the expansion of coils all overestimated the force of repulsion and yielded fully stretched macromolecules, which was not confirmed from simple experiments. Shneior Lifson and Aharon started to work on a model which, they felt, though it would not be useful to determine dimensions, would be useful for calculating potentiometric titration, ion-binding, conductance, osmotic pressure, and so forth. This model states that, for relatively short distances along the chain it is possible to assume a rod-like shape and calculate the distribution of counter-ions surrounding a cylinder with equally spaced oppositely charged co-ions. This was seen by Shneior to correspond closely to the distribution of electrons around a glowing wire, a solution of the differential equation of which was given on page so and so in a standard mathematical text [18]. This model of polyelectrolyte behavior known as the cell model sounded very interesting and, in one form or another, has maintained itself to the present day [19].

An exciting event, maybe commonplace today, but probably a first then for our young country and Institute, was a letter Aharon received in 1950 or 1951 from Ray Fuoss, from Yale University, New Haven, who had read our initial papers, and thought they were highly exciting in this new field of polyelectrolytes, for which Fuoss had coined the name. Fuoss was a well-known electrochemist, a student of Charlie Kraus, who had pioneered research at Brown University, Providence, from around the turn of the century, working on organic electrolytes. When Lars Onsager first came to the U.S. he had worked at Brown and he and Fuoss wrote important articles on the theory of electrolyte conductance. Now Onsager, Fuoss, Kirkwood, Herbert Harned, Benton Owen, a young Julian Sturtevant and outstanding others, were all in the Yale Chemistry Department, a Mecca for electrolyte research. Fuoss was also well known for his work on polymers, so it was no wonder he was a 'par excellence' polyelectrolyte type. Ray and Ann Fuoss came to Israel by boat from Egypt, where Ray was visiting a former student, via Cyprus. Aharon, Shneior and myself drove to Haifa port to pick them up. There were very few cars in Israel at that time, and Aharon had a small one. He was so excited that he had to show the Fuosses the magnificent view from Mount Carmel to which we drove up in the small car, including the terrified Fuosses and their luggage, at breakneck speed. Finally, another detour was made to show Tel Aviv where we had cake and coffee at Café Kassit on Dizengoff street, in the middle of the summer heat - there was no airconditioning in cars, houses or cafes at that time - before we dropped the exhausted Fuosses at the small guest house in Rehovot. We became good friends after that and Ann, in particular, worked hard for Israel and lectured for Wizo and Hadassah. One evening after a long discussion at the Gat Rimon Bar in Tel Aviv, and a large amount of scotch-on-the-rocks, I was invited to a Sterling Chemistry Fellowship at Yale; I believe I became also one of the first postdocs from Israel in the U.S.

While at the Institute Aharon had coopted Ray on his research with Shneior on the rodlike polyelectrolyte model and their joint paper appeared [20], at about the same time as a similar paper by Turner Alfrey, Paul Berg and Herbert Morawetz [21]. The Fuosses were given a big farewell party and Aharon said that, now that we know and like each other so well, everybody from our Department would be welcome to stay with the Fuosses in New Haven. I am not sure that they themselves felt that way, though they had become close to us, to Aharon, to Pnina and to Ernst and his wife, Hani Bergmann.

Our sons, Shai, the first child to be born on the Weizmann Institute 'Shikun' (housing), where we had a nice three-room apartment, on September 18, 1949, and Danny, born on December 24, 1950 were still very young when I was scheduled to leave for Yale, in February of 1952. The Fuosses convinced Nutzi and myself that I should go by myself for 11 months, as she and the boys would be better off, surrounded by friends, in our comfortable apartment on campus, and it would be more difficult to adjust in New Haven under much more awkward circumstances. We did agree to this and I went by myself though, on second thoughts, maybe Ray had a selfish view in this matter, or maybe he did not like babies.

Yale University: electrolytes (1952–1953)

It was such an unusual event in those distant days that, when I was to leave for my stay at Yale, together with our neighbor and friend the organic chemist, David Ginzburg who was going with his family to spend a year at Harvard, practically the whole Weizmann Institute came to the diminutive airport at Lod to have a final coffee and cake together and say good-bye. I took leave from my family and friends, and we were on our way. There was only first class in propellor-driven Constellations in those days and KLM took good care of us. At the relatively low height, low speed and luxurious seating the Greek Islands and then the Swiss Alps practically at eye level were unforgettable on the eight-hour trip to Amsterdam, not to forget my first acquaintance with Dutch 'Oude Genever'. Spotless and well-to-do Amsterdam, almost fully recovered from the destruction caused by the war, was in stark contrast to austerity-ridden Israel, still grappling for its existence. I stopped for a few days to meet Anjut, Aunt Rahels elder daughter, her husband Ward Messer, journalist and editor of *Het Vrije Volk*, who had been in the resistance during the war and helped in the rescue of many victims of Nazi persecution, their children and a number of more distant relatives. Ward also drove me over the icy highway to Utrecht were I visited Overbeek, the well-known colloid chemist in the Van 't Hoff University.

We proceeded to New York a few days later stopping in Prestwick near Glasgow and in Halifax, Nova Scotia, after a close view of the Greenland iceberg scenery. At Idlewild, now Kennedy, airport I was received by the Fuosses who, after I went through an extensive and intensive customs and visa examination, drove me to New Haven for first exposure to the American way of life.

While working with Aharon I became aware of the broad aspects of science, and the big questions still unsolved, with less insistence on detail and precision, whereas Fuoss taught me to ask precise questions to solve problems with accuracy. I owe them both my ability now to handle these two aspects with enough flexibility to remain on a straight path of progress. In my year at Yale I would study the conductance of 'bolaform' electrolytes, which Ray had been studying for a while. Bolaform electrolytes are chain-like structures with two charged end-groups. The compounds

$$Me_3N^+(CH_2)_2\,X\,CO(CH_2)_nCO\,X\,(CH_2)_2N^+Me_3$$

show curare-like activity when X=O. Physiological activity of the bisquaternary salts is related to chain length. Succinylcholine (n=2) binds to the acetylcholine receptor, and causes depolarization of the end plate. Unlike the esters, the corresponding amides (X=NH) show no curare-like activity for n=0 to n=6; instead, they prolong the duration of the block of neuromuscular transmission produced by certain bis-esters. We could show, from an analysis by conductance of the ratio of the first and second counter-ion dissociation constant, for the same value of *n*, that the amides are hydrogen-bonded, and therefore the amides and esters differ in configuration. Charge separation rather than maximum chain length is the significant variable leading to curare-like activity [22].

We had previously shown that hydrogen bonding could affect the behavior of simple macromolecule coils in solution. Dissociation of carboxylic acids also affects the stability and folding of proteins. With increasing concentration, the dissociation constants of simple carboxylic acids calculated from conductance measurements show pronounced deviations from the extrapolated values, though the limiting Debye–Hückel activity coefficients should still hold at these concentrations. Dimerization of carboxylic acids, a consequence of hydrogen bonding modulated by increasing Van der Waals interactions between monomers with increasing chain length, provides a plausible explanation for obtaining constant values of the ionization constants in the range of concentrations examined [23].

My first assignment at Yale was to calibrate a set of analytical weights, a process which took about two weeks. They had been calibrated a year ago, by a previous postdoc, but if you wanted to be certain of the validity of your concentrations you should be sure that your weights were reliable. Once this hurdle was overcome the flow of research was even. I was also introduced into the mystique of baseball on the slope descending from the Sterling Chemistry lab to Whitney Avenue, on which the Molecular Biology Building, a discipline then nonexistent, is now situated. The lectures of Kirkwood were clear and instructive. Onsager lectured by speaking to himself, often turning to the blackboard on which he scribbled things nobody could see, and once in a while he broke out in a small private laughter. On a personal level Kirkwood, though Chairman of the Department, was accessible and friendly, but hardly anyone went to consult with Onsager. A young lady from Israel, Bruria Kaufmann, had done it, and the result was a clear exposition of the two-dimensional Ising model in electrolyte solutions, which had been previously demonstrated by Onsager in a form difficult to grasp by mere mortals [24]. Of Kirkwood, who has developed definitive presentations of complex statistical mechanical and hydrodynamic problems, the story went round that when both he and Peter Debye were at Cornell, his office was right on top of Debye's. Debye would point to the ceiling in his room, saying:

> 'Why is he worrying upstairs and walking up and down? I have solved this problem many years ago.'

Kirkwood was trying to understand electrophoresis of proteins; this was before the days of gel electrophoresis, in collaboration with Jonathan Singer, then a young assistant professor. Terrell Hill, with whom I established a deep friendship based on scientific discussions, human relations and, somewhat later, the great game of tennis, was on leave from the Navy Laboratory in Bethesda, writing his book on *Statistical Mechanics* [25]. He went on to write his classic text *Statistical Thermodynamics* [26] a few years later. I had long conversations about the future of our globe with Harned who, with Owen, also in chemistry at Yale, were authors of the electrolyte bible [27]. Sturtevant was already developing his calorimeter for the study of electrolyte solutions. Among the students I remember Marshal Fixman, Irvin Oppenheim, Bob Zwanzig, Bernie Coleman, Peter Geiduschek.

One day it was given to me to meet the still active grand old man of electrolyte research who was Ray's teacher, and whose publications on nonaqueous electrolytes in the American Chemical Journal, the predecessor of the Journal of the American Chemical Society, went back before the turn of the century [28]. Charlie Kraus was going to Seattle to receive a medal and Fuoss was chaperoning him there. Kraus never took a plane and they were going by train. Kraus had two suitcases, one of which contained exclusively bottles of Irish whiskey. No wonder Ann Fuoss was not happy. In the great hall of the New Haven railroad station, which has been demolished since, I also went to hear and be excited by Adlai Stevenson, then unsuccessfully competing with Eisenhower for the U.S. presidency. Ernst Bergmann who had broken his relationship with Weizmann and, unfortunately, also with the Weizmann Institute, came to work for a few weeks in the chemistry laboratory at Yale. He had been invited by Ray, and introduced me to a piece of Americana more sophisticated than the hamburger or the milkshake, namely the old-fashioned strawberry shortcake, in a drugstore just opposite the old railroad station.

The Yale Chemistry Department had strong connections with the Rockefeller Institute in New York; this is before the days of the Rockefeller University, and every year the Yale electrolyte wizards would meet with their counterparts at Rockefeller: Lewis Longsworth, who had designed original electrophoresis and diffusion equipment occupying a large room. Theo Shedlovsky, expert in electrolyte conductance, Duncan McInnes and others. McInnes had retired and in the freedom of his retirement was redetermining the Faraday constant, the product of Avogadro's constant and the elementary charge.

It was my good fortune, in this year at Yale, to meet and appreciate almost the whole spectrum of scientists young and old, experts in polymers and in electrolytes, and moving into the field of biology. A visit to Boston with David Ginzburg, working at Harvard, included Ephraim Katchalsky, spending a year with Edwin Cohn and John Edsall, producers of the long-lasting opus on *Proteins, Amino Acids and Peptides* [29]. I met Henri Benoit away in Paul Doty's lab from Charles Sadron's CRM (Centre de Recherches sur les Macromolecules) in Strasbourg, which I was to visit on my way home. Although many of the outstanding team assembled by Mark in the Polymer Institute in Brooklyn Poly had spread into industry and academe, a regular one-day symposium lecture series I once attended, attracted speakers of the caliber of Bruno Zimm, Walter Stockmayer, Turner Alfrey and others. I was pleased to meet the crystallographer, Isidore Fankuchen, who had worked on TMV, both structure and spontaneous orientation of the rodlike tobacco mosaic virus, with John Desmond Bernal, in England. A few years later Mrs. Fankuchen, on a visit to the Weizmann Institute, told us the story that, during their stay in Bernal's lab she and Bernal's secretary had founded the Society of women who had not gone to bed with the great man. Mrs. Fankuchen was president, and the secretary was the treasurer. Following their return to the U.S., Mrs. Fankuchen once received a telegram: 'Congratulations, you are now both president and treasurer!'

In the summer of 1952 I went to my first Gordon Conference on Polymers in New London, New Hampshire. Ray Fuoss had recommended me to Paul Flory, the conference chairman, to speak about polyelectrolytes. In those days the format of Gordon Conferences was two lectures in the morning, one in the evening, and the rest of the time discussion. No short talks, no communications. In this baptism of fire I overcame my fears and spoke before Flory, Stockmayer, John Ferry, Mark and many others. Another first in my budding scientific career was a Faraday Society discussion meeting in Toronto on the reactions of free radicals, organized by E.W.R. Steacie and attended by many outstanding physical chemists. It is amazing how Toronto has since changed from a boring English provincial town, with drinking places with separate entrances for men and women, with shutters coming down Saturday noon bringing life in the city to a complete standstill until Monday morning, to an attractive modern metropolis. I was coming by Greyhound bus all the way from New Haven to Toronto, reading Hemingway's *The Old Man and the Sea* which had just appeared, and stopping in various prearranged places on my way. To visit Art Bueche and Bruno Zimm at General Electric I went through a traumatic experience by lodging overnight in the Albany YMCA, which someone had recommended me to do. It was filled with outcasts and drunks, a picture which affected me deeply. Strangely enough, a year ago, I saw a picture with Jack Nicholson and Meryl Streep, *Ironweed*, which depicted exactly the same scene of human degradation and lost struggle for escape, in Albany. For the continuation of my trip I had written to Flory in Ithaca and was very disappointed when, upon arriving, his secretary in the Cornell Chemistry Department gave me a letter in which Paul apologized deeply that he had to leave unexpectedly, at short notice. I was lost for a moment and then realized that Peter Debye was at Cornell.

In my mind, at that time, Debye was a titanic figure who had established as early as 1912 the theory of specific heats, in 1915 the laws of scattering of x-rays and in 1923, with E. Hückel, the dependence of the activity of electrolytes on electrolyte concentration [30]. He had walked into a lecture by J. Chandra Ghosh who had proposed a dependence of the activity on the cube root of the concentration. After some thought Debye concluded that his was not likely and proposed a square-root dependence, which turned out to be true. One of his maxims was never to try to solve a problem, the correct answer to which could not be predicted in a simple process of thought. How many researchers grapple their whole life with one insolvable problem, without ever reaching an answer. Nowadays one can of course use a computer and spin out an unlimited number of numbers and beautiful graphics, which may be related only very distantly to the solution of the problem at hand. Scientists of the intellect of Debye do not appear very often on the scene of human creativity. I had in my mind a story I had recently heard about Debye that he was snoozing through a lecture, with eyes closed, while a desparate young postdoc was frequently interspersing his lecture with the cry, 'is it clear?', pointing at Debye. At the end of the lecture Debye slowly got on his feet and said: 'It is clear but it is wrong.'

Having no other option I nevertheless collected all my strength, and asked to be received by Debye. The meeting became a surprisingly pleasant encounter, which I have remembered since. Debye was working with his son Peter P. and together they had designed an instrument to measure light scattering of polymers. Debye himself had, during the war, shown how molecular weights can be determined absolutely from the assessment of the angular dependence of the scattering light. Now Debye was all excited about experimental details, trying to convince me that his instrument was much better than any other instrument used at that time.

I stayed for a few more days in Ithaca in a charming guesthouse in the woodlands close to the University, eating wholesome American meals of fish and steaks, downtown in the College Inn. Harold Scheraga was already at Cornell at that time, but I did not know him. I was so pleased to rest in Ithaca that I decided to skip my trip to Rochester, to visit Maurice Huggins, who, working independently of Flory, became famous for the Flory–Huggins equation in polymer solutions. Years later I met Huggins who scolded me for not informing him about the cancellation of my visit.

If the reduced specific viscosity of polymer η_{sp}/c is plotted against concentration c, a positive slope is observed, the intercept at vanishing concentration is called the intrinsic viscosity $[\eta]$ and is a measure of molecular size. The positive slope is a measure of interactions between macromolecules and is given by the product $K[\eta]$. K was established by Huggins and carries his name.

In polyelectrolytes, in the absence of added salts, the viscosity behavior is totally different, and η_{sp}/c increases sharply with decreasing c. Fuoss, who liked to express data in linear plots noted that a linear plot was obtained by plotting the reciprocal of η_{sp}/c against $\sqrt{c}$, and identified the intercept with the reciprocal of $[\eta]$ [31]. My own feeling was that, in the absence of added salt the viscosity had no relation to molecular shape and the increase of η_{sp}/c with decreasing c was due to an increase of intermolecular electrostatic interaction. I believed there should be a downturn of η_{sp}/c at very low polymer concentrations when the ionic strength became comparable to the ionic strength due to ions stubbornly remaining in the purest water, including OH^- and H^+ ions. Also, I believed that small amounts of electrolyte would eliminate the 'polyelectrolyte' effect. To convince Ray some experiments had to be done, preferably in a shear viscosimeter, at very low rates of shear. I knew that such an instrument had been built in Strasbourg by Georges Vallet and wrote to Charles Sadron, who generously invited me to stop in Strasbourg on my way home. Therefore, on my return trip to Israel, in February 1953 I took with me a sample of quaternized polyvinyl pyridine, a polyelectrolyte being studied in Ray's lab, and proceeded to Strasbourg, after a brief first stop in the Paris of my dreams.

The CRM had not yet moved into its own quarters and occupied the basement at the University, 3, rue de l'Université. I was staying in the modest Pension Elisa, which had one bathroom for three floors of rooms, unfortunately permanently occupied by a long-term visitor from the U.S. with wife and kids. There were public baths not far away. Strasbourg was still in a drab post-war condition and had not yet assumed the glitter of the Capital of the Council of Europe. Still, the 'choucroûte' on Place Gutenberg was quite satisfactory. As most Alsatians spoke the local dialect I tried to speak German, but was met by total silence. Luckily my French was quite good.

My time was rather limited so I was quite upset by the fact that the lab was closed: it was 'mi-carême' holiday, and electricity repairs were planned. The whole lab had rented a chalet in the Vosges and, with boss and cook, left for a skiing holiday to which I was invited as well. A miracle then occurred, which I had not expected.

Vallet was not there anymore and I was introduced to Jean Pouyet to help me in my experiments. Jean, with whom I then forged a life-long friendship, agreed to leave in the evenings, to drive in his old 'deux chevaux' to Strasbourg over icy roads, to work with me all night on the planned experiments. In the evenings electricity was available. The Couette viscometer, in its classical form consists of a stationary cylinder, suspended by a galvanometer quartz wire, to which a mirror is attached, surrounded by a rotating outer cylinder, with the liquid to be examined in the gap in-between. It is a very delicate instrument and requires great artistry for successful operation. To cut a long story short, we succeeded to prove the hypothesis I had put forward that a tiny amount of salt, of the order of 0.1 mM, restored the viscosity behavior nearly to that of nonionic polymers. Changes of conformation and coil expansion were milder than expected. Our paper was published and had a funny title because the polymer turned out to be only 60.8% and not 100% quaternized [32]. Physicists have in recent years entered the polymer and polyelectrolyte field. They use fancy new language but their results have not gone far beyond of what we knew in the earlier days. They love Fuoss's empirical polyelectrolyte viscosity equation and believe from their scaling relations that polyelectrolytes expand to full length. I have stated my views on polyelectrolyte expansion some time ago [33]. Currently we are is undertaking in our laboratory some total intensity angular dependence light-scattering experiments with salts of polystyrene sulfonic acid to establish the true expansion process of polyelectrolytes. It is not easy to do this as experiments are very difficult at low polyelectrolyte, low salt concentrations.

My visit at Sadron's laboratory lasted less than two weeks, though all my friends believe I had spent a longer time there. At that time I first heard about DNA which was intensively studied in Strasbourg as a function of salt concentration, pH and other parameters, by viscosity, streaming birefringence and light scattering. They were truly amazed that it was impossible to rationalize their data in terms of what was known from coiling polymers. Only after the Watson–Crick paper was published in *Nature* [34] later the same year, did their results make sense in the context of the very stiff double-helical Watson–Crick helix. On my way home I decided to design an effective low-shear viscosimeter for a study of both polyelectrolyte and nucleic acid solutions. We often came back to Strasbourg for meetings, discussions and just to enjoy visiting old friends. Almost the whole laboratory, with Sadron in charge, came to attend the IUPAC Polymer Symposium in Rehovot at the Weizmann Institute in 1956 and travelled by sea on the ss Herzl. They enjoyed the trip very much but did not realize that it was the Passover week and only 'Mazzoth' was available on board, no bread. Still, we remained good friends and Jean Pouyet made an official film of the meeting. In 1966, Jean came with his family to Rehovot to work with us on glutamate dehydrogenase, using mostly the ultracentrifuge.

From Strasbourg my trip went back via Amsterdam to pick up the KLM flight home, after a short delay. When I entered Ward's house he said: 'There has been a terrific flood in Holland, in particular in Zeeland (where he was born). I am leaving for Zeeland, will you come with me?' I said, yes, and we both drove for a few days to Zeeland; we had to go through Belgium, and then cross by ferry from Breskens to Vlissingen, as all other land routes were blocked, and I have a photograph of myself taken next to the big Willemsdorp ferry boat, sitting inland on swampy 'dry' land. Through my family relations and friends, I felt close to the Dutch people struggling for a bit of extra land, and this was just one brief episode in the centuries' old struggle between nature and man.

Back in Rehovot, surrounded by my family which had patiently awaited my return, I was now ready to strike out on my own in the study of biology by the application of physical means.

Weizmann Institute: nucleic acids (1953–1958)

The scale of things must have changed because upon my return from Yale I was approved by Benjamin Bloch, Administrative Director of the Weizmann Institute, the generous sum of US $1000.- (one thousand) and proceeded to gather components for building a Shedlovsky conductance bridge which served to determine the conductance of polyelectrolyte solutions, reduced as a result of strong interaction of counter-ions, surrounding and attracted to the oppositely charged macromolecules. I established a simple empirical relationship, which fitted well into the Manning counter-ion condensation model to be developed later. To avoid electrode polarization corrections which were a nuisance in low-concentration polyelectrolyte solutions, I developed a differential bridge circuit with identical unequally spaced conductance electrodes [35]. It was thus not necessary to eliminate electrode polarization by coating with platinum black, which also led to significant errors due to adsorption.

Another, somewhat more ambitious project was the construction of a Couette viscometer, covering a wide range of rates of shears, without using the delicate galvanometer suspension wire. I was fortunate to collaborate with Ephraim Heini Frei, then Head of our Electronics Department, on the original design of our precision rotation viscometer with electrostatic restoring torque, first reported at the founding session of the Israel Physical Society and published in March 1954, and dedicated to Albert Einstein on his 75th birthday [36]. The viscometer consisted of a freely floating inner cylinder, centered almost without friction by a single-point suspension, surrounded by a rotating cylinder driven by a wide-range planetary gear box. The restoring force required to prevent rotation of the inner cylinder was provided electrostatically, yielding precise measurement as well. At the Institute there was still a free flow of information and interchange at lectures, coffee breaks and social occasions and I received valuable advice from my physicist friend Saul Meiboom, who later became one of the early pioneers in NMR research. With the help of this instrument we could determine viscosities of polyelectrolytes, correctly extrapolated to zero rates of shear [37]. Though, from basic hydrodynamical principles, the dependence of the viscosity should be with even powers in the rate of shear only, experimentalists working with Poiseuille capillary viscometers were merrily extrapolating in a linear fashion. We could show convincingly that the first term in the expansion was indeed quadratic. From Jean Pouyet I received a sample of the best quality protein-free calf-thymus DNA, and from Arthur Peacocke a sample of salmon-sperm DNA. The molecular weights of these materials were not well known in those days and were grossly underestimated at around 6×10^6 g/mol. I will come back to this point later. I could show that, unlike coiling single-chain polyelectrolytes, DNA behaves like a stiff coil, its size only mildly depending on ionic strength, in line with the newly available Watson–Crick double-helical structure [38].

My good friend from the days of 'Hemed', and colleague, Uriel (Uri) Littauer, just back from his postdoctoral research with Arthur Kornberg in Saint Louis, came to me with a problem which proved to be of great interest. Ribonucleic acid, RNA, in distinction to DNA, is much more delicate, and prone to attack by indigenous enzymes. Therefore, no good RNA preparations were until then available for proper study. Uri had just succeeded, by using phenol extraction and other tricks, to obtain what looked like excellent preparations from *Escherichia coli.* It turned out that none of the best analytical phenol available was good enough for the purpose but had to be redistilled just before use. Sedimentation in the ultracentrifuge yielded three boundaries. The slower moving component (which, unfortunately, we did not then investigate!) was separated from the faster one by ammonium sulphate precipitation from a phenol-saturated water solution. From sedimentation, viscosity and light-scattering data the molecular weight was found to be about one million. The RNA preparations in their viscosity behavior closely resembled coiling synthetic polyelectrolytes, and behaved quite unlike DNA. The potentiometric titration curve showed some hysteresis. We also performed birefringence of flow measurements and found the sign of birefringence in neutral aqueous solutions of RNA to be positive as opposed to DNA which shows negative birefringence of flow. The positive birefringence observed for RNA could be compared with the measurements of Rosalind Franklin who found a positive contribution to birefringence of flow in TMV suspensions, which disappeared upon removal of the RNA constituent from the virus particle. The small birefringence exhibited by the

RNA preparations in pure water solution also completely disappears upon addition of salt, and no orientation is observed. In solutions of DNA considerable birefringence and orientation persist at neutral pH even in the highest concentrations of NaCl that have been investigated. We therefore discussed the viscosity behavior, birefringence of flow and potentiometric titration data in terms of a single-coil contractile model [39].

I have in front of me a letter in Hebrew from Uriel to me, dated 14 October 1958, from which I will translate excerpts:

> 'The Congress in Vienna was very interesting, 2000 lectures were announced and 6000 participants attended. It required a special effort to meet anyone in particular. I was lucky that my lecture on our work took place the first day and therefore many attended. We were lucky in the context of our conclusions on the single strand and we almost missed the boat. This concept reverberated throughout the lectures of some of the virus people and some other lectures, though without convincing physical proofs . . . Our lecture was well received, as well as our work on rat liver RNA. (J.A.V.) Butler asked whether we assume that the two low molecular weight components of the RNA are products of degradation of the higher molecular weight product – this is a possibility but the sedimentation curves do not support it. (Paul) Doty reported similar conclusions relating to rat liver RNA. After the lecture I met Doty, (Alex) Rich, (Francis) Crick and (Vittorio) Luzzati for lunch. Doty believes as we do that RNA is made of single fibers, but he believes this is an aggregate formed by a number of short fibers, and the proof for this, in his mind, is that heating with urea to high temperatures leads to molecular changes. This is a possibility which we should check. Crick also had a very interesting proposal which was based on the work of (Heinz) Fraenkel-Conrat on TMV-RNA, namely that the single RNA fiber can, under certain circumstances, fold onto itself, maybe not in as sophisticated form as TMV-RNA but sufficient to give some secondary structure characteristics, this might explain the reversible hysteresis curve obtained by potentiometric titration. Zillig from Germany has isolated three nucleic acids from *E.coli* which relates well to what we have found. He was very interested in our work and invited me to lecture in München but I refused, because I somehow still feel incapable to go to Germany*.
>
> I gave some of our material to Rich, and also to (Aaron) Klug, Director of Rosalind Franklin's laboratory, let us see what will happen. The lecture of Dave (Elson) was very well received and I enjoyed listening to it . . .'.

---

* We both now have strongly overlapping relations with Wolfram Zillig and of course also with Dieter Oesterhelt from the same laboratory and they both successfully participated in a Rothschild School on Modern Aspects of Halophilism, which I organized this year.

At the time I received this letter I was working in the Mellon Institute in Pittsburgh on the theory of multicomponent systems. Uri continued to work on the RNA problem with R.A. Cox, visiting our laboratory. Since we published our first papers we became very friendly with Aleksandr Spirin in Moscow, who referred to them frequently and exchanges with us reprints and New Year cards ever since. We occasionally meet at international congresses. Rosalind Franklin once came to our laboratory. She came from a distinguished British Jewish family and was related to the wife of one of my colleagues. Aharon had met her in Aaron's (Klug) laboratory and they had planned to do with my help some x-ray studies on bromine substituted polyelectrolytes, for greater contrast. When Rosalind died, these experiments were never done.

I was, after my return from New Haven, trying very hard to obtain some well defined polyelectrolytes. One such idea was to transform a polycarboxylic acid into the corresponding amine and to obtain thereby two polyelectrolytes with opposite charge but the same degree of polymerization and size distribution. They would be studied separately, and then the interactions between them. This could be done by the Schmidt reaction, acting with hydrazoic acid on carbonium ions formed in concentrated sulfuric acid solution. We found that very good yields were obtained by using 100% sulfuric acid obtained by the addition of sulfur trioxide to concentrated sulfuric acid. I was very happy to have the advice of Arieh Berger, an extraordinary personal friend and wizard as far as chemical reactions were concerned, with whose help we found that, in the successful transformation of benzoic acid to aniline all the reaction intermediates could be successfully followed [40]. In 1974 I published a paper with Dori Cwikel on polyvinyl-*para*-aminostyrene from polyvinyl-*para*-benzoic acid by the Schmidt reaction in pure sulfuric acid in the Arieh Berger Memorial issue of the *Israel J. Chem.* [41].

I made another attempt to study a simple well-defined polyelectrolyte and settled on polyvinylsulfonic acid, prepared by polymerization of the monomer, vinylsulfonic acid, the synthesis of which was first described by Kohler in 1897 [42].

With Ram Mohan, who came from India, we had some very interesting results. Aqueous solutions of polyvinylsulfonic acid and its salts with various monovalent cations, separate into two liquid phases at high concentrations of added monovalent electrolytes, the phase separation depending on temperature, polymer and added electrolyte concentration. The phenomenon is highly specific with respect to the monovalent electrolytes investigated. In the alkali halide series the order of specificity is $\mathrm{NaCl} < \mathrm{KCl} > \mathrm{RbCl}$ and $\mathrm{KI} > \mathrm{KBr} > \mathrm{KCl}$; no phase separation occurs with HCl, LiCl, CsCl and NH_4Cl. Partial separation of cations in mixtures containing the chlorides of both respective cations was achieved. From potentiometric and conductimetric titrations polyvinylsulfonic acid was shown to behave as a fully ionized polyelectrolyte. The exciting observation was that extremely interesting specificities were observed, though this macromolecule was mostly a sulfonic acid ion and the backbone hardly contained any structure-forming, or interacting groups. Specific effects with alkali cations were shown, from viscosity and conductivity measurements, to exist also in dilute solutions of the polymer, both in the absence and in the presence of added monovalent electrolytes. They are correlated to the order of specificity of the same cations in phase separation [43].

In 1958 I went to work for two years at the Mellon Institute and the analysis of the behavior of polyvinylsulfonic acid solutions led me, with Ed Casassa, to the analysis of the behavior of multicomponent solutions. More about this in the next section.

Another activity which gave me much pleasure and from which I learned a lot was the establishment of a laboratory for the study of plastics in Israel, on which I reported to *Mada* in 1957 [44]. We were fortunate to be joined, soon after the establishment of the Polymer Department, by an immigrant engineer from Czechoslovakia, Saul Gassner, who had great expertise with polymer fabrication and an unusual knack for the production of complex miniature equipment from plastic (polyethylene, nylon, plexiglas) components. He thus built many laboratory research instruments but also a water meter with noncorrugating plastic components, which was commercially produced. With very simple means he produced plastic thin tubing, syringes, complex plastic manifold taps, long before they became available for medical uses and standard items in the laboratory. Citrus fruit, a mainstay of our agricultural exports, was wrapped in treated paper sheets in a costly process. Gassner developed a process for generating an extremely thin invisible plastic film letting air through, but preventing loss of water by evaporation. This process eventually became a standard process in the packing of export fruit. In those days our modest plastics industry was slowly expanding without much expertise for manufacture or testing. It was therefore decided, with the generous help of the U.S. 'Point 4' program to set up a plastics laboratory at the Weizmann Institute for small-scale manufacturing and testing of plastic products. With the collaboration of Saul and the background support of the basic studies of the Polymer Department I then proceeded to establish the Plastics Research Laboratory, also containing a full documentation service, for anyone wishing to go into the manufacture of a new product. The laboratory was well equipped with moulding presses, injection molding machines, ovens, extruders, a universal testing machine, and a valuable collection of plastic materials of all kinds, assembled by Saul. When David Vofsi joined the Weizmann Institute he took charge of the laboratory, which eventually grew into the fully fledged independent Plastics Research Department.

The family of the Weizmann Institute was still a wonderful institution in those days and everybody knew each other, scientists, administrators and technical personnel. I have already mentioned that my broad background in science dated back to these early days. We were in particular befriended with the Ginzburgs: David, Hemda and their two daughters, Nini and Montzi. When Montzi, age six, returned to Rehovot after her visit to the U.S. in our joint trip in 1952, she cried out happily, upon being offered a coke: 'Pepsicolale, you so good . . .' Montzi is now an excellent, active, Oxford graduate, neurobiologist. They had come to Israel from New York in 1949; David's father had been a Hebrew writer in New York, and his uncle, still living then, a mathematician. David had been to high school in Palestine when he met his Israeli wife-to-be. David had synthesized a precursor of morphine, and later established a modern Chemistry Department in Haifa. He was one of our best chemists.

When, in 1954, Nutzi and myself left on our first joint trip abroad to the Milan/Torino polymer congress already mentioned, David and Hemda aided in supervising the well-being of our little boys. I

To which Neurath replied that what we had seen did not count, since **he** had not shown it, and proceeded to present a long exposé, well into the luncheon hour.

Many of us young scientists lived in the 'Shikun', on the Institute's grounds, in apartments that became available to us, because many of the foreign scientists originally scheduled to come to the Institute had, for some reason, changed their minds. In our apartment house, Neveh Weizmann 7, above us lived Joe Gillis, a Cambridge mathematician, originally from Sunderland up North, who greatly helped Aharon in mathematical derivations, in the early stages of our work. Joe was a bachelor who owned the only car in the 'Shikun' at that time, a small British postwar creation. This came in handy, because it was Joe who drove Nutzi to the maternity hospital, to deliver Shai. With Joe and Anushka Weizmann, organic chemist and Chaim Weizmann's younger sister, we drove to Tel Aviv in May of 1949, to watch the first Israeli Independence parade, which did not parade because the narrow Tel Aviv streets were so crowded with proud Israelis; there were very few cars in those days. Anushka was a conscientious organic chemist who always made sure, at lectures, that the valency of carbon atoms chalked on the blackboard had the correct value of four. She once walked into a lecture by Herman Mark, who was a perfect gentleman. When she entered he stopped his talk, walked up to her through the lecture hall, and gave her a big kiss. Ruzicka, apparently, was a little rougher. One day, when hearing that Anushka had studied at the ETH, he asked her in what years, to which she blushed and did not reply. Whereupon he said: 'Was it in the new building, or was it in the old building?' It seems the 'new' building had been completed some time in the late years of the nineteenth century.

From Joe I learnt that, when you were asked: 'Do you know a famous eccentric mathematician?' the reference is always to Paul Erdös. One day Erdös, who was visiting and stayed in the guest house of the 'Shikun', started a big row because he had locked himself in, lost the key, and could not come down. He finally climbed down the drainpipe from the second floor.

The day Joe came to tell us about his forthcoming wedding to Olga, a young poet from the South African 'Veldt', we were very would like to mention one more remembrance of the trip to Italy. When Nutzi went during one of the sessions, which did not concern her, to visit the Brera Gallery in Milan, she met a nice old gentleman, doing the same. He told her he was also attending the meeting. I asked her: 'Do you remember his name?' She said: 'Yes, I think it was Staudinger.'

It will not be possible to repeat all the names of colleagues already mentioned in one context or another. Michael and Margalit Sela were close friends. Margalit was exceptionally gifted and the right hand of Meyer Weisgal, who gave her a wonderful dinner party in the newly founded Accadia Hotel in Herzlia, when Michael and family left for a post-doctorate with Chris Anfinsen at the NIH in Bethesda, Michael told me that his mentor, Ephraim Katchalsky, had written to Hans Neurath in Seattle who, for some reason, had not answered. Michael himself therefore wrote to Chris, who was not yet very well known in those days. Michael was the first Israeli visitor at the NIH, and a large number followed in his footsteps. Can you imagine the course of Israeli life sciences if Michael had gone to Seattle? I must, at this point tell a little story about Hans Neurath, whom I have since met and come to know well. I was, in 1970, attending the IUB Meeting in Interlaken, Switzerland. I was quite new to the scene of biochemical meetings, but both Nutzi and myself enjoyed seeing many of our good friends, Flossie and Chris Anfinsen, Eva and Herb Sober, Margalit and Michael Sela, Inga and Bill Harrington. Bill asked me to chair a morning session, because the announced chairman had not arrived. The sessions were running quite late and he asked me to be a little strict with the speakers. The last speaker in my session was Hans Neurath. The speaker before him, whose name I do not remember, got into trouble because, instead of his own slides, Hans' slides appeared on the screen. It took a while to rectify this, and when he had finished we were really overtime. In introducing Neurath I made a little comment hinting to this, which he apparently resented. I said:

> 'Well, we are a little late, but we have already seen Dr. Neurath's slides, so we might be able to catch up and finish on time.'

happy. We all went to the wedding party in the Wizo Club in Tel Aviv. We and they are still happy today, and after moving through a few apartments and houses now live again under one roof in the magnificient House Markus for elderly scientists, just off the Weizmann Institute campus. Opposite Joe in Neveh Weizmann 7 lived Gerhard and Anita Schmidt; we sometimes played bridge with them and listened to Gerhard's music. Gerhard, the son of a well-known German chemistry professor, had taken his degree with Dorothy Hodgkin in London, and created solid-state chemistry and modern crystallography of small molecules at the Institute. Gerhard later became Scientific Director of the Institute. I still remember about ten young ladies working hard on their Frieden calculators, before the arrival of the computer, to evaluate the Fourier transforms. I also remember the day when the IBM perforated-card reader arrived.

Computers were introduced into the Weizmann Institute and, for that matter, into Israel a long time before anyone even thought about it, by Chaim Pekeris, geophysicist and applied mathematician from Boston who, with his wife Leah, also lived in a house on the 'Shikun'. Pekeris brought over Gerry Estrin who had participated on the construction of John von Neumann's first computer in Princeton to build a similar machine, later called 'Weizac', in Rehovot. We became very friendly with Gerry and Thelma, though I clashed mildly with Gerry in our precision workshop, as the construction of his magnetic drum memory took precedence over my low-shear viscometer gearbox. Pekeris was starting geophysical exploration for oil in Israel, and when oil was found in Heletz – it later turned out not to have been so much – we drank wine to honor the occasion and the Weizac played the Israeli national anthem, *Hatikva.* Pekeris went on to build bigger and faster computers, Golem I and Golem II, but when I asked one of the U.S. engineers, who had been brought to Israel by Pekeris, whether it would one day be possible to carry a small computer in your pocket, he said, 'Don't be ridiculous!'. I cannot really talk about everybody who stayed with us in the 'Shikun' then, I can only give a taste about days gone by. In 1950, there was a big snow (it snows about once every 50 years in Rehovot); nobody worked for three days, and we just threw snow balls at each other. Israel Dostrovsky, who had done outstanding work in physical organic chemistry with Edward Hughes and Christopher Ingold in England, explored the Negev desert with our friend Cornel Laub (Lahav) for minerals which Israel was lacking, and discovered phosphates, uranium, copper and others. Daphna Dostrovsky, painter and poet, was not too happy with progress in our ancient country and one day asked me, why Jews are so strange: Bethlehem is one single beautiful word and you insist to call it Beth Lechem (meaning House of Bread in Hebrew). She was very unhappy when the dirt road leading to the 'Shikun' from the main road was replaced by an asphalt highway, with buildings springing up on both sides. In the old days we could still hear the sound of hyenas barking in the night, but this slowly disappeared as the orange groves moved further south.

Joe Jaffe had come from Manchester where he had specialized in spectroscopy with Samuel Tolansky. From him I heard about Ronchi gratings, and even went to a lecture by the old man, visiting the Institute. Joe, who was a cousin of Joe Gillis, built a big revolutionary infra-red spectrophotometer, which was set up in our new Physics Building before the final walls were closed, so big was the instrument. Joe later decided to contribute to the development of applied science and established the first high-tech industry in Israel. He has since contributed in many ways to the development of applied science in this country. He and his wife Hali live in the artists' quarter, in the ancient harbor of Jaffa, and also enjoy sailing in the Mediterranean Sea. When they were still living on campus in 1956, Joe's mother was visiting from Manchester, during the IUPAC polymer symposium in Rehovot that year. We then had a visitor, Arthur Peacocke, originally also from Manchester, who became well known in DNA studies, but also in theological aspects of science and the impact of science on religion, and vice versa. One evening we visited the Jaffes and Arthur was introduced to Joe's mother as a Mancunian. She said:

> 'Peacocke, Peacocke, never heard this name; tell me, what was your name before?'

About Meyer Weisgal much can be said, and much is said in his autobiography, *So Far*, and it is not necessary to repeat here all the stories and wisdom he has generated [45]. He had collaborated with Weizmann for many years and was the superb architect and now President of the Institute. Weisgal wanted the Institute to be the best and the most beautiful in the world. He would not tolerate clothing hanging out on the balconies of our apartments and would raise hell if he saw anything of the kind. In his autobiography he relates the story that our younger son once asked his mother, upon being scolded by her: 'Who do you think you are, do you think you are Weisgal?' For our trip to the U.S.A., in 1958, we bought sports jackets for our boys, who till then never owned or wore jackets or ties; the newly acquired garments were immediately called Weisgals because Meyer was the only properly dressed person in those days on campus. I cannot resist telling two stories of Meyer, though you will find much more in his autobiography, including the more serious aspects of his important endeavor. He was telling us how, on one of his frequent money-raising tours, he stopped in London, at the Dorchester as usual, and went out for a small evening stroll around the hotel. Thereupon he was accosted by a prostitute who kept insisting that he was Ben Gurion: there was a certain resemblance in the shape of their white hair, and she was offering her favors for free. He says it was not easy to extricate himself. When he told the story to David Ben Gurion, the prime minister got very excited and kept asking: 'Was she Jewish? Was she Jewish?'

Meyer Weisgal was strongly attracted to the theater. In his youth, he had brought Max Reinhardt to the U.S. with an immense production about the story of the Jewish people, *The Eternal Road*, music by Kurt Weill and book by Franz Werfel, and it was even whispered that, somehow, he was trying to run science and the Weizmann Institute as a big theatrical production. He had many friends in the theater and one evening the whole 'haute volée' of the Israeli theater was at his home at a huge party – I do not remember in honor of what – including the great aging 'grande dame' of our stage, Hanna Rovina. In an adjoining room, Meyer was telling the following story. Three young archeologists had discovered a new pharaonic grave in the Egyptian desert. In the darkness of the grave, barely lit by the assistants' torches, a pharao-mummy raises himself slightly, rubs his eyes, and asks the first archeologist: Where are you from?' To the answer, Paris, the mummy says, never heard of it and treats the archeologist from London in the same way. But, when the third archeologist says, 'I am from Jerusalem', a big smile and grin covers the face of the mummy, saying: 'Ah, Jerusalem, ah, Rovina . . .'

The age of innocence came to an end roughly after the tenth anniversary of the State of Israel, in the spring of 1958. That night we went to Tel Aviv and danced all night in the streets and in the homes of our friends. We had grown up together and shared a great deal. Robert Robinson collected flowers on the road from Jerusalem to Rehovot and told us about the pigments they contained; Niels Bohr probed into the depth of human knowledge, Linus Pauling carried a toilet chain in his pocket to demonstrate polypeptide folding; Max Perutz told us about hemoglobin structure; from Abraham Pais we heard about Einstein; Felix Bloch introduced us to the new technique of nuclear magnetic resonance; Harold Jeffreys discoursed on the structure of the universe. And so on, and so forth. Great musicians played on campus and were entertained along with political leaders of the day in the President's residence, the walls of which were decorated with the best in modern art. I remember Jean Pierre Rampal, Jascha Heifetz, Isaac Stern on the one hand and Golda Meir, Abba Eban, Moshe Dayan, on the other. Also, only in Israel would participants of a scientific congress be received by the President of the State. In 1956 we successfully organized the IUPAC polymer congress in Rehovot and Jerusalem with strong international participation, and in 1957 we were proud to watch the Israeli flag displayed at the corresponding congress in Prague. That was the year of the Sputnik. I remember asking Mikhail Volkenstein from the Soviet Union whether we should learn more Russian, and he advised me to learn eating with chopsticks. I decided the time was ripe for extending my activities temporarily on an international level and in 1958 secured a two-year appointment, offered by Paul Flory, Director of the Mellon Institute, in Pittsburgh.

Mellon Institute: multicomponent systems (1958–1960)

The Mellon Institute in Pittsburgh was founded in 1913. Its purpose was to provide experimental facilities for the performance of industrial research, for which most industrial companies were not equipped at that time. Thus, to perform research in a given area a company would rent well equipped laboratory space from the Institute and would additionally have access to all the specialized research facilities, workshops and outstanding library provided by the Mellon Institute. This arrangement worked very well for years but obviously became counterproductive in our own times when many industrial firms, large and small, undertook their own research. Indeed, the Mellon Institute can be credited with having catalyzed a sizeable fraction of present-day industrial research in the U.S. Later, when the impact of the Institute was slipping, it was decided to use its still outstanding facilities and resources to create a first-class institute for basic research. Paul Flory was called in to direct this transition, and it was at that time that I went to work there on my leave of absence from the Weizmann Institute. It was an outstanding experience, meeting as well the excellent young people Paul had brought in, and I went home with a renewed feeling of exhilaration. Unfortunately, for a variety of reasons, this magnificent venture lost its momentum, and Flory left soon afterwards to take up a professorship at Stanford. It is a pity because there were hardly any other generously supported basic research institutes available at that time. The Mellon Institute merged with Carnegie University, to become The Carnegie–Mellon University. Having intimately known both the Weizmann and the Mellon Institutes, it seems to me that the now very much in the fore Hughes Foundation may have found the key for a successful amalgamation of research with teaching, guaranteeing a continuous through-flow of young intellect. The Weizmann Institute does have a school for graduate research, but I am afraid the emphasis is more on research than on teaching and the undergraduate fountain of youth is sadly lacking.

In the summer of 1958, this time all together, our family took the ss Maasdam from Rotterdam to New York after a vacation in Holland and a visit to the World Fair at Brussels, where eight-year-old Danny amazed the visitors in the Israeli pavilion by reading out loud from the Dead Sea scrolls, and I took a picture of the boys near the exhibit and model of our Couette viscometer displayed in the science pavilion. From New York we took the train to Pittsburgh (my, how things have changed!). Pittsburgh was quite a good place to live in, was then at the stage of total rebirth, including music in the 'Syria Mosque' and some art and theatre. The boys were of course about to discover the addictive attraction of baseball, with Roberto Clemente and the Pittsburgh Pirates approaching their much celebrated winning of the World Series over the New York Yankees. I never really understood American football, but baseball was another story.

The Mellon Institute was a beautiful building in pure Greek style, surrounded by huge magnificent columns. Not far from it were the 42-storey, imitation Gothic, Pittsburgh University main building and the Heinz Chapel, a faithful copy of the Sainte Chapelle on the Ile de la Cité in Paris.

Uri Littauer had told me, if you go to Pittsburgh, you must look up a young friend of mine, Gary Felsenfeld, who had recently moved from the NIH, into the Biophysics Department of the University of Pittsburgh, headed by Max Lauffer, just opposite the Mellon Institute. Upon calling Gary he immediately offered us the use of their apartment until we found our own accommodation as he, his wife Naomi, and their beautifully dark, recently born baby, Sarah, were leaving for a few weeks. This spontaneous gesture of friendship to an unknown couple with two dangerous young brats, and many other things to come, sealed a deep friendship between our two families, lasting ever since. Gary and Naomi had only been married a year ago and Nutzi was so anxious Shai and Danny would ruin their beautiful wedding-gift couch, that, within three days of moving in with the Felsenfelds, we moved out into our abode for the coming two years, in Alderson Street on Squirrell Hill. I could reach the Mellon Institute by walking through Schenley Park, in which Carnegie University was situated. We bought a $90 1949 vintage Chevrolet and our boys, intently watching the 1959 cars with their big wings, declared that, when we would come back in 1969, that is what we should be buying. I bought the car from a Swiss postdoc who demonstrated its viability to a novice in these arts; it was our first car, but it seems that between sale and delivery a few vital parts had been exchanged for less reliable ones. I was given the address of a Jewish mechanic, Murray, who advised me it was worth caring for this car, as all used cars have troubles which must be continuously attended to. In the two years in Pittsburgh I hardly even saw his face, as he was always lying flat under a car he was fixing, and thus his philosophy emerged under these rather strained circumstances. I should have taken notes to publish a book concerning this horizontal approach to modern philosophy. Our blue Chevvy became a close companion of our lives. It lasted two years, the length of our stay in Pittsburgh. On our last day we packed our valises – our large baggage had been sent separately – and we prepared to drive to New York and sell the car there, before flying home. On a clear Sunday, on the empty Pennsylvania Parkway, a car passed us and I noticed a strange smell. It turned out to be our car which smelled strangely, and it came to a final, well-deserved stop. We were towed into a gas station, sold the car for the junk heap for a nominal sum, and called Saul Meiboom, who very kindly picked us up in this middle of nowhere from his home near Bell Labs in New Jersey. Sic transit gloria mundi. The real moment of glory of our car was much earlier, close to the beginning of our joint careers, when we drove to New York to visit Arthur, Mathilda and Daphna Krim. Arthur, president of United Artists, had recently married beautiful Mathilda, who had been a colleague scientist at the Weizmann Institute. Nutzi had flown in a bad ice storm to New York for the wedding, and reached a peak in her career by singing a duet with Harry Belafonte.

Anyway, when we reached Sutton Place, where the Krims were staying, the car was gently moved by an attendant into the apartment garage and, I believe, almost completely overhauled. I think they had never seen a creature of this type. After that it ran like a charm. With Mathilda and Daphna we went in great style on a few days' trip on the United Artists luxury cruise boat 'Aladdin's Lamp', symbolized by a flag with a movie strip emerging from the oil lamp's spout, around Manhattan, and into Long Island sound. When we docked in Cold Spring Harbor, NY, and had a luxurious lunch in a dock-side restaurant, though the corned beef on board was much more delicious, I now reached a peak in my own career, because the restaurant owner thought I was the owner of the boat. I must now go back to the main trend of my story.

Work started at the Mellon Institute by being officially received by the Institute's chairman, General Matthew Ridgway of World War II Far Eastern fame, together with a number of other new workers. When six or seven of us were introduced to him in his office, he immediately talked to us as if he had known us for ever. It was an amazing exercise of having prepared himself very carefully for this meeting. Of course, we never saw him again. My lab at Mellon was in the department devoted to Polymer Research, in a broad sense, enthusiastically directed by Tom Fox. Hershel Markovitz and Don Plazek, a student of John Ferry, were engaged in studies of rheology and creep. Bernie Coleman, whom I knew from Yale, was engaged in a similar vein, but in a more basic fashion. He was promoting the science of continuum mechanics, closely collaborating with Walter Noll of the Mathematics Department of Carnegie University. The high priest of this science was Clifford Truesdell, who believed in the magnificent pattern of scientific thinking evolved in the 17th and 18th centuries by Isaac Newton, Gottfried Wilhelm Leibnitz and others. When lecturing, Truesdell was dressed in a beautiful robe. Noll's lectures always started with the fundamental statement: 'Take a body B . . .'. I was honored by being invited at evening gatherings of the continuum mechanics elite, at which large amounts of dark German beer were consumed, of which they exclaimed: all beer, no molecules. Ed Casassa and Guy Berry were busily engaged in obtaining reliable ultracentrifuge and light-scattering data of well-characterized polymer samples to advance the theories of Flory, Zimm and Stockmayer with respect to polymer solutions. Ed, who was coming from MIT, was extremely well versed in George Scatchard's thermodynamic theories concerning proteins, which turned out to be of great advantage when we tried to understand the behavior of multicomponent solutions.

Across the road, in the Pittsburgh University Biophysics Dept., was Felsenfeld, as I have already mentioned. When one day Bob Hexter redecorated his apartment, the wall painting was done by the crew of Marshal Fixman, Bob Hexter, Gary Felsenfeld, and myself.

The whole operation was very entertaining and Norma Hexter kept serving us a broad range of drinks.

I should mention Henry Frank, at Pittsburgh University, whose theory of water structure represented a milestone in the development of this important field, and I also remember lecturing visits by Manfred Eigen and Charlie Tanford. Flory organized a meeting, attended by a broad range of outstanding physical chemists to cell physiologists, to discuss aspects of muscular contraction which I summarized in 1960 with Dick Podolsky and Bill Carroll in *Science* [46]. I remember Aharon Katchalsky delivering one of his masterful lectures in which complicated mathematical expressions became simpler and simpler, until Aharon triumphantly finished by exclaiming; 'Lo and behold, A=B.' Whereupon Marshall Fixman, who was sitting next to me, bent to my ear saying: 'Well, indeed, he has rederived the Gibbs–Duhem reciprocity equation.'

I very much enjoyed my interaction with Ed Casassa which extended beyond our quest for an understanding and simple formulation of the thermodynamics of multicomponent systems, including ionized components: Ed and his wife Ethel, a scientist herself, played Beethoven's Trio in B flat major No. 11, with our ten-year-old Shai playing the clarinet part. The analysis of multicomponent systems can be viewed on a number of levels, all aimed at simplifying the analysis of experimental data. George Scatchard defined components including proteins and simple electrolytes, and J.J. Hermans introduced the concept of 'isoionic dilution', in which dilutions were made at constant ionic strength, including in its definition the counter-ions of the charged polyelectrolyte. It turned out that the simplest and most straightforward concept was to assume that, in a process involving a change in concentration, dilution is performed not at constant concentration of the low molecular weight components, but rather at constant chemical potential of the low molecular weight components. In principle many experiments, in one way or another, lead to the analysis of an osmotic pressure experiment, in which low molecular weight components diffuse through a semipermeable membrane, while the high molecular weight component is restricted to one side of the membrane only. If a macromolecular component is a polyelectrolyte with dissociating charges, then the additionally dissociating charges push some small electrolyte components into the polyelectrolyte-free solvent, again to maintain constant chemical potential of small ionic components. This is known as the Donnan effect. The most interesting conclusion in the case of the osmotic pressure is that the number concentration of polymeric ions is correctly determined, in the limit of vanishing polymer concentration, and the polymer ions are correctly counted in this colligative procedure. We have often emphasized, and I will only mention it here briefly, that it is not molecular weights which are determined, but rather the number of polymer particles in solution, from which molecular weights are determined in agreement with whatever concentration units are used to relate to number concentration. All so-called molecular weight determination methods, such as ultracentrifugation, light and x-ray scattering relate to osmotic pressure in one way or another, and are therefore subject to the same considerations. Thus, in ultracentrifugation, the sedimentation field generated in the ultracentrifuge leads to change in macromolecular concentration, which is countered by the derivative of the osmotic pressure with respect to concentration. In a simple two-component system, whether a material sediments, floats upwards, or stabilizes in a band, is determined by the value of the Archimedes buoyancy term, $1-\bar{v}\rho^\circ$, including partial specific volume, $\bar{v}$, and solvent density, ρ°. In a charge-carrying multicomponent system the buoyancy term becomes more complicated because of the phenomenon of 'preferential interaction', which has created a great deal of confusion. We could show that, whereas in a simple two-component non-ionic system the Archimedes buoyancy term is simply the derivative of solution density with macromolecular concentration, the density increment, in a multicomponent system it is simply the same quantity at a constant chemical potential of all small diffusible components. Formulated in this way, equations can be reduced to a very simple form. A similar conclusion had also been reached by Agienus Vrij, a student of Overbeek's in Holland, but this was not known to us until much later, since Vrij's work was only published in a thesis, I believe in Dutch.

How are the scattering of light, of x-rays and more recently also of neutrons, related to this general scheme? According to Einstein we observe scattering of electromagnetic radiation (light, x-ray, neutrons) because, if we observe adjoining small volumes in solution by light, x-ray or neutrons, their refractive index, electron density, or neutron scattering density are different due to fluctuations in the concentration of the various components. Concentration fluctuations obviously depend on osmotic pressure variations, and this is how scattering relates to osmotic pressure derivatives and to the strength of the contrast-variation terms, similar to the density increments in sedimentation. It was important to point out that this approach correctly yields the number concentration of macromolecular polyelectrolyte particles in multicomponent systems, from which molecular weights may be derived, yet it provides no information about salt binding or hydration. Information concerning salt binding and hydration can be satisfactorily derived from a study of the density increments themselves, and interesting complementarities arise when some of these methods are used in conjunction with each other.

A detailed summary of my work with Ed Casassa on the *Thermodynamic Analysis of Multicomponent Solutions* was presented in 1964 [47]. In this work we considered an unlimited number of components and went to great lengths to precisely develop such quantities as molecular weight averages and virial expansions. The Editors thought the presentation was awe-inspiring but maybe also a little frightening and suggested an explanatory summary, which I thought we should call, *A Guide to the Perplexed*, following Maimonides. Finally it was just called *Summary*. Our opus was much quoted, but I am not sure it was much read. Ed told me one day he received a letter from *Current Contents* that it was a citation classic and he should send in the necessary summary. He never did, so we did not get into the books. We continued to work on various aspects of nucleic acid and protein research in Rehovot, and I published in 1976 a book, *Biological Macromolecules and Polyelectrolytes in Solution*, at Clarendon Press, Oxford [48], in which I discussed various advances and applications, reducing all derivations to the much simpler situation of three components only: (1) water, (2) the macromolecule and, (3) a salt or nonionic low molecular weight component, for instance, a sugar. In 1981, I published an extensive article on forward scattering of light, x-rays and neutrons [49], which was mainly meant to point out the extremely interesting advantages arising from the use of contrast variation and neutron scattering by itself or in conjunction with ultracentrifugation, towards the study of hydration or salt binding in nucleic acids or in extreme halophilic proteins. Finally, as mentioned earlier, I have recently written a detailed review [9] which summarizes the development and achievements, if any, of my scientific contribution and thought.

Weizmann Institute and NIH: nucleic acids and proteins (1960–1973)

The period between our return to Israel in 1960 and the 'Yom Kippur' war in 1973, can be called the golden age of our lives. We were young and active, enjoyed our children and friends, and believed in the success and justice of our young nation. The 1967 Six-Day War, though a traumatic experience, led to a powerful feeling of strength. Very few then realized that the problems the war had created, and which were not solved, would come back at us with tremendous disruptive strength. The world had recovered from World War II, prosperity was widespread in Western Europe and in the U.S., money for research was easily available, and the big problems of misery, disease, colonial wars, Viet Nam, corruption, drugs, AIDS, totalitarian regimes rising and collapsing, ruin of the environment, religious fanaticism, and so on and so forth, had not yet manifested themselves.

When I came back, the Polymer Department was about the same. Here I must pay credit to our magnificent secretaries, who have followed us through dark and through light moments. In the old days the success of the secretary was much more a work of art than it is nowadays, in the era of the computer and the xerox machine. Everything, including papers and theses were carefully typed with carbon copies; the old Gestetner copyer was not very effective, and corrections had to be artistically introduced by corrective fluids, displacing the rubber eraser. Equations were of course drawn in by hand. I am now cleaning out my old papers and get terribly upset, when looking for some interesting relic of a long-gone age, that the copy has faded beyond recognition. Of course, in addition to their technical prowess, we valued our secretaries for their human attributes, which are as important today as ever, and cannot be replaced by word processors or copying machines. Thus, when I came back to Rehovot, the secretary of our youth, Lola Pinto, had left and was replaced by our close friend for many years, Ruth Goldstein. Ruth, fresh from literary studies in college, with hardly any knowledge of secretarial duties, was hired by Aharon on the promise of the great excitement this job would provide. Aharon was travelling a great deal and in his absence Alex Silberberg was running the mechanics of the place. Ruth, not being very busy at one time and I being away temporarily, was instructed by Alex to go through my private records and reorganize all my letters, reprints and files, in whatever manner she would think best. When I then came back it took me a long time to find what I was looking for, and I often wondered what made Alex give this instruction. An outstanding help over many years and until her retirement was Sylvia, the rock of Gibraltar – this happened to be her family name – who had immigrated with her family from the U.S. But the dearest of the dear, Herut Sharan, a child of Yemenite immigrants, started work in the Institute many years ago as a cleaner, and has now raised herself to qualify for the award of the Institute's outstanding worker. Her strength does not only consist in the technical innovations, which she has all mastered, but also in the human qualities which, in our society, become rarer and rarer as time goes by. As Herut is now typing this into the Apple, and probably also blushing, I will only say that it is people like her who make our efforts worthwhile.

In those days the Weizmann Institute was still run efficiently with a small administration. The scientific director was an outstanding young physicist, Amos de Shalit, whose intense interest in furthering our research was unsurpassed. I remember that one day, when I came to see him with our chief engineer, Abraham Dines, to debate the cost of installation of our newly acquired Model E ultracentrifuge, he took one good look at the figures proposed and told

Dines: 'You will build it for half the sum shown here', which was eventually done. Another powerful instrument we acquired at the time for our light-scattering studies was a SOFICA photometer, originally developed in Sadron's lab in Strasbourg. This was still before the days of the laser, but a long way from the Pulfrich visual split-field photometer, barely ten years earlier.

As regards the specter of ever-growing administrations, apparently inevitable, I have heard a story about a recent event at our Institute. The present head of our Administration, now called a Vice-President, brings in a powerful commercial training outfit to raise our administrators to a high level. In one of the exercises a task force is asked to work on the following problem: the Institute goes through a terrible crisis and must temporarily fire most of its staff. The question is, who should be fired. The correct answer of course is, the scientists should be fired and the administration kept, as it is relatively easy to rebuild a new scientific body, but it is almost impossible to rebuild a successful lost administration!

In 1963 a festive moment was the opening of the Ullman Life Sciences building. I remember going to Jerusalem to pick up John Kendrew and John Edsall who had come from Jordan via the Mandelbaum gate. After the official opening ceremonies there was to be a lecture by Jacques Monod in the huge Wix auditorium. Meyer Weisgal, being afraid that the audience would not be large enough, instructed secretaries and gardeners to attend the lecture. I shall never forget the moment when, while Monod was lecturing on the major problems of biology, an old Yemenite gardener, who probably had heard enough, got up from his seat and, rather than leaving by the back door, quietly came forward and slowly walked across the stage to the front exit, in the eyes of a speechless lecturer and audience.

Scientists are often boring and show little interest in things beyond their narrow scientific pursuit, such as intellectual activities or the arts. This statement should not be generalized as I proudly own a real Ef Racker painting and have often listened to excellent chamber music recitals at Gordon Conferences or similar meetings. Unusual practitioners of the trade of science have crossed my path. Whereas the scientist today usually travels with a box of slides or transparencies vastly exceeding the number he can show in the time alotted to him, elegant performers, such as Bob Woodward for instance, travelled with a magnificent wooden box of color chalks, that enabled him to create his fascinating story on a huge blackboard in front of the eyes of a spellbound audience, not to disappear practically unexplained, as when one slide or transparency succeeds the next in a mad rush of keeping up with the time. My friend Bernie Coleman used to say that scientists are like troubadours of the past, and they should go from city to city singing their wares. Eccentricity is also a desirable quality and I remember Franz Grün, from Basel, a co-worker of Werner Kuhn, who, visiting us, cycled to the Dead Sea, carrying an alarm clock, 'Wecker', attached to a long string, in his bosom. When once asked by some passers by, not far from the Neveh Yaakov asylum, what time it was, he slowly extracted his clock, whereupon the frightened questioners fled at top speed.

But above all the qualities one mostly looks for in the cross-section of people one runs into in a lifetime of endeavor, are characteristics associated with true nobility of mind and action. One of my best friends is Bill Carroll, a Harvard Ph.D. and for many years engaged in competent research at the NIH. One day Bill decided that what he was doing, many other scientists were also doing, sometimes better and sometimes less so, and that we should use our abilities and the intellectual benefits we have acquired from a benevolent Society, to aid vast strata of the population who have not benefitted in similar fashion, and thereby to start a process of, hopefully, catalytic redemption. Bill resigned his job from the NIH and went to teach in junior high school in the terror-ridden black ghetto of Washington, the proud capital of the U.S., for 15 years. To get to his school, he left every day by bus before six in the morning, feeling that cars should not be used to pollute the atmosphere. Bill is now retired and, I am afraid, disappointed that his efforts have not led to the dramatic results he had expected. I collaborated with Bill on devising a beautiful method for sulfonating monodisperse polystyrene samples to *para*-polystyrene sulfonic acid, a material we are still using to derive some unknown basic laws of polyelectrolyte behavior [50]. Bill and Bunny and three of their children spent a wonderful year at the Institute in 1962. Their son Bruce considerably raised the level of basketball during their stay in Rehovot. We had a big fright when they left because we knew that, on their way back in July 1963 they were driving through Yougoslavia and had planned to stop in Skopje. We could not establish contact with them but later found out that they had left one day before the big earthquake.

A tremendous personality, who became a close friend, was Gordon Tomkins, head of the NIH laboratory, where I spent a year in 1965. Gordy died a young man a few years later in 1973, the result of an unfortunate operation. He was an outstanding biologist with a very deep intuition and understanding of the basic phenomena of life. He was not only interested in his own work but spent much of his time in the corridors on the second or third floor of the famous Building 2 of the NIH, listening to and discussing the work of his colleagues. He combined a classical Jewish humor, which he expressed rather more warmly than Woody Allen, for instance, with a precise universal intellect. Unlike many of our colleagues, he wanted to know for the sake of knowing and not for the sake of pushing himself up on the ladder of success. He was therefore always prepared to help, without ever expecting gratitude in return. The same was true with respect to private matters where his offers of help were unlimited. Gordy was an accomplished jazz musician and went to play on a variety of instruments with outstanding Negro professional groups in the District. His wife, Millicent, was a very good singer and painter, and I spent unforgettable moments at their home, when they were making music with their two daughters in Chevy Chase, and later in Mill Valley, when Gordy was a professor at the University of California Medical School in San Francisco. With Gordy we solved on my 1965/66 visit to the NIH the problem of the subunit structure and linear association of bovine liver glutamate dehydrogenase [51], which had been incorrectly understood previously. Here, application of the theory of multicomponents solutions came in handy.

In those days I met a few more scientists in the biophysical sciences, which raised my faith in the human race, as expressed in the 'homo sapiens' species, combining the virtues of outstanding scientific achievements with the gift of friendship and humanity.

Arieh Berger, John Schellman, Pete von Hippel, Ken van Holde, Howard Schachman, Bill Harrington, David Davies, Terrel Hill, easily qualify for this status. With Walter Gratzer I have enjoyed a wonderful exchange in matters of science, music, the English country side, the theater, literature, the arts, good food and drink and, above all, relaxed positive human relationship. I admire his writings and have enjoyed the occasions when we could exchange views concerning literary forays. I remember in particular the days when in the mid-seventies, stopping in London, I joined Walter and Joan Morgan for business lunches in small pubs near his lab in Drury Lane, where they were planning future issues of the recently started, and later immensely successful, *Trends in Biochemical Sciences* (TIBS). At one of these occasions I conceived the idea and was encouraged to write a leading article for TIBS [52]. As far as I can recall Walter never received or asked for any written credit in this enterprise. If I am not mistaken, he is also responsible for the creation of *News and Views* in *Nature*, a new style in modern scientific journalism, and I have a complete set of his biting book reviews. Occasionally, on my way to the U.S., I would visit Walter and his wife, Hannah Gould, in London, as a brief interlude. Though we now live in the era of the compact disk, I might pick up some interesting 50p LP records in places he knows in Soho.

Research in the lab was now devoted to a study of light scattering, refractive index and pressure dependence of refractive index of water, heavy water and other liquids, in collaboration with my students Gerry Cohen and Emil Reisler. We could show that the Einstein-Smoluchowsky fluctuation theory for the scattering of light was correct for unusually structured liquids such as water and heavy water, and derived a very precise refractive index law for these liquids, predicting that at very high pressures their properties should revert to those of simple liquids [53]. We had come across a claim by Vittorio Luzzati that caesium and sodium DNA might have different structures, as expressed in deviations from the Watson-Crick B-form helical periodicity, and Gerry could show, again by application of multicomponent solution theory to small-angle x-ray scattering, that the structures were identical [54]. Gerry determined a large number of density increments and partial specific volumes over a broad range of salt concentrations, and we could derive DNA hydration values which are in good agreement with values derived today by high-resolution x-ray crystallography of various oligomers [55]. We were rather less successful with the precise determination of DNA flexibility, the so-called persistence length, with DNA fragment samples then available, but could do much better when well defined and uniform plasmid DNA samples became available. The sum total of these data, which have successfully withstood the test of time, are now appearing in *Landolt-Börnstein, New Series, Biophysics* [56]. I had previously reviewed hydrodynamic and thermodynamic studies of DNA solutions [57]. Emil Reisler continued the glutamate dehydrogenase studies initiated in Bethesda and we also benefitted from the electron microscopy and image reconstruction expertise which Bob Josephs had picked up at the MRC in Cambridge. We did report our work on subunits to superstructures, assembly of glutamate dehydrogenase, in the 23rd Mosbach Colloquium on protein–protein interactions in 1972, organized by Rainer Jaenicke, co-editor of the present volume, with Ernst Helmreich [58].

In other activities in our laboratory Allen Minton and Emil analyzed the temperature and density dependence of the refractive index of pure liquids [59] and Jamie Godfrey studied equilibrium sedimentation and light scattering of carefully prepared DNA fractions [60, 61]. This was a last heroic effort using classical technologies, before the laser and equally sized DNA plasmids and restriction fragments became available. Jamie went on to collaborate with Van Moudrianakis at Johns Hopkins on nucleosome structure and has recently developed a very interesting approach for measuring the equilibrium electrophoresis diffusion pattern of charged macromolecules [62]. I should also mention Nobuhisa Imai from Nagoya who successfully faced the deep cultural clashes involved in adaptation to an entirely new medium, and also determined the activity of KCl in potassium vinylsulfonate solutions using cation-sensitive glass electrodes [63]. Both he and his wife Eiko remained our close friends.

In my year, 1965, at the NIH with Gary Felsenfeld we studied the phase diagram of poly (A) at neutral pH in high salt and could determine single strand polynucleotide conformations at various temperatures in ideal, so called theta, or Flory, solvents, by light scattering and ultracentrifugaton [64].

Life is not all work and a favorite activity of top grade scientists at NIH is year-round indoor tennis [65]. I have enjoyed playing with Terrel Hill, David Davies, Bill Eaton, Ed Rall, Mort Lipsett, George Khoury, Mones Berman, Tsuchiya Takashi, Bob Rubin, Ira Pastan and others. Sometimes the game was enjoyed more, and at other times and with other partners it was taken more seriously. When we came to Bethesda in 1965 we rented an apartment at the Linden Hill Apartment Hotel with its tennis courts, which have since supported this important aspect of research activity. Later, the hotel building and courts were razed, to make room for a new housing development. I do hope the blow will not be total and alternative courts will be found.

After-hours' activities in Rehovot included one phone call from Mrs. Vera Weizmann's secretary: would I come to dinner and a game of bridge. They apparently needed a fourth hand, and my wife was not included. I declined the dinner invitation, but said I could come later, for the game. Mrs. Weizmann's partner was Prof. Isaac Berenblum, our famous cancer scientist, and mine was his wife Doris. We played in the old Weizmann library, in the magnificent villa on top of the hill designed by the Jewish German refugee architect Mendelsohn, surrounded by unbelievable, historical mementos. Mrs. Weizmann, at one time was displeased by her partner, turned to him sharply and said: 'Professor Berenblum, if you run your department the way you play bridge, you cannot be doing very well.'

I will not discuss in detail the Six-Day War in 1967 which, apart from leading to boundless euphoria, the feeling that Israel was now one of the big powers and the possibility to go skiing on Mount Hermon without going abroad, did not lead to a solution of the conflict which has come so strongly to the fore within the last two years. Our boys were too young for the war in 1967 but enlisted soon thereafter in the 'Nahal' and parachute fighting units. They were both wounded in the War of Attrition during their military service, but fortunately not seriously. On the completion of their military service Shai and Danny entered the Hebrew University to study physics and medicine, respectively. My army service still continued in the form of reserve service and, for a number of years, I joined the Educational Corps, to lecture frontline soldiers. In the following I reproduce a short article I published in the Jerusalem Post in June 1974.

'We set out from Rehovot at seven a.m. The bus from Tel Aviv to Kiryat Shmona leaves at eight, winding northwards out of the metropolis past the pleasant Sharon coastal plain, the starkly beautiful Ara valley, the lush Emek, the lovely Sea of Galilee and on to upper Galilee. At eleven-thirty a.m. we are in the army canteen at Rosh Pina; after a short wait, a little van arrives to take the assembled lecturers to central camp past the bridge of the Daughters of Jacob over the muddy Jordan River into the Golan.

We are assigned to various army units holding the new line, and from now on frustrations compound; we may well spend the next 24 hours lying on army cots, listening to the same tired pop music for the umpteenth time, rereading articles we would ordinarily never look at in yesterday's evening paper. The logistics of moving us to the frontline are complicated and besides we are not the highest priority assignment, and unnecessary risks should not be taken. Just before the ceasefire a well-known young lecturer (in literature) from Tel Aviv University was killed by an old Syrian shell, on a similar assignment.

We are equipped with steel helmets and travel in fast jeeps or armoured cars. The going is rough; now and then, frightened animals make an appearance, groups of donkeys, set 'free' by the course of events, picturesque storks, an occasional hare. As the moonless night falls, an amazing starlit sky emerges. I myself have never before seen so many stars.

Eventually the outpost is reached. Our reception is warm; the men seem eager to listen to a messenger from the world outside. A faint light is provided by oil lamps though sometimes the pale glimmer of an electric bulb is available. Throughout the proceedings communication instruments keep humming and an occasional warning comes through. At a barely noticeable signal, a group of young men reach for automatic rifles and file silently into the night to return when the alert is over.

I introduce myself, mention my place of work, try to explain what we are doing – and why. Sometimes the audience is 'high-brow' – all high school graduates – and then the level of discussion is more technical. At other times, a more popular stance is appropriate. But the level of interest and intelligence is universally high. I speak about molecules; large molecules, macromolecules or polymers, describe their chainlike structure, the many shapes they can assume, the way they may fold into specific well defined conformations. I refer to the impact of science and technology on man since the Industrial Revolution, in particular since World War II. I talk of the changing environment, of ecology and pollution, of the energy problem and the inability of our society to adjust to a new world. I mention the cataclysm that faces us if we do not heed the warnings.

Questions are sometimes slow to come but they eventually develop. Not all the questions are of equal interest to everyone. Remember, this is not a classroom. But enough questions are asked for a good discussion to ensue. How will Israel solve its energy problems? What about solar energy and the problems associated with atomic energy? Are we about to solve the problem of cancer? Can plastics replace metals? And many more until speakers and audience alike are exhausted. After the talk, there are private conversations. Always there are two or three prospective chemists, physicists or biologists in the audience; they want advice in preparation for their discharge from the armed services. We talk in an atmosphere that is serious - but relaxed.

Then it is time for a simple substantial meal, served at tables, unlike the lining-up with mess-tins of my days with the British Army in World War II. Either I will spend the night here or I will proceed to another outpost - if transport is available. We give between one to three talks a day and after four days I find myself exhausted. Intellectually, it is hard to recreate the same sense of excitement each time. These are not dry presentations of facts and theory designed for fellow-scientists or research students. Unless I can imbue these talks with some special excitement I might as well not talk. Physically, I am exhausted too and grimly envy the twenty-year-olds their stamina and endurance.

After a week or so the 'tour of duty' draws to its end. Whoever has been privileged to spend a few days with Israel's young people cannot but believe in the future of this country. I return - on a bus crammed with Army personnel going home for leave - straight to Tel Aviv. We stop briefly for refreshments at Tiberias and spend five minutes lolling in an easy chair near the blue waters of the ancient lake watching girls in bikinis, tanned tourists and other sun worshippers, all seemingly oblivious of the drama being played out on the hills up North.'

When Milton Kesker graduated from Columbia University in 1949 studying with the well-known colloid chemist Victor LaMer, shortly after the end of the war, the opportunities for employment were limited and in particular industrial employment for young Jewish Ph.D.s. Colloid chemistry had been called the world of neglected dimensions by Wolfgang Ostwald, and was still rather neglected, though I believe it represents an extremely important dimension in technology and biology as well. Milton, the young graduate, went to Clarkson College in Potsdam, NY, and founded an outstanding Colloid Institute in the Chemistry Department, on the Raquette river, close to the Canadian US border, in the middle of nowhere. I am told that, when they once were looking for a young biochemist to join the faculty, the latter asked: 'How many hundreds of miles away is the nearest biochemist?' This situation has changed now. Milton became a world authority on light scattering, in particular Mie scattering from large particles, and atmospheric scattering. He has written a definitive textbook on the topic, and he never left the place. Much more, he has attracted there a group of excellent scientists who succeeded in putting Potsdam on the map. It is Clarkson University now; the adjoining New York State University Campus has a quartet-in-residence and their ice-hockey team competes with Cornell University. Potsdam has very long and cold winters, though Montreal, which is not much warmer, is only one hour away by car. I came to Potsdam for the first time in 1962 for the Interdiciplinary Conference of Electromagnetic Scattering (ICES), attended by Peter Debye, Victor LaMer, Karol Mysels, J. Th. Overbeck, Anton Peterlin, Henryk van de Hulst, Otto Kratky, J.J. Hermans, John Hearst and many others [66]. I was a visiting Professor at Clarkson in 1970, lectured on multicomponent solutions, and wrote my light-scattering review [67]. The cold outdoors in Potsdam is matched by the warmth indoors, in particular of my colleagues originally from Croatia, world-famous pillars of Colloid Science, Egon Matijevic and Joza Kratohvil, and their wives Bozica and Stanka. Egon will always remember his visit to Old Jaffa one day, when some young people mistook him for Salvador Dali. The total lack of offering of the type provided by Times Square or Broadway has led to the development of home culture, including fine wines, foods, and paintings, difficult to locate elsewhere. I went back to Potsdam a number of times, to lecture or for American Chemical Society Colloid Division Meetings. In the early days you could take a train, or fly to Potsdam or Massena by a tiny plane, but now it is a four-hour bus ride from Syracuse airport.

Shai and Hanna were married in Rehovot on July 2, 1970, and went back to their studies of physics and 'Kabbalah', respectively. They are still happily married, their oldest son Ido was born in 1974 and he is a dedicated jazz trombonist, and proud of his sisters, Tamar, 1981, Anat, 1982, and Hagit, 1988.

On May 1972 Aharon was pointlessly murdered along with other passengers by a young Japanese terrorist in the entrance hall of Lydda Airport, coming back from a scientific meeting in Göttingen. It was a sad moment when I eulogized Aharon at the open grave in

Rehovot cemetery in a section set apart for Weizmann Institute scientists and, unfortunately, already partially filled.

I became Head of the Polymer Department, which later split to form the Membrane Research Department, with some of the members of our old group, Ora Kedem, Israel Miller and Roy Caplan leading the pack. The year after, in 1973, I became a Visiting Fogarty Scholar-in-Residence at the NIH. I spent the time to start writing what would eventually become my book, already mentioned, on *Biological Macromolecules and Polyelectrolytes in Solution* [48]. Fogarty Scholars still lived then in Stone House, the original mansion of the estate on which the NIH was built, magnificently restored to accommodate Fogarty Scholars and their wives in separate suites. Dining was in an elegant common dining room, other meals were self-service in a joint kitchen-cafetaria. The set-up was abandoned soon after our stay there, because it created logistic difficulties of one kind or another. The 'Kibbutz' principle of living together does not necessarily apply to individualistic scientists. The Scholars now live in small apartments of their own in Building 20, under the loving care of Ophelia Harding, and still use Stone House for their offices, library and social functions. Of the Scholars in attendance at my time I recall William Rushton of Cambridge University, Albert Sabin following his somewhat shortened term of President of the Weizmann Institute, Osamu Hayaishi, Derek Brown, Charles Davidson and Margaret Mead.

When Aharon died, his young coworker Eberhard Neumann, who had come to us for a number of years from Germany inspired by Manfred Eigen, decided to return to Göttingen. With Eberhard was a young student, Moshe Mevarech, who had just about started his graduate studies. They were beginning to isolate enzymes from halobacteria. Halobacteria had been discovered in the Dead Sea many years ago by Beni Volcani, studying for his Ph.D. at the Hebrew University in the late thirties, and are extremely interesting having adapted to a more than saturated internal salt concentration, mostly KCl. Aharon became interested in this topic for some reason, wanting to study hysteresis, or biological memory, by drastic lowering and raising the salt concentration. I was interested in the topic, because one of the two enzymes isolated, glutamate dehydrogenase along with malate dehydrogenase, had been studied by us from beef liver, and it was certainly interesting to compare its properties with the properties of the halophilic enzyme. Moshe then became my student. We have now for many years studied the halophilic malate dehydrogenase: it is a dimer like its nonhalophilic counterpart, whereas the glutamate dehydrogenase is a tetramer, compared to the hexameric bovine liver enzyme [68,69]. Moshe completed his Ph.D. in due time, went to Chicago to expand his education into molecular biology and is back now at Tel Aviv University studying manifold aspects of halobacteria. I am happy to say that we have an ongoing successful collaboration in this field which is assuming increasing worldwide interest. With Eberhard we have maintained a continuous friendly relationship and, following his suggestion and Manfred's invitation, I became a frequent attendant of the Eigen 'Winterseminar', a yearly gathering, first in Zuoz, and now in Klosters, promoting the ancient 'Mens sana in corpore sano', and much more than that. More of all of this will come in the next section.

Current activities: chromatin and extreme halophiles (1973 – 1989)

Though maintaining close levels of traditional interactions with science in the U.S.A., my involvement in European science, and science on a worldwide level increased after 1973. In 1972 I was elected member of EMBO, the European Molecular Biology Organization. EMBO had been founded in 1963 by a group of European molecular biologists, including Israeli participants, to promote the development and evolution of molecular biology in Europe, and so to strengthen European science, which had lost a lot of impetus and initiative to the U.S.A. following the access of Hitler to power in 1933. EMBO succeeded in raising support, first from the Volkswagen Foundation, and then from 13 European governments, including Israel. It became a very successful tool for scientific exchange through long- and short-term scholarships, workshops and courses, and in due time also established an active laboratory in Heidelberg and the successful EMBO Journal. From 1976 through 1979 I served on the EMBO Course Committee, which enabled me to meet and establish friendships with a number of outstanding European molecular biologists, and to learn a great deal about science evaluation and organization. Members of the committee at one time or another, during my tenure were Werner Arber, Giorgio Bernardi, Ricardo Miledi, Klaus Weber, John Paul, Giorgio Semenza, John Subak-Sharpe, Franco Celada, Dieter von Wettstein, Harald zur Hausen and Anne MacLaren. But in particular I remember the skill with which John Tooze was acting as secretary of the meetings, making it easy and efficient for ourselves to reach the right decisions. John, as executive secretary of EMBO, was handling a number of committees, as well as the fellowship funds. He was doing all this with the help of one or maybe two secretaries only, and, unlike other administrative bodies known to me, the size of this one never increased. Altogether the ability of European scientists to get together at short notice, or to exchange knowledge and be taught at courses, proved to be a tremendous boost to a renascent European science. I myself organized two EMBO basic schools on Modern Analysis of Biological Structures, with my good friend Maurizio Brunori, which were held in Pavia in 1981 and then again in 1986. Pavia is a remarkable Italian University city and maintains a relaxed old time provincial atmosphere, as the tourists visiting the famous 'Certosa di Pavia' monastery by-pass the city. We were shown the bench where Napoleon sat while listening to Alessandro Volta lecturing. The meetings were held in the ancient 'Collegio Ghislieri' and we could split in the afternoons into tutorial study groups shaded by big old trees, in the college garden. Sandro Coda, Head of the Crystallography Department, was responsible for smooth running of the school, including visits to Milan, Cremona and Mantua. Both teachers and students performed extremely well. I recall in particular Ueli Aebi, Eraldo Antonini, Wolfgang Baumeister, Emilia Chiancone, Jan Drenth, Bill Eaton, Stuart Edelstein, Jürgen Engel, Sture Forsén, Stanley Gill, Robert Huber, Bernard Jacrot, Edward Kellenberger, Michael Levitt, Shneior Lifson, Alfonso Liquori, Brian Matthews, Max Perutz, Heinrich Stuhrmann, Attila Szabo, Gregorio Weber, and others, at one or both schools. I have been privileged to be invited by Maurizio to a meeting in Caprarola, North of Rome, in

1980, one of the meetings known as the LaCura conferences, organized by the well known Rome protein group centered around Jeffreys Wyman and Eraldo Antonini, combining a high level of science with a promise, given to the host monks, to finish the wine in the monastery's cellar - for free - as space had to be provided for the new wine coming in that year. Which reminds me of a story during the war, in Italy, that yesterday's brandy was gone, and to-day's was still warm. We got very hungry one evening at Caprarola and Maurizio broke into the kitchen and cooked a large and tasty midnight 'pasta' for the believers in allostery and hemoglobin. Maurizio and myself are Vice-Presidents now of IUPAB, the International Union of Pure and Applied Biophysics, and we may well clash in friendly contest, competing for the next Presidency! Other activities, centered around the European Molecular Biology Laboratory in Heidelberg with outstations established in Grenoble and Hamburg, related to participation in the Instrumentation Policy Planning Committee chaired by Leo de Maeyer. John Kendrew was Director General of the Laboratory then and I was very keen on the neutron and x-ray scattering developments originally due to Heinrich Stuhrmann. I have been collaborating on a very broad basis, for a number of years, with my good friend Joe Zaccai from the Institut Laue Langevin in Grenoble, in our studies on the structure of halophilic enzymes by neutron scattering, ultracentrifugation, and other methods.

A very pleasant way to do science is associated with Manfred Eigen and his group. Meetings organized by Manfred are always exciting and intellectually stimulating. Music mingles with hypercycles and I remember an opening session in 1978 in Göttingen, prior to the Israeli/German Minerva meeting in Braunlage in the Harz, at which Manfred played the Mozart piano concerto in A major KV 414, with a professional chamber orchestra. When I first met a very young Manfred at the Faraday Society Meeting in Oxford in 1957, he was the darling and hope of chemistry and the chemists. When he later decided to study in what ways physics and chemistry can be used to explore the laws of biology, he gathered an outstanding group of young scientists at Schloss Elmau in Bavaria for an intensive discussion, and the process has continued unabated.

The 'Winterseminar' is more mellow and, in the following I present thoughts expressed by myself on the occasion of the 20th seminar in 1985:

'The second half of the month of January has, for many of us, become a source of great pleasure and satisfaction, keenly expected with anticipation. What must have started out as a wintery laboratory outing, maybe with a little mix-in of science, has now graduated to a fully fledged scientific 'palaver', a wonderful spiritual collision between current and aspiring members of Manfred's laboratory, former students and associates, distinguished guests and friends, interested in the scientific, philosophical, mystical and mythical aspects of the secrets of life and its origins - or maybe just keen to be together and have a lot of fun.

Though I only came in Zuoz to number nine, I feel very much the oldtimer. Zuoz, in a way, though less streamlined than Klosters, was more romantic: the old hotel, the waiter bringing in a tray of sherry glasses while the meeting had almost started, Connie Kaufmann playing relaxed songs or dance music in the basement bar, Jack Cowan demolishing the unfortunate 'morituri' trying to block his ever ascending table-tennis career, the bowling game on the ice-rink, the traditional 'rodel' night outing. I vividly remember when Niels Jerne presented a masterful 'aperçu' of immunology, culminating in his ideas about networks, for which he was this year honored; Manfred suggested that continuous discussion could now proceed, extending into the week, about the problems raised. Jerne, on the other hand, suggested that discussion be completed that same evening as, not tolerating the snow and the cold, he was moving the next day to sunny Italy, with companion and dog.

At the tenth seminar I received two plastic remembrance medals, the second one, unfortunately, to commemorate a compound ankle fracture acquired on the Diavolezza slopes. Yet, if you have to be hospitalized, I recommend the Dr. Bernard Klinik in St. Moritz, with its youthful doctors and nurses, wine list, selected foods, and an atmosphere totally divorced from what one would usually expect from a hospital. And when, following five hours of work, the chief surgeon said I would both ski and play tennis again, he was absolutely right.

Klosters offers a little of everything to participants and companions. The most famous ski slopes in the world within easy access, swimming pool, crosscountry slopes, long walking tours for the less adventurous. The 'Bündnerstube' provides strong attraction to discussions of scientific and general nature into the late hours of the night. An unforgettable year for Nutzi and myself was four years ago when we brought our then six-year-old grandson Ido along to glimpse a completely new world for him. A highlight of the yearly visit to Klosters is the invitation to Hans and Dorothée Dutler's to their 17th century farmhouse, tucked away in crisp deep snow in the nearby hillside. Here wine and conversation flow easily and 'Bündnerfleisch' and 'Appenzeller Käs' keep body and soul together.

I was originally assigned by Ruthild Winkler to discuss the weather and the

'Winterseminar'. This proved to be a very difficult undertaking and I have instead provided a personal glimpse into the past. The weather is not really the important problem at the 'Winterseminar'. Manfred, Hans Frauenfelder, Eberhard Neumann, Hans Zachau, Benno Hess, and all the other skiing devotees, ski in every weather. They are daredevils when it comes to skiing, and of course to talking about it. Only once, some years ago, alpine skiing was not possible and we went crosscountry skiing into the hilly woods of Davos. It must have been very painful for the masters of the deep gorges to move around on ridiculously narrow cross-country skis, on a well trodden path. Yet the restaurant at the top of the hill compensated by providing excellent fare. And both 'Veltliner' and conversation were not lacking. Whether with too little or too much snow or frost or thaw there is always something worthwhile doing at the 'Winterseminar'. Science and human friendships engendered by it unite all in a common bond. As long as the fire twinkles in the fireplace and the wine sparkles in the glasses, the show will go on.

Besides, what do we know about the weather? Nothing much, even with the new powerful computers. I am reminded about the old Yemenite Jew in Jerusalem sitting on a doorstep and, according to popular belief, always predicting the weather correctly. When asked how he was doing it the answer was that he was always predicting the opposite of what the learned meteorologist professor from the Hebrew University was announcing. I hope our science is in better shape.'

In 1978 I went to Kyoto, Japan to attend the 6th International Biophysics Congress of the IUPAB and was elected member of its council. The first congress had taken place in Stockholm in 1963 and Aharon was a founder of the new union. With Israel Pecht and the secretarial support of Ruth Goldstein and the Aharon Katzir Center, we organized a very successful congress in 1987, in Jerusalem. Service on the IUPAB council has again allowed me to interact in the service of science, and for our own pleasure, I must admit, with such good friends as Bernard Pullman, Setsuro Ebashi, Kurt Wüthrich, Claude Hélène, Jozef Tigyi, Lee Peachey, Maurizio Brunori, Archie Wada, Sergio Estrada, Leshek Wierchowski, Leo de Maeyer, Richard Keynes, Werner Reichardt and many others.

My happy relationship with Bernard and Alberte Pullman has expressed itself in the organization of three very successful Edmond de Rothschild Schools, in Molecular Biophysics on topics close to our own work which, I believe, have benefitted science both here and worldwide.

Our first school on the structure of chromatin was held in 1979 at Kibbutz Ayelet Hashahar in Galilee and the second one on the structure of DNA and of chromatin in 1985, at Kibbutz Neve Ilan, in the Jerusalem mountains. Teachers included Richard Axel, Morton Bradbury, Howard Cedar, Pierre Chambon, Gary Felsenfeld, Irving Isenberg, Aharon Klug, Roger Kornberg, Ruth Sperling, Edward Trifonov, Alex Varshavsky, Harold Weintraub, Abe Worcel, Donald Coffey, Fritz Cramer, Colyn Crane-Robinson, Richard Dickerson, Martin Gellert, Hannah Gould, Claude Hélène, Ulrich Laemmli, Marc Leng, David Lilley, Tim Richmond, Zippora Shakked, Paul Ts'o, Ken van Holde, Pete von Hippel, Jim Wang, Jonathan Widom, and others. The Tel Aviv Chamber Players, Menahem Breuer, violin, Rafael Frenkel, violin, Rachel Kam, viola, Marcel Bergman, cello and Eli Eban, clarinet, played Mozart's clarinet Quintet in A Major, KV 581, and other pieces, in the large living room of the Weizmann House. I also would like to recall the happy marriage of Joseph Parello and Ruth Gjerset, who met at the 1979 school.

The most recent 1989 school on modern aspects of halophilism took place in Neve Ilan and teachers included Mordecai Avron, Roy Caplan, Pat Dennis, Rainer Jaenicke, Martin Kessel, Janos Lanyi, Moshe Mevarech, Yasuo Mukohata, Knud Nierhaus, Dieter Oesterhelt, Felicitas Pfeiffer, Walter Stoeckenius, Manfred Sumper, Joel Sussman, Hans Günther Wittmann, Ada Yonath, Joe Zaccai and Wolfram Zillig. Arieh Nissenbaum lectured on the Dead Sea. A special treat was a jazz concert by the big band of the Rimon School for Jazz and Modern Music of Ramat Hasharon, conducted by Amikam Kimelman and Robin Palmquist. Our grandson Ido played the trombone.

Research in this period was centered around the study of nucleic acids on the one hand, and halophilic enzymes, on the other. Doug Jolly, coming from Glasgow, developed reliable technologies for the use of angular dependence of laser light scattering and the study of correlation functions [70], whereas Michael Reich, coming from Melbourne, obtained his doctorate by setting up small-angle x-ray scattering technology with linear and two-dimensional position-sensitive detectors, built by Reuven de Roos, for the study of proteins, nucleic acids and chromatin [71]. Gerrit Voordouw was the first to use monodisperse plasmids, studying supercoiled, relaxed and linearized ColE1 DNA [72]. Later on Gerrit extended his work to histone–DNA interactions and could show that nucleosomes adsorb additional histone cores [73]. Juan Ausio and later Otto Greulich studied conformational transitions in nucleosomes and attempted to advance the difficult topic of chromatin transitions between the lower-order and higher-order forms [74–76]. Nina Borochov was mostly concerned with DNA flexibility [77,78] and Dalia Seger made significant progress towards the isolation of the chicken β-globin gene in chromatin form, work still unpublished, in collaboration with Joanne Nicoll and Gary Felsenfeld.

In the study of enzymes from halobacteria Wolfgang Leicht characterized the halophilic glutamate dehydrogenase from *Halobacterium marismortui* [79], and Moshe Werber contributed to most studies in this field in the laboratory [80], and in particular concentrated with Moshe Mevarech on the isolation and characterization of the halophilic ferredoxin [81]. Shlomo Pundak, in his Ph.D. thesis, could determine by light scattering and ultracentrifugation the large amounts of water and salt bound to the halophilic enzymes, and showed that these unusual binding properties are lost following enzyme denaturation by lowering concentration of salt [82,83]. Much of this work was extended in recent years by Ellen Wachtel, using x-ray scattering, and, in particular, Joe Zaccai who, with the use of neutron scattering and imaginative thought, considerably advanced our ideas concerning the unusual structures of halophilic enzymes [84–86]. Structural studies of halophilic proteins, ribosomes, and organelles of bacteria adapted to extreme salt concentrations were summarized by Ellen and myself [87]. We have, more recently, with the support of the ONR, and jointly with Moshe Mevarech, extended our work into molecular biology aspects, in particular the study of ribosomal RNA genes in *H. marismortui*. A fruitful collaboration was set up with Pat Dennis, from Vancouver, who was a visiting Professor in 1987, in our laboratory [88]. This work is still in progress.

I will finish this chapter with a few brief personal notes. In 1981 Arieh Durst removed a cancer of my colon in a successful operation. My son Danny, who was at the time resident in radiology at Hadassah Hospital in Jerusalem was of great moral and practical help. I am grateful to Bill and Lee Abramovitz for establishing the chair of Macromolecular Biophysics of which I am the present incumbent. Bill was a great friend of Israel and did much to promote science and industry. I headed the Mazer Center for Structural Biology from its inception in 1980 till 1986, which did much to promote this field of studies at the Weizmann Institute. I was greatly saddened by not receiving sufficient support from the Institute's local authorities to expand structural biology significantly, a major field of modern biology research, although enthusiastic young scientists were available and strong support from the Scientific Advisory Committee had been given. In 1982 and 1983 I was chairman of the Council of Professors at the Institute, and Head of the Tenure Committee of Twelve, an important yet traumatic assignment in the extreme. For the last 12 years I have been acting on the Editorial Advisory Board of the *European Journal of Biochemistry*, a rewarding experience allowing me to explore science evolution in depth. In 1986, on a lecturing visit to Bucharest I was made an honorary member of the Roumanian Biophysical Society, which gave me much pleasure considering that the basis of my education had been cast in Roumania, before the terrors of World War II. And finally, on June 14, 1984, Danny and Lauren were married in the garden of our home, on campus in Rehovot. They have two charming daughters Ravit, 1985, and Adva, 1987, and Danny practices radiology in Haifa.

Current thoughts (1989)

Yesterday we celebrated with children and grandchildren 50 years of my coming to Palestine, in September of 1939. Tomorrow is 'Yom Kippur', the Day of Atonement and Reckoning, memorial of grievous loss to our young nation. The morning radio broadcast reports the daily incidents resulting from the 'Intifada' and the bickerings inside our so-called National Unity Government. We live in a small country facing serious social, economical and political trouble, but seem uncapable to approach their solution in a constructive way. Honest contributions towards attempting talking to the Palestinians, in the hope of solving the conflict between us, are punished within the context of a law expressing a serious breach of individual freedom in a democratic society. On the other hand, unjustifiable actions against the people in rebellion, escape punishment. These are problems which are difficult to solve, but there are tragedies and losses in human life which can be easily avoided by activation of self-control. Buses carrying school children drive too fast on wet roads, and overturn out of control. Private cars collide frontally and entire families are wiped out. The Israeli driver dashes madly, but unlike his Italian counterpart, with a total lack of driving skill, consideration and experience. When I first came to this country a story went round that, when a murder was first committed in Tel Aviv, our national poet, Chaim Nahman Bialik, was pleased to proclaim that we had now finally matured and become like all other well established nations in this world though, when I first came to Tel Aviv 1939, locking one's front door was an act still unheard off. We have unfortunately, in the intervening years acquired many of the negative attributes of the well established nations in this world, whose survival does not depend on a continuous struggle with an inhospitable environment. The modern rebuilding of the ancient land of Israel was based on hard work, draining of swamps, elimination of disease, sharing of limited resources, creation of vital industries, settlements, successful public health and educational systems accessible to all, a defence army, created and existing for this purpose only. The ideal which kept us going through years of darkness and persecution has unfortunately weakened. It was natural for a young country, fighting for survival, to receive support from friendly outside sources. Now, more than 40 years after the establishment of the State, we still cannot survive without the immense support of the U.S.A., other friendly nations, as well as the Jewish people. Yet a tremendous amount of the good-will earned following the Holocaust and in our formative years will not last for ever and is sure to wane. A vote in the UN General Assembly sponsored by the Arabs to condemn our actions is almost unanimously against us, the only country still voting for us being the U.S.A. By no means do I justify the action of the Arab nations. They are blessed with immense resources and territories, yet they have made no efforts to solve the problems of misery and suffering pervading their own territories. They are dominated by religious fanaticism, and the death sentence on Salman Rushdie, for the crime of having written a searching book, is a good point in case. Hundreds and thousands of innocent people are killed by airplane bombs, in the war-torn Lebanon and elsewhere. I do not believe in Arab intentions to solve the Israeli–Palestinian conflict in terms of what one could consider as a compromise acceptable to both sides. I believe the only solution acceptable to the Arabs is total elimination of the Jewish State from an area which they consider their own. Anwar Sadat maybe thought otherwise but had to pay with his life for it.

Survival of Israel in a hostile Arab World, while maintaining the hope that there will arise a mellowing of the conflict now gripping both partners, is a long process which must be based on the recovery of principles which made the establishment of Israel possible. A basic achievement in the 'renaissance' of Jews as a nation was the principle of Jewish labor in the fields and in the factories. Unfortunately now, since the War of 1967 vast areas of activity have been taken over by cheap Arab labor. The symbol of the Jewish worker was short-lived. Manufacture in many areas of industrial activity has ceased and we have become another Levantine importing community. Every day we read about another bankruptcy and the once prosperous 'kibbutzim' and 'moshavim' are going broke. While unemployment is rising in many of our cities in the Galilee and in the South, we read in the papers about enormous salaries being paid to officials of companies under public control. The Labour Unions insist on the maintenance of agreements which lose all meaning in relation to a population refusing to work. Soldiers finishing their three-year army stint leave the country for an elongated drift throughout the world, sometimes ending up as furniture movers in the Bronx or as hostesses in Tokyo. Jews allowed to leave the Soviet Union prefer waiting for months in a transit camp in Italy for emigration to the U.S.A., rather than joining us in the Promised Land.

I am afraid these are hard words and the justification for putting them in here is that these are not ordinary times and we cannot easily split our work from the circumstances which surround us. It is unfortunate that man's preoccupation is more heavily engaged in conflicts relating man-to-man, rather than in solving the major problems also created in our own times by man's activity on earth, namely the consequences of the uses of nuclear energy, the spread of drugs, provision of food for a growing population, tampering with the environment, and so forth.

Rehovot, September 1989

REFERENCES

1 G. von Rezzori, Ein Hermelin in Tschernopol, Rowohlt, Hamburg, 1966.
2 I. Chalfen, Paul Celan, Insel, Frankfurt, 1979.
3 O. Manning, The Levant Trilogy, Weidenfeld and Nicolson, London, 1980.
4 P. Levy, If Not Now, When?, Simon and Schuster, New York, 1985.
5 R.M. Fuoss and G.I. Cathers, J. Polymer Sci., 2 (1947) 12–15.
6 W. Kuhn, O. Künzle and A. Katchalsky, Helv. Chim. Acta, 31 (1948) 1994–2037.
7 J.J. Hermans and J. Th.G. Overbeek, Rec. Trav. Chim., 67 (1948) 761–776.
8 G. Jones and R.C. Josephs, J. Am. Chem. Soc., 50 (1928) 1049–1092.
9 H. Eisenberg, Eur. J. Biochem., 187 (1990) 7–22.
10 A. Katchalsky and P. Spitnik, J. Polymer Sci., 2 (1947) 432–44.
11 W. Kuhn, Experientia, 5 (1949) 318.
12 A. Katchalsky, Experientia, 5 (1949) 319.
13 W. Kuhn, B. Hargitay, A. Katchalsky and H. Eisenberg, Nature, 165 (1950) 514–516.
14 J. Riseman and J.G. Kirkwood, J. Am. Chem. Soc., 70 (1948) 2820–2823.
15 A. Katchalsky and H. Eisenberg, Nature, 166 (1950) 267.
16 M.M. Hunt, New Yorker, September 13 (1958), 48–71; September 20 (1958) 46–79.
17 H. Morawetz, Polymers: The Origins and Growth of a Science, Wiley, New York, 1985.
18 S. Lifson, in M. Balaban (Ed.), Molecular Structure and Dynamics, International Services, Rehovot, 1980, pp. 213–243.
19 M. Mandel, in H.F. Mark, N.M. Bikales, C.G. Overberger, G. Menges and J.J. Kroschwitz (Eds.), Encyclopedia of Polymer Science and Engineering, Vol. 11, 2nd ed., Wiley, New York, 1988, pp. 739–829.
20 R.M. Fuoss, A. Katchalsky and S. Lifson, Proc. Natl. Acad. Sci. USA, 37 (1951) 579–589.
21 T. Alfrey Jr., P.W. Berg and H. Morawetz, J. Polymer Sci., 7 (1951) 543–547.
22 H. Eisenberg and R.M. Fuoss, J. Am. Chem. Soc., 75 (1953) 2914–2917.
23 A. Katchalsky, H. Eisenberg and S. Lifson, J. Am. Chem. Soc., 73 (1951) 5889–5890.

24 B. Kaufmann and L. Onsager, Phys. Rev., 76 (1949) 1244–1252.
25 T.L. Hill, Statistical Mechanics, McGraw-Hill, New York, 1956.
26 T.L. Hill, An Introduction to Statistical Thermodynamics, Addison-Wesley, Reading, 1960.
27 H.S. Harned and B.B. Owen, The Physical Chemistry of Electrolytic Solutions, 2nd ed., Reinhold, New York, 1950.
28 E.C. Franklin and C.A. Kraus, Am. Chem. J., 20 (1898) 820–836.
29 E.J. Cohn and J.T. Edsall, Proteins, Amino Acids and Peptides, Reinhold, New York, 1943.
30 Peter J. N. Debye, Collected Papers, Interscience, New York, 1954.
31 R.M. Fuoss, Disc. Faraday Soc., 11 (1951) 125–134.
32 H. Eisenberg and J. Pouyet, J. Polymer Sci., 13 (1954) 85–91.
33 H. Eisenberg, Biophys. Chem., 7 (1977) 3–13.
34 J.D. Watson and F.H.C. Crick, Nature, 171 (1953) 737.
35 H. Eisenberg, J. Polymer Sci., 30 (1958) 47–66.
36 H. Eisenberg and E.H. Frei, Bull. Res. Counc. Israel, 3 (1954) 442–443.
37 H. Eisenberg, J. Polymer Sci., 23 (1957) 579–599.
38 H. Eisenberg, J. Polymer Sci., 25 (1957) 257–271.
39 U.Z. Littauer and H. Eisenberg, Biochim. Biophys. Acta, 32 (1959) 320–337.
40 A. Berger and H. Eisenberg, Bull. Res. Counc. Israel, 4 (1954) 97–98.
41 D. Cwikel and H. Eisenberg, Israel J. Chem., 12 (1974) 35–46.
42 E.P. Kohler, Am. Chem. J. 20 (1898) 680–695.
43 H. Eisenberg and G. R. Mohan, J. Phys. Chem., 63 (1959) 671–680.
44 H. Eisenberg, Mada, 2 (1957) 49–50.
45 Meyer Weisgal., . . . So Far, Weidenfeld and Nicolson, London, 1971.
46 W.R. Carroll, H. Eisenberg and R.J. Podolsky, Science, 132 (1960) 475–478.
47 E.F. Casassa and H. Eisenberg, Adv. Prot. Chem., 19 (1964) 287–395.
48 H. Eisenberg, Biological Macromolecules and Polyelectrolytes in Solution, Clarendon Press, Oxford, 1976.
49 H. Eisenberg, Quart. Rev. Biophys., 14 (1981) 141–172.
50 W.R. Carroll and H. Eisenberg, J. Polymer Sci. A2/4 (1966) 599–610.
51 H. Eisenberg and G.M. Tomkins, J. Mol. Biol. 31 (1968) 37–49.
52 H. Eisenberg, Trends Biochem. Sci., 1 (1976) N121–N122.
53 H. Eisenberg, J. Chem. Phys., 43 (1965) 3887–3892.
54 H. Eisenberg and G. Cohen, J. Mol. Biol. 37 (1968) 355–362; erratum, ibid. 42 (1969) 607.
55 G. Cohen and H. Eisenberg, Biopolymer, 6 (1968) 1077–1100.
56 H. Eisenberg, in Biophysics - Nucleic Acids, Landolt-Börnstein New Series, Vol. 7, 1c, Springer-Verlag, Berlin, 1990, in press.
57 H. Eisenberg, in P.O.P. Ts'o (Ed), Basic Principles in Nucleic Acid Chemistry, Vol. 2, Academic Press, New York, 1974, pp. 171–264.
58 R. Josephs, H. Eisenberg and E. Reisler, in R. Jaenicke and E. Helmreich (Eds.), Protein–Protein Interactions, Springer-Verlag, Berlin, 1972, pp. 57–89.
59 E. Reisler, H. Eisenberg and A.P. Minton, J. Chem. Soc. Faraday Trans. II, 68

(1972) 1001-1015.
60 J.E. Godfrey, Biophys. Chem., 5 (1976) 285-299.
61 J.E. Godfrey and H. Eisenberg, Biophys. Chem., 5 (1976) 301-318.
62 J.E. Godfrey, Proc. Natl. Acad. Sci. USA, 86 (1989) 4479-4483.
63 N. Imai and H. Eisenberg, J. Chem. Phys., 40 (1966) 130-136.
64 H. Eisenberg and G. Felsenfeld, J. Mol. Biol., 30 (1967) 17-37.
65 L. Stein, Cell Biophys., 11 (1987) 17-18.
66 M. Kerker, J. Coll. Interface Sci., 105 (1985) 290-296.
67 H. Eisenberg, in G.L. Cantoni and D.R. Davies (Eds.), Procedures in Nucleic Acid Research, Vol. 2, Harper and Row, New York, 1971, pp. 137-175.
68 M. Mevarech, E. Neumann and H. Eisenberg, Biochemistry, 16 (1977) 3781-3785.
69 M. Mevarech and E. Neumann, Biochemistry, 16 (1977) 3786-3792.
70 D. Jolly and H. Eisenberg, Biopolymers, 15 (1976) 61-95.
71 M. Reich, Z. Kam and H. Eisenberg, Biochemistry, 21 (1982) 5189-5195.
72 G. Voordouw, Z. Kam, N. Borochov and H. Eisenberg, Biophys. Chem., 8 (1978) 171-189.
73 G. Voordouw and H. Eisenberg, Nature, 273 (1978) 446-448.
74 J. Ausio, N. Borochov, D. Seger and H. Eisenberg, J. Mol. Biol., 177 (1984) 373-398.
75 K.O. Greulich, J. Ausio and H. Eisenberg, J. Mol. Biol., 186 (1985) 167-173.
76 K.O. Greulich, E. Wachtel, J. Ausio, D. Seger and H. Eisenberg, J. Mol. Biol., 193 (1987) 709-721.
77 Z. Kam, N. Borochov and H. Eisenberg, Biopolymers, 20 (1981) 2671-2690.
78 N. Borochov and H. Eisenberg, Biopolymers, 23 (1984) 1757-1769.
79 W. Leicht, M.M. Werber and H. Eisenberg, Biochemistry, 17 (1978) 4004-4010.
80 M. Mevarech, W. Leicht and M.M. Werber, Biochemistry, 15 (1976) 2383-2387.
81 M.M. Werber and M. Mevarech, Arch. Biochem. Biophys., 186 (1978) 60-65.
82 S. Pundak and H. Eisenberg, Eur. J. Biochem., 118 (1981) 463-470.
83 S. Pundak, H. Aloni and H. Eisenberg, Eur. J. Biochem., 118 (1981) 471-477.
84 G. Zaccai, E. Wachtel and H. Eisenberg, J. Mol. Biol., 190 (1986) 97-106.
85 G. Zaccai, G.J. Bunick and H. Eisenberg, J. Mol. Biol., 192 (1986) 155-157.
86 G. Zaccai, F. Cendrin, Y. Haik, N. Borochov and H. Eisenberg, J. Mol. Biol., 208 (1989) 491-500.
87 H. Eisenberg and E.J. Wachtel, Annu. Rev. Biophys. Biophys. Chem., 16 (1987) 69-92.
88 M. Mevarech, S. Hirsch-Twizer, S. Goldman, H. Eisenberg, E. Yakobson and P.P. Dennis, J. Bacteriol., 171 (1989) 3479-3485.

G. Semenza and R. Jaenicke (Eds.) Selected Topics in the History of Biochemistry: Personal Recollections, III. (Comprehensive Biochemistry Vol. 37) © *1990 Elsevier Science Publishers BV (Biomedical Division)*

Chapter 9

'The Highest Grade of This Clarifying Activity Has No Limit' – Confucius

CHEN-LU TSOU

National Laboratory of Biomacromolecules, Institute of Biophysics, Academia Sinica, Beijing, 100080 (China) Fax 86-1-256-5689

Confucius said: *'At fifteen I wanted to learn. At thirty I had a foundation. At forty, a certitude. At fifty, knew the orders of heaven. At sixty, was ready to listen to them. At seventy, could follow my own heart's desire without overstepping the t-square' [1].*

From chemistry to biochemistry – days in Cambridge with David Keilin

Although I studied chemistry in the National Southwest Associated University in Kunming in the interior of China during World War II, I had always been attracted by the mystery of life's processes and believed firmly that the best approach to solve these mysteries would be a chemical approach. My university years ended with a brief spell in the army until the end of the War in 1945. After two years' work in organic chemistry I finally won a keenly contested examination for a scholarship to do postgraduate study in England: I naturally selected biochemistry as my subject. I was very much disappointed when I was sent, in autumn 1947, to the Department of Chemistry at the University of Birmingham which no doubt was an excellent department, especially in the field of carbohydrates, with the Nobel laureate Sir Norman Haworth at its head. Fortunately, before I left China, Ying-Lai Wang had just returned from Cambridge to take up a Professorship at the National Central University at Nanking. Armed with a recommendation letter from Wang to Prof. David Keilin, I went to Cambridge for an interview and eventually moved to Cambridge after only a few months working on the 'sweetest of all sugars, inulin'. Coincidentally, I was later destined to spend a large amount of my life's work on a protein with a very similar name, insulin. However, my background and training in organic and physical chemistry did me much good with my work on proteins and enzymes.

I started my scientific career as a Ph. D. student with the late Prof. D. Keilin in the beginning of 1948 in the Molteno Institute of Parasitology. I believe very few biochemists today know the Molteno Institute and may wonder why I should go to an Institute of Parasitology to study biochemistry. David Keilin was very well known in both fields and as the Director of the Molteno Institute made numerous important contributions to both parasitology and biochemistry and was in fact the founder of the discipline biochemical parasitology. In those days Keilin's fame as the discoverer of the cytochromes made the Molteno Institute the mecca for biochemists working on respiratory enzymes. Among the people who had worked at the Molteno were such well known names as Emil Smith, David Green, Bruno Straub, Britton Chance, Max Perutz, John Kendrew, E. C. Slater and many others.

I was given the problem to study the structure–function relationship of cytochrome *c* by proteolytic digestion of the protein molecule which must be an early attempt for such an approach [2–4]. In those years, we used the Keilin–Hartree heart-muscle preparation as our enzyme source. The heart-muscle mince had to be washed thoroughly and squeezed firmly through muslin several times to get rid of all the myoglobin attached to the muscle. I thought it might be easier to remove the myoglobin by washing with salt solutions, but to my surprise, not only myoglobin but also cytochrome *c* was removed. I was later able to show that cytochrome *c* can be easily extracted from heart-muscle mince by salt solutions and can be put back again under appropriate conditions. There are important qualitative as

Plate 14. Chen-Lu Tsou.

well as quantitative differences of cytochrome *c* in the bound and the free states [5]. I also noticed that, no matter how thoroughly I extracted the heart-muscle mince with salt solutions, there was always some pigment left which absorbed near 550 nm in the reduced state. It was cytochrome c_1 of course, as we all know now [6], however, I missed it when it was sitting right there before my eyes. Working on succinate dehydrogenase I showed that, contrary to a very popular suggestion at that time [7], it is not identical with cytochrome *b* [8].

Professor Keilin made his great contributions to science with very limited equipments. Apart from the hand spectroscope with which the professor made his discovery of the cytochromes, people at the Molteno had at their disposal only the Barcroft manometers, Thunberg tubes, and an old swing-out centrifuge. A later addition of a Beckman DU spectrophotometer was a luxury. A refrigerated centrifuge (MSE) came only a few weeks before I finished work on my Ph. D. thesis. For centrifugation in the cold, we had to put a few ice cubes made from distilled water to each of the centrifuge tubes when dilution was no objection. The professor told us a story that once he had a rich professor from a rich country as a visitor who asked his advice on what to work on with all kinds of fancy equipments available and enough money to buy anything you can think of. I am sorry to say that this kind of thing does not happen only in rich countries. In recent years, many Chinese laboratories had generous loans from the World Bank to buy scientific instruments and I had the same experience more than once with my visitors asking not only what to do but also what to buy with all the money made available to them! In China, as anywhere else, it is not always those with the best equipment who are doing the best work.

Among those who are engaged in fundamental research in China today, we are fortunate in being generously supported by Academia Sinica and during the last few years when hard currency was indeed very hard to come by for us, we had a Grant from the National Institutes of Health of the U.S. so that we are now moderately well equipped for the work we are doing. Nevertheless, I still ask my graduate students to make their own enzymes and to synthesize whatever reagents that are good for their training; above everything I make them always remember that worthwhile contributions in fundamental research can be made from very limited resources.

I got into the bad habit of smoking when I was very young, and in England those days cigarettes were difficult to obtain: one had to know the tobacconist and get the cigarettes 'under the counter'. I therefore learnt to smoke a pipe as pipe tobacco was easier to obtain. Unfortunately, my pipe-smoking technique was not very good and someone in the Molteno once remarked that my pipe seemed to work most of the time in anaerobic conditions. It is for this reason that I kept the pilot flame of a gas burner always on so as to light my pipe whenever it went anaerobic which happened quite frequently. One day Prof. Keilin walked into my room in the corner of the third floor and saw the lighted pilot flame of the burner. He asked me to turn it off and use matches for lighting my pipe; he said that if I always left it on I might regret it if I had an accident.

It was a few days after, when I went home after work, my wife, who was then working in the Goldsmith Laboratory of Metallurgy downstairs from the Chemistry Department at Pembroke, told me that Prof. A. R. Todd dismissed a graduate student because of an accident. The student had been doing a distillation the previous evening and forgot to turn off the cooling water when he finished the experiment and left for home. Because of the increased water pressure after working hours when everyone turned their taps off, the connection between the rubber tubing and the condenser burst and his room, which was right above Prof. Todd's office, was flooded. The water dripped through the floor and soiled some important papers. This made Prof. Todd very angry and he dismissed the student on the spot when he came to his office in the morning. After my wife told me this story, we had our dinner, talked a little and then went to bed. It happened that I could not go to sleep and sometime around midnight it suddenly occurred to me that I left the pilot flame of the gas burner in my lab on. The more I thought of it the more I was sure that I did not turn it off when I left. I hurriedly put on my mackintosh over my pajamas and rode my bicycle to the lab. I had to climb over the locked iron gate of the Dawning premises, went into my lab on the top floor of the Molteno and found to my great relief that the pilot flame was off after all.

Recently, something very similar to the accident in Prof. Todd's lab happened to one of my graduate students. Although this room was not directly above my office, he did make a mess of a few laboratories. However, I had no say in what to do with him; the Security Officer of the Institute came, inspected the damage, talked to my student and asked him to compose an essay of 'self-criticism' on the accident. Depending on how well he did with this essay, he may get away with the accident, pay a fine or be dismissed as a graduate student. In case the security officer decided on a fine, as his supervisor, I should pay 80% of the fine for 'failure to teach the student to observe the security regulations'! It is only common sense that, since I have more money than the poor student, I should pay the lion's share of the fine. There you have the difference in Chinese and Western philosophies. Fortunately for him and for me, he apparently did his essay well enough so as to satisfy the Security Officer that he had learnt the lesson and further punishment was not necessary.

Early years in Shanghai

I came back to China in 1951 and, accepting an invitation from Y.-L. Wang, I joined the Shanghai Institute of Biochemistry of Academia Sinica. I was made a full professor a few years later at a relatively young age. As I also accepted the invitation to do part-time teaching at Fudan University of Shanghai, I found that my bad habit of smoking a pipe was helpful in making me look more dignified before the students.

Modern Biochemistry was introduced to China in the 1920s when H. Wu returned from the U.S. and established the first biochemical laboratory in China in the Peking Union Medical College. Biochemists of older generations might still remember the method developed by Folin and Wu for sugar determination which bore their names and appeared in many textbooks for experimental physiological chemistry. Wu was also known for his contribution to protein denaturation which was envisaged as a process involving 'change from the regular arrangement of a rigid structure (of a protein) to the irregular, diffuse arrangement of the flexible open chain' [9] and thus laid the foundation of our present concept of the unfolding of protein molecules. Apart from protein denaturation, immunochemistry and clinical analysis, Wu's group was also mainly responsible for early nutritional research in China which was the major activity of many biochemical laboratories at that time not only in medical colleges, but also in science faculties of Chinese universities.

Research activity was interrupted when the War broke out in 1937, and serious fundamental research was not possible until the founding of the People's Republic in 1949. During the period 1949–1966, the most important center for basic biochemical research in China was the Institute of Biochemistry of Academia Sinica in Shanghai, founded in 1958 with Prof. Y.-L. Wang as Director. It was originally part of the Institute of Physiology and Biochemistry which had three divisions: Physiology, Biochemistry and Chemistry of Natural Products which all later became separate Institutes. For some years, the majority of the more significant research papers came from the Institute of Biochemistry, which had five research laboratories: proteins, enzymes, nucleic acids, metabolic studies, and radiation biochemistry. Shortly after I joined this Institute, I was made the head of the enzyme division.

My training with Prof. Keilin in Cambridge made it much easier for me to start working with a research group of my own in Shanghai. The Institute of Biochemistry of Academia Sinica had then similar equipments as the Molteno except that we had to use a model A stone mill (A for ancient, this kind of hand-operated stone mill must have been in use in China for making flour from wheat since time immemorial) instead of a mechanical mortar for grinding up heart-muscle mince with sand to make Keilin–Hartree preparations of succinic dehydrogenase-cytochrome oxidase. During our work on the purification of succinic dehydrogenase, we used the same trick of adding ice cubes made from distilled water to the centrifuge tubes to keep them cool in some of the steps where dilution did not matter.

With Prof. Keilin, I made all the enzymes that I used including the proteases for the study on the effect of proteolytic digestion of cytochrome *c*. One day I was discussing with him my work on the reactions of cyanide with cytochrome *c* and succinic dehydrogenase and he suggested that I try the effect of methyl carbylamine on the respiratory enzymes; it is after all similar to cyanide in structure. It was not available commercially and, as I had some training in organic chemistry, I tried to make it myself. On account of its smell, Prof. Keilin suggested that it would be better to work with as little material as was convenient. The Molteno was no chemical lab and a hood was not available for such experiments. I therefore worked on the roof which served not only as a cold room for Bill (E. C. Slater, in spite of his first name Edward, he has always been known as Bill), but also as a hood for me. Prof. Keilin commended me upon my dexterity for making the final fractional distillation step with 0.5 ml of material collecting fractions with a drop each and determining the boiling point of each of the fractions. However, I was an unwelcome person for some time because of the smell which was retained in my lab coat for quite a while. Joan Keilin also used this reagent to study its reactions with haems but fortunately for her, and unfortunately for the others working in the same room, she temporarily had no sense of smell precisely at the time when she was working on this problem.

In my early years in Shanghai, it was apparent that my training in organic chemistry and the habit of making myself some of the things I needed stood me in good stead. In those early days, we had to synthesize even dimercaptopropanol ourselves [10]. For our work on the resynthesis of insulin from its chains, we made sodium tetrathionate for the splitting of the insulin into the *S*-thiosulfonate derivatives of the chains.

Collaborating with Wang and Wang, I continued my interest in succinate dehydrogenase and purified this enzyme using the method of butanol extraction first suggested by R. K. Morton whom I had known in Cambridge and was deeply impressed by his success in the extraction of a number of membrane-bound enzymes with butanol which was dubbed Mortonol at that time. I was very much saddened when I heard that Morton was killed in an accident of acetone explosion while drying acetone powder in his laboratory in Adelaide.

The successful purification of this enzyme [11] depended very much on a convenient and reliable assay method. It was not known at that time that dyes commonly used for the assay of the enzyme in particulate preparations (membrane-bound) such as methylene blue or 2,6-dichlorophenolindophenol do not react directly with dehydrogenase and require another factor so as to accept the electrons from the dehydrogenase. I suspected this when working on this enzyme in Cambridge and we hit on the correct idea of using potassium ferricyanide as the artificial acceptor. Relying too much on my Cambridge experience, I suggested to include substrate and cyanide in the extraction of this enzyme, the substrate was to protect the enzyme from being inactivated and cyanide to prevent the exhaustion of the substrate during extraction; as I had shown previously it did no harm to the dehydrogenase in the presence of the substrate. Little did I realize that the added cyanide made the enzyme no longer reconstitutable with the particulate enzyme preparation of the respiratory chain then known as the Keilin–Hartree preparation. This was demonstrated by Keilin and King a little later [12].

The excitement came when it was found that this enzyme has as its prosthetic group an FAD molecule covalently linked to the protein moiety of the enzyme [13]. It was at that time very difficult for us to attend international meetings, especially when they were held in a Western country. Y.-L. Wang was the only one of our group who went to the Third International Congress of Biochemistry in Brussels and presented our results. Coincidentally, T. P. Singer was also there to present his paper on the purification of succinate dehydrogenase. Although the purification procedures were necessarily different, the final preparations obtained by the two groups were remarkably similar [14]. The method of Wang, Tsou and Wang modified by various authors is still commonly used for the preparation of this important enzyme.

During my early years in Shanghai, I had the good fortune to get help of my young colleague, Dr. C. Y. Wu, in my studies on respiratory enzymes [15]. Unfortunately, he was to fall a victim of the 'Cultural Revolution' while still in his thirties.

Total synthesis of insulin

The year 1958 was a memorable one in Chinese history. Mao Zedong's idea of a great leap forward caught the imagination of hundreds of millions of Chinese people who, understandably, were eager to shake off the legacy of over a century of poverty and backwardness and the scientists, who can be at times very unscientific in nonscientific matters, were no exception. When everyone else was achieving wonders, what could we poor biochemists do? In the Shanghai Institute of Biochemistry, we had many heated discussions on the great problem we were going to tackle; then suddenly someone hit on the idea of synthesizing a protein. Considering that at least the amino acid sequence of one protein, insulin, was then known (the sequence of myoglobin was to come a little later) and an active fragment of ACTH 26 amino acid residues long had already been synthesized, the idea appeared logical. In fact, apart from insulin, we had no other choice. It could have been myself who first made the proposal, but during those days when many people often spoke at the same time, the original proposal could have come from a dozen different people. Anyway, the idea seemed to be the only proposal acceptable to both the authorities and the scientists and the fact that China was not producing any amino acid except monosodium glutamate at that time was not even considered a problem. The idea of synthesizing a protein molecule which had been described by such a classic of Marxist literature as the *Dialectics of Nature* by F. Engels as the mode of existence of life soon caught the imagination of many Chinese scientists, chemists and biochemists especially, and we at the Shanghai Institute of Biochemistry had to dissuade quite a number of enthusiastic biochemists in many institutions either to join our group or to start a project of their own on the synthesis of a protein.

The 'great leap forward' of 1958 made havoc of Chinese economy and has since been admitted by most people in China as a great failure. Many people have also since voiced their doubts whether it was worthwhile to spend so much effort on the total synthesis of a small protein. Nevertheless, apart from the synthesis itself we have now established our amino acid industry and built up good research teams on peptide synthesis. I have myself never lost my interest in this small protein. Because of the success in the synthesis of insulin, when almost every other serious scientific research was taboo during those difficult years of the 'Cultural Revolution', the attempt to solve the crystal structure of insulin was allowed to carry on and thus formed the basis of a very good research group on x-ray crystallography of proteins today in the Institute of Biophysics in Beijing.

Thus, I had to put aside my interest in enzymes for the time being and joined the group in the Institute of Biochemistry working towards a first ever total synthesis of a protein, bovine insulin. My responsibility was to find out whether the A and B chains of insulin, after reduction and separation, could be joined again by oxidation to form the native hormone. At that time, this was a key to the successful total synthesis of the protein as there was no way to protect the respective thiols by different protecting groups so that these can be deprotected sequentially so as to form the desired disulfides one after another. I had the good fortune of being assisted by a group of talented young people including Y.-S. Zhang, Y.-C. Du, G.-J. Xu and Z.-X. Lu who are now all in responsible positions: Zhang the Head of a National Laboratory and Du a Vice-President of the Shanghai Branch of the Chinese Academy of Sciences.

When we started to look up the literature, the prospect of successfully joining the reduced chains to form native insulin with a reasonable yield did not appear at all promising. Not only had many people tried and failed, some even reported that with partially reduced and inactivated insulin, oxidation led to further decrease in activity. With a model peptide, Rydon reported that, when two peptide chains each containing two thiol groups were oxidized together, the peptides were most likely to be joined in an anti-parallel way [16]. Undaunted by these reports, we started on our way to the preparation of the separated A and B chains and the joining of them to form, at least some, native insulin again. It must be said that in my mid-thirties I was the oldest of the group and according to Confucius [1] 'should have a foundation'; I was more apprehensive of the outlook of our job and took the literature results more seriously than did my younger colleagues who, as the ancient Chinese saying goes, were 'young calves not afraid of the tiger'.

We hit on the correct idea of avoiding oxidizing agents and carrying out the reoxidation of the reduced chains by air slowly at a low temperature. Within a fairly short time, good results were coming and soon we could obtain 10% yield of insulin activity on the oxidation of the reduced chains. We took meticulous care to make sure that the separated chains were not contaminated with each other. The crude product was then subjected to purification and under the conditions for the crystallization of insulin, indeed crystals indistinguishable from those of authentic insulin and with biological activity identical to that of the native hormone were finally obtained. Furthermore, finger printing of the pepsin hydrolysate by twodimensional paper chromatography and electrophoresis showed the identity of the resynthesized insulin with the native hormone.

Looking back now, our success depended on three crucial conditions, the use of the stable thiosulfonated derivatives for the separation of the chains before reduction, slow air oxidation at alkaline pH 10.6, instead of using oxidation agents or catalysts to accelerate the oxidation at neutral pH and above all, a low temperature of 4°C. This was in the autumn of 1959 and at that time, because of the 'great leap forward', no scientific journal in China was published, including *Scientia Sinica* which was then the only Chinese journal that published articles in western languages. In that period, no one was publishing abroad. It was only in 1960, when a paper by Dixon and Wardlaw appeared in *Nature* [17] reporting resynthesis of insulin by reoxidation of the reduced chains with a yield of 1–2%, that we decided to publish our results in two papers in *Scientia Sinica*, the first one in the first issue when this journal resumed publication in 1961 [18,19].

We continued to explore the conditions for the resynthesis so as to improve the yield and soon increased the yield to 30–50% based on the mouse convulsion assay. Later we found that with A chain in excess, the yield calculated on the basis of B chain can reach 70% [20]. This paved the way for the eventual total synthesis of this protein which was a collaborative effort. The Shanghai Institute of Biochemistry, led by C.-I. Niu and Y.-T. Kung, was responsible for the synthesis of the B chain, as well as the joining of the chains [20,21]; workers in the Shanghai Institute of Organic Chemistry, led by Y. Wang and the Chemistry Department of Peking University, led by C.-Y. Hsin were jointly responsible for the synthesis of the A chain [22]. It was a great moment indeed when the synthetic product was proved to have the same activity in the mouse convulsion test as native insulin. It was 1965 and the final test was made in the small hours of the day, but everyone concerned was there, either in another room waiting anxiously for the results, or fortunate enough to be present (only a few directly responsible for the final test were admitted) to watch the mice as they suffered with the first ever chemically synthesized protein. After purification, the totally synthetic protein had the same biological activity and was again proved to be identical with the native hormone by finger printing.

A preliminary report was published in 1965 in *Scientia Sinica* and the full papers in a special issue of *Kexue Tungbao* (*Science Bulletin*) in 1966 [23]. When Y.-L. Wang and myself were invited to speak at a FEBS Meeting in Warsaw in the Spring of 1966, we decided to speak on the total synthesis of insulin and Y.-T. Kung who played an important role in the actual chemical synthesis went with us to present our results for which we were given exactly ten minutes before an audience in a packed room. We did submit our abstract well in advance and I had thought that we would be given more time. However, to be fair to our hosts, I was invited to a buffet dinner party attended by a limited number of people who, according to my old friend Prof. Bruno Straub, now the President of Hungary, had contributed important papers to the Meeting. Little did we know then that we were to face the 'Cultural Revolution' only two months after we went home.

Sir John Kendrew came to visit China in the early days of the 'Cultural Revolution' and told us that the total synthesis of insulin was in the BBC television news in the golden hour which must have been watched by millions and definitely was the best known Chinese scientific achievement to the public in the U.K. When Sir John gave a lecture at the Shanghai Institute of Biochemistry, I was asked to translate for him but I was instructed at the same time not to tell him who I was. I had known Sir John since my Cambridge days when he came occasionally to the Molteno Institute to see Prof. Keilin. Sir John did not recognize me as I must have changed very much during the 15 years since I left Cambridge. Furthermore, in a shabby Mao uniform as becoming a poor interpreter, I undoubtedly looked very different from the Cambridge student he knew. Nevertheless, he did remark on my English and asked whether I had been abroad and upon receiving a negative answer asked where I learned to speak English so well. I had to answer 'in school' and then he said 'you must have been to a very good school'.

A. Tiselius came toward the end of 1966 and by that time my political standing had deteriorated to such an extent that I was not even allowed to be anywhere near him during his visit. As Tiselius was then the Chairman of the Nobel Committee for Chemistry of the Swedish Royal Academy of Sciences, it was thought naturally that he was probably looking for a possible Nobel Prize candidate. People who had very little to do with the effort either elbowed themselves or were pushed by the authorities to the forefront but the Prize was not to be. I was subsequently told that Tiselius was very much interested in the synthesis of insulin and when he was asked to comment on the explosion of the first Chinese atom bomb which happened at about the same time, he remarked:

> 'The making of an atom bomb you can learn from the textbooks but not the synthesis of insulin'.

I do not remember the exact date, but it must have been in the early years of our effort in this project, when we still had an exchange programme with scientists from the Soviet Union, that academician Oparin came to visit China. He was probably best known for his book on the *Origin of Life* and was then the Director of the Bach Institute of Biochemistry of the Soviet Academy. When he was in the Shanghai Institute, Prof. Wang told him about what we were doing, which was kept secret for our friends from the 'capitalist countries'. He was of course incredulous and said:

> 'We will send people to learn from you when you succeed to synthesize a protein'.

We did so in 1965, but probably due to the much worsened Sino-Soviet relations by that time, he did not send anyone.

During the years of the 'Cultural Revolution', although genuine scientific efforts were looked upon with strong disapproval if not outright prohibition, it did not prevent those in power to make use of the fact that the total synthesis of insulin was popularly very well known in the West. In 1973, when it had become possible to exchange visits with the U.S. under official auspices and a group of biochemists was being organized to go there, Prof. Y.-L. Wang, nominally still the Director of the Shanghai Institute of Biochemistry, was asked to be the Deputy Leader of the Group and a young man of the same Institute who joined the insulin project only shortly before its completion was made the Leader of the Group for political reasons which were obvious to everyone at that time. My old friend Tsoo E. King, then Professor of Biochemistry at the State University of New York at Albany, later told me that the visit was a great success and during the tour of the group around biochemical laboratories in the U.S., all members were loaded with reprints except the Leader who apparently was not familiar with either biochemistry or English.

The successful resynthesis of insulin from its chains was confirmed in different laboratories which were also engaged in the synthesis of insulin [24,25] and a modified method was later used in the commercial production of human insulin from A and B chains made separately by the recombinant DNA technique [26]. The good resynthesis yield of insulin from A and B chains led us to the suggestion that the native insulin structure is the most stable among all its isomers [18,20]. However, in spite of all the above, neither the fact that insulin can be obtained from its chains in good yield nor the related suggestion of the stability of the insulin structure seems to have been generally accepted [27–29]. This is probably due partly to the low yield reported in the final step for the total chemical synthesis of insulin and the fact that previous workers [30] failed to obtain the native hormone from 'scrambled' insulin with randomly formed disulfide bonds through thiol-disulfide exchange reactions catalysed by protein disulfide isomerase, under similar conditions, that proinsulin can be obtained in 25% yield from the scrambled molecule.

Indeed, I was asked frequently why the yield was so low during the final step of the total synthesis of insulin from its chains. This was not surprising after all as the synthetic chains were usually heavily protected, only slightly soluble in most solvents and thus difficult to purify. It was only logical to join up the chains first, remove all the protecting groups and then purify the final product. The fact that the A and B chains of correct sequences can single out each other and then be joined together through the native disulfide bonds to form insulin at all in the presence of a very large excess of so many different peptides of similar sequences indicates strongly the ability and specificity of the chains to recognize and pair alongside each other correctly. Incidentally, the treatment by metallic sodium in liquid ammonia to remove all the protecting groups also did some damage to the chains to diminish the final yield.

Back to enzymes

With the first hurdle taken for my work on the total synthesis of insulin and the tumult of the 'great leap forward' gradually quieting down, I was able to go back in 1961 to what has always been my primary interest, the enzymes. Partial proteolytic digestion to produce an active fragment of an enzyme was very much in vogue at that time, like site-directed mutagenesis today, but had met with very little success. With my experience on cytochrome c [2–4], I suspected that the active site might spread over different segments of the primary sequence and the prospect for obtaining an active fragment considerably reduced in size compared to the native enzyme may not be very bright. Chemical modification studies would be a more profitable approach to map the active sites of enzymes.

Looking up the literature, I was surprised to find that, although chemical modification of proteins had been extensively used to ascertain the amino acid residues which are essential for the activities of proteins, no quantitative treatment of the results obtained was available for the determination of the number of relevant residues which are indeed required for activity. The paper by Ray and Koshland in 1961 on this subject was a timely publication [31]. However, their method was based on the comparison of the first-order rate constants for the modification and inactivation reactions and does not apply when the modification reaction is not first-order or too fast for the course of reaction to be conveniently followed.

At that time, there already existed in the literature a fairly large number of papers providing data for both the extents of modification and the loss of activity. Not infrequently, when a number of groups of the same kind, X, was modified by a certain reagent with complete loss of activity, these groups were all considered to be essential for the activity of the enzyme. It appeared evident to me that, if the number of essential groups among those modified is i, the modification of any one of them would lead to complete inactivation and the random modification of a fraction, x, of these groups would result in an overall fractional activity remaining of x^i. A statistical treatment of the quantitative relation between the extents of modification of the side-chain functional groups and the inactivation of the protein concerned was published in 1962 in the English edition of *Scientia Sinica* [32]. In this paper, equations relating modification and inactivation were derived for different cases and a graphical method was proposed for the determination of the number of essential residues among those modified by suitable plots. Taking the simplest case of all susceptible groups of the same kind are modified with the same rate, a plot of the ith root of the fractional activity remaining, a, to the extent of modification should give a straight line when the number of essential groups among those modified is i.

$$a=x^i \text{ or } \sqrt[i]{a} = x \qquad (1)$$

In that paper, the existing data in the literature were treated accordingly so as to provide definitive conclusions on the number of essential groups among those modified. Fig. 6 gives an example taken from the 1962 paper showing that only 2 among the 28 carboxyl groups of ovomucoid trypsin inhibitor esterified are in fact essential for its activity. The hypothetical curves for one and three essential groups are also given for comparison.

Unfortunately, as it must be admitted that very few people ever bothered to look up *Scientia Sinica*, for many years this paper remained practically unknown. Freedman and Radda [33] and Paterson and Knowles [34], both in Oxford at that time, were among the first not only to read the paper but also to use the method. However, because of the interruption of my scientific activities, their papers came to my notice only after the 'Cultural Revolution'.

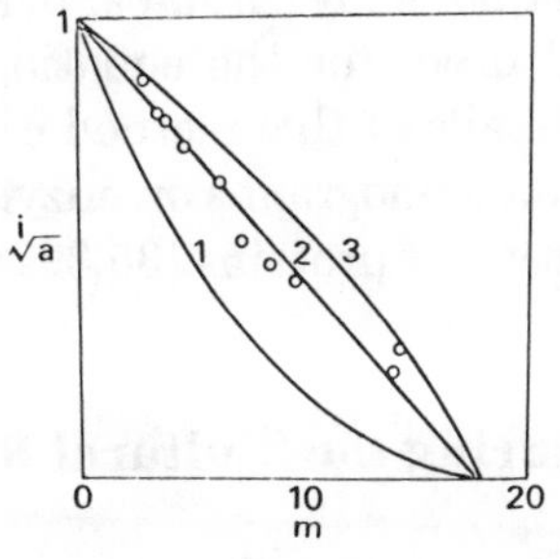

Fig. 6. Number of essential groups in pepsin modified by acetylation as determined with a Tsou plot. The straight line with $i = 2$ (curve 2) shows the acetylation of two essential amino groups. The hypothetical curves for 1(1) and 3(3) essential groups are also given for comparison. (For source of experimental data and other details, see [30].)

The method has gradually become fairly well cited in the biochemical literature and the plot as that shown in Fig. 6 has since become known as the Tsou plot. The above method has the advantage of simplicity over the kinetic method of Ray and Koshland and is applicable when the modification reaction is either too fast or not first-order where Ray and Koshland's method cannot be used. According to the *Science Citation Index*, it is the most cited paper ever published in *Scientia Sinica* and accounted for over one tenth of all the citations to that journal, but of course it is nothing compared with the citations of Lowry's method for the determination of proteins. During one of my recent visits to the Biochemistry Department of Cambridge, Richard Perham told me that if one took Vol. 11 of *Scientia Sinica* from the shelves of the University library, one could easily locate this particular paper without remembering the page number by looking at the blackened pages as a result of thumb marks. When I was introduced to Robert Feeney of Davies, California, during the FASEB Meeting at San Francisco in 1983, he surprised me by saying: 'So you are the man who owes me money'! He explained that because he cited this article in his book [35] on chemical modification, he had to pay for copying and mailing that paper as requests were sent to him instead of to me.

Theoretical aspects of the underlying principles have since been discussed and further explored by a number of authors [36,37] and a rigorous mathematical proof for the equations derived was published recently [38]. Details of this method can now be found in a number of textbooks and monographs on enzymes, enzyme kinetics and chemical modification of proteins [35,39–41].

Lost years during the 'Cultural Revolution'

Although the so-called 'Cultural Revolution' in China officially started May 16, 1966, it was plain for everyone to see that the storm was already gathering during the summer of 1965 when my first two papers on the *Kinetics of irreversible modification of enzyme activity* appeared in Chinese in *Biochimica Biophysica Acta Sinica* [42,43] which, according to my old friend E. C. Slater, is simply 'BBA China'. In those days practically no Chinese authors ever published outside China especially in the Western world and *Scientia Sinica* was the only home journal published in western languages, mostly English. At that time, the common practice was to publish our papers in Chinese first in 'BBA China' and upon the recommendation of the editors of that journal, the paper was then translated into English and submitted to *Scientia Sinica* for publication. I was about to translate into English these two articles which had taken up a sizable portion of my thoughts in that period and publish them in *Scientia Sinica* but alas, this was not to be and I had to wait for 17 years to come back to this problem and have the essence of these two articles published as an Appendix to a paper in *Biochemistry* in 1982 [44].

Serious fundamental research came to a complete standstill from 1966 to 1976 during the 'Cultural Revolution' – a revolution which, if anything, was anti-cultural. The study of insulin structure by x-ray crystallography and the effort to synthesize a transfer RNA chemically were among the very few exceptions. These projects were made possible because of the extremely wide publicity in the Western world of the total synthesis of insulin completed in 1965 and published in full early in 1966 at the time that the clouds of the 'Cultural Revolution' were already gathering. During this period, most intellectuals in China had to abandon their work and the scientists were no exception. Many were sent to do manual labour, in a farm or in a factory, especially in the early days of the tumult. When it was all over and I was able to go to the Western world for the first time since my return to China in 1951, I was frequently asked by many friends how did I fare during the 'Cultural Revolution'. I did no better and no worse than the average senior Chinese scientist. Like the others, I had to do some manual labour both in a farm and in a factory. I was recommended to be a Fogarty Scholar-in-Residence at the National Institute of Health (NIH) of the U.S. in 1985 and asked to submit a publication list. People at the Fogarty International Center of the NIH were apparently puzzled by a gap of over ten years during that period. They asked my old friend Bert L. Vallee whether he knew what I was doing then. Bert later told me that his first impulse was to tell them that I was doing 'Agricultural Biochemistry', but on second thought he was a little apprehensive that some people may not take this as a joke and believed I could be singularly nonproductive during these years.

I moved from the Shanghai Institute of Biochemistry to Beijing in the middle of the 'Cultural Revolution' in 1970. However, when I arrived at the Institute of Biophysics, I was put in an office together with four other people and no laboratory space was given to me which was neither surprising nor mattered very much as, with a few exceptions, no one was then doing serious work anyway. A laboratory was subsequently given to me in a most unexpected way.

Following Nixon's visit to China in 1972, my old friend Emil Smith of the University of California at Los Angeles led the first delegation of American scientists to visit the Peoples' Republic in 1973. Before he came, he asked specifically to visit my 'laboratory'. This message was transmitted from diplomatic channels through the headquarters of Academia Sinica to the Director and Communist Party Secretary of the Institute of Biophysics and they knew that I did not have one. They had no other alternative but to give me a 'laboratory' in a hurry, only a few weeks before the arrival of Emil. I had to borrow benches and glassware to decorate the room, I hoped, adequately. The day when Emil came to see my 'laboratory', all the reagent bottles contained pure tap water! Nevertheless, I was very pleased that I could keep that laboratory after Emil's visit but had to return the lab benches. It was a few years later that I could furnish the room and began to do a little work.

Out of his kindness, when asked by my American friends on his return to the U.S. about how I fared, Emil always answered I was doing all right. However, when I saw Emil a few years later and told him the above story, he said 'of course, I saw at first glance that you weren't doing anything'! Emil also told me that this was the second time that this kind of thing happened to him. The first time was when he visited Sweden after the War and asked to see B. Malmstrom who was then serving in the Swedish Army. Malmstrom was given a temporary leave to go back to his lab to show Emil around and was allowed to quit the army permanently after Emil's visit.

The 'Cultural Revolution' ended officially in October, 1976 after which scientific life in China gradually returned to normal and it took a couple of years before the resurgence of scientific activities in all fields became apparent with the universities taking an increasing share of the fundamental research activities. During this period another center for biochemical research emerged in the Institute of Biophysics of Academia Sinica in Beijing.

I was able to resume my research work at the age of 53, an age at which, according to Confucius, I should 'know the orders of heaven'. However, some help from heaven was indeed required as it took some time for me to get things moving again. For one thing, I had practically no equipment whatsoever. I picked rabbit muscle D-glyceraldehyde-3-phosphate dehydrogenase to start with for the only reason that it can be easily purified and crystallized in good quantity with glass jars and funnels only. Moreover, the supply of reagents was much worse than it was in the early 1950s. For instance, we had to recrystallize ammonium sulfate as the local product available at that time was no better than that used as fertilizers. We obtained distilled water from the local bath house and made our own redistilled water.

Considering the above circumstances, it was perhaps not surprising that it took two years before my first paper after the 'Cultural Revolution' appeared in 1979 [45]. Since I returned to China from England in 1951, political tumult occurred frequently and each time it lasted a few years. Mao once said that it was necessary to have a 'Cultural Revolution' every seven or eight years and on that some intellectuals remarked: 'Unfortunately each time it also lasted seven or eight years.' This was nearly true if one includes all large-scale political movements as mini cultural revolutions. As a consequence, of the period 1951–1977, only less than one third could be devoted to serious research. Nevertheless, as far as interruptions are concerned, it improved greatly from 1977 on and we had a continuous stretch of over ten years to do serious research. For getting anything worthwhile, such an uninterrupted stretch of time is absolutely essential even under difficult conditions.

After the 'Cultural Revolution', I published my papers in *Scientia Sinica* and other Chinese journals under a different name, which must be somewhat perplexing to my friends abroad. Pingying was introduced to transliterate Chinese characters into the roman alphabet and by decree of the State Council during the 'Cultural Revolution', all Chinese names had to be transliterated with Pingying. Thus Peking became Beijing and Chen-Lu Tsou became Zou

Chenglu because the family name comes first in Chinese. No matter how you sign your name in your paper, the editorial offices of all Chinese journals always change it automatically into Pingying with the family name first. Very recently, taking full advantage of my position as the Deputy Editor-in-Chief of the journal *Chinese Science Bulletin*, I finally persuaded the Head of the Editorial Office of this journal to publish one of my invited review articles under the name of Tsou, Chen-Lu, followed by the name in Pingying in parentheses. It appeared to me to be a convincing argument which, I hoped, had persuaded the Head of the Editorial Office that for the attraction of both authors and readers, it is not the best policy for a journal to change an established scientist into a nobody with his name in Pingying. I carefully checked this on the page proof but, alas, when it finally appeared [46], I found my name again as ZOU Chenglu followed by Tsou, Chen-Lu in parentheses instead! Abstract journals in the West, unfortunately, never take notice of the names of Chinese authors in parentheses.

On my passport, my name appears, of course, in Pingying. Anthony Linnane kindly invited me in the early 1980s for a lecture tour in Australia and provided me with a round trip air ticket from Hongkong to Melbourne. When I was to pick up that ticket at the Cathay Pacific Office in Hongkong, I had a little difficulty in proving that the passport bearer was the same person the ticket was for. This was several years ago and people are now used to the two different ways of transliterating Chinese names. It is a nuisance to find the names of a number of Chinese authors including myself to appear in two different places in *Chemical Abstracts* and in *Science Citation Index*. Undoubtedly, thanks to this practice, I am receiving twice as much junk mail as most others.

Ever since the founding of the Peoples' Republic, it was particularly difficult for people engaged in basic research, as we are under constant pressure to produce something which can be easily converted to hard cash. Sometimes it was in the form of direct pressure from above and recently, in the more subtle form, of withholding financial support. It is necessary for us to persuade those who hold the purse string that fundamental research is worth the money in the long run. It always takes some time to convince them and unfortunately they do not remain in office long enough and we often have to repeat the same argument all over again to convince new people who come into power.

Since the end of the 'Cultural Revolution' we have had an enormous increase in the exchange with the outside world. It was during the 11th International Congress of Biochemistry in Toronto that, with Y.-L. Wang, I took part in the negotiations leading to the joint discovery that a formula could be found for both the Chinese Biochemical Society and the Biochemical Society located in Taipei, Taiwan, to join the International Union of Biochemistry. For this I wish to take the opportunity to thank W. Whelan (General Secretary) and E. C. Slater (Treasurer) of the IUB, T.-B. Lo and J.-C. Su of the Biochemical Society in Taipei for their patience in finding a formula that not only both the Chinese biochemical societies can remain in the IUB at the same time but also paved the way for other societies to join the respective international unions. Since then Chinese scientists from both sides of the Strait have participated side by side in numerous international meetings.

Almost all of my senior colleagues have worked for some time abroad and I myself go to meetings and other visits several times each year so that one of my younger colleagues once remarked that I go abroad more often than to downtown Beijing. It was in Toronto that I saw Bert Vallee again whom I had known since the First International Congress of Biochemistry in Cambridge. This led to frequent mutual visits and my appointment as a visiting Professor in Harvard Medical School in 1981–1982. No less important is the fact that we are now not only allowed but actually encouraged to publish outside China. As far as I know, I was probably the first to publish abroad in 1979 with a letter to *Nature* [45], exactly 30 years after my first ever paper appeared in that journal [2]. In recent years, I am publishing most of my papers in Western journals. As suggested by Bill (E.C.) Slater, I was invited to join the Editorial Board of *Biochimica Biophysica Acta* in 1979 and began a long association with that journal. I am listed as C.-L. Tsou and not Zou Chenglu. Bill remarked that the BBA had only limited space and regretted that it was not possible to list both my names with an explanation of the Pingying system. Since then, I have served on the

Editorial or Advisory Board of a number of Western journals including *Analytical Biochemistry, Biochemistry* and the *FASEB Journal*.

Kinetics of irreversible modification of enzyme activity

After the 'Cultural Revolution', when it became possible for me to do some fundamental research again, I had to begin with something which not only would produce quick results but also require the very minimum of funds and lab equipment. I started with D-glyceraldehyde-3-phosphate dehydrogenase for the simple reason that it can be purified and crystallized without using a centrifuge which we did not have at that time. This enzyme proved to be very obliging in producing a fluorescent derivative when the carboxymethylated enzyme was irradiated in the presence of NAD^+ [45,47]. Nevertheless, as conditions in my lab gradually improved, I was soon able to continue my old interests in a number of different subjects.

Enzyme inhibition has always been an important field of study owing not only to its usefulness in providing information on such fundamental aspects of biochemical problems as enzymatic catalysis and metabolic pathways but also to its implications in pharmacological and toxicological problems. In most textbooks on enzyme kinetics, chapters on enzyme inhibition have been devoted almost entirely to reversible inhibition with barely a passing mention on irreversible inhibition which recent developments have shown to be at least equally important.

This was the problem which occupied a great deal of my thoughts before the 'Cultural Revolution' resulting in two papers published in Chinese [42,43]. I was translating these two papers into English for publication in *Scientia Sinica* and had one of my graduate students beginning to do some experimental studies when it was abruptly interrupted and all the efforts had to be abandoned in spite of the fact that the initial results of the experimental studies had shown very satisfactory agreement with the equations derived. I had to wait 17 years before I was finally able to continue where I had left off and had a different graduate student to finish the experiments started before the 'Cultural Revolution'. The results were published in a Western journal [44] in 1982 with the two papers in Chinese highly condensed as an appendix to this paper. During the 'Cultural Revolution' I was very much interested to read in the second edition of the well known textbook on enzyme kinetics by Laidler and Bunting [48], published in 1973, a very similar treatment on the kinetics of enzyme inactivation during denaturation by guanidine HCl and urea. The equations presented by these authors are identical to some of those given in my two papers in Chinese [42,43].

The conventional method for the determination of the rate constants for the irreversible modification of enzyme activity is to take aliquots from an incubation mixture of the enzyme with the modifier at definite time intervals and assay for the activity. This method is not only laborious and time consuming but also not applicable to fast reactions with a halftime of, say, a few seconds. In my systematic study on the kinetics of irreversible modification of enzyme activity in 1965 published in Chinese, the kinetics of irreversible inhibition during the substrate reaction was considered. The production of product, P, in the presence of both substrate and the inactivator, Y, has been shown to be:

$$[\mathrm{P}] = \frac{v}{A[Y]}\,(1 - \mathrm{e}^{-A[Y]t}) \tag{2}$$

where v is the reaction rate in absence of the inhibitor and A the apparent modification rate constant. From the above, it can be shown that the asymptote of a plot of [P] against t gives the value of $[\mathrm{P}]_\infty$, the concentration of product formed at time infinity as shown in Fig. 7 [49].

$$[\mathrm{P}]_\infty = v/A[Y] \tag{3}$$

A semilogarithmic plot of Eqns (2)-(3) gives a series of straight lines for different [Y], as shown in the inset of Fig. 7.

$$\log([\mathrm{P}]_\infty - [\mathrm{P}]) = \log[\mathrm{P}]_\infty - 0.43\mathrm{A}[Y]t \tag{4}$$

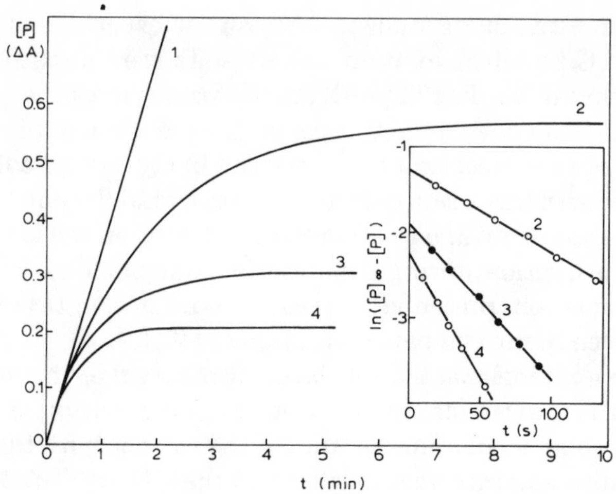

Fig. 7. Substrate reaction of creatine kinase in the presence of an inactivator. The reaction was followed by absorbance change of thymol blue 597 nm, at 25°C due to proton generation. Concentrations of the inactivator, 5,5′-dithiobis(2-nitrobenzoic acid) were, 0, 1.5, 3, 4.5 μM, respectively, for curves 1 to 4. The inset shows semi-logarithmic plots of lines 2–4 according to Eqn. (4). (Reproduced with permission from Biochemistry, 27, 5095. Copyright American Chemical Society, 1988.)

Since [Y] is known and v can be separately determined in experiments in the absence of the inhibitor, the apparent rate constant, A, is easily obtained directly either from Eqn. (3) or (4). The rate constants thus obtained for a number of known reactions agree satisfactorily with those obtained by the conventional method.

Thus, instead of taking several aliquots from an incubation mixture of the inhibitor and the enzyme at time intervals and assay for the activity remaining, the apparent rate constant for the irreversible modification of enzyme activity can now be obtained in one experiment. In the present approach, with a stopped-flow apparatus, a first-order rate constant in the order of 10 s^{-1} can be easily obtained. With the conventional method, the determination of fast-inhibition rate constant would require a stopped-flow machine and have the reaction quenched at different intervals. Both reversible and irreversible inhibitions are considered in a general scheme in the recent approach and the concept of substrate competition has been shown to be applicable to both reversible and irreversible inhibitions. The effect of substrate on the apparent equilibrium constant for the binding of the inhibitor to the enzyme has been routinely used to determine whether the inhibitor is competitive, noncompetitive or uncompetitive for reversible inhibitions; it is the effect on the apparent rate constant for the binding of the enzyme with the inhibitor which can be used for irreversible inhibitions.

Since the appearance of the paper in *Biochemistry* [44] a few years ago, this method has been employed in a number of different laboratories and produced useful results not only for inhibition kinetics [50–55] but also for activation kinetics [56]. Because of its simplicity, it is particularly useful for the determination of rate constants when a large number of inhibitors are to be compared [50]. A review article was published in 1985 [57] and when it was suggested that the derivation of the equations was to be included as an Appendix, I gladly complied. The resulting article contained most of the more important points of the two papers in Chinese. After the publication of this review, the above approach has been further shown to be applicable to enzyme reactions involving two substrates [49] and to slow but reversible inhibitions [58].

Insulin again

The job on insulin was given to me in 1958 to see whether the insulin chains can be put together again to form native insulin, but since then I have always kept an interest on this small but fascinating protein. Because of the successful resynthesis of insulin from its chains with a reasonably good yield, we proposed that the two chains are capable of recognizing each other and the insulin structure is the most stable among all its isomers. The former suggestion implies that the A and B chains contain in themselves enough structural information for them to be joined correctly. However, for a long time neither the fact that insulin can be resynthesized from its chains, nor the above proposition were accepted in the international literature, as can be read in a number of review articles [27–29] of that period and textbooks up to the present time [59–62]. In a much-used textbook published in 1982 [59], it was given as a problem to students to contemplate why all the other disulfide-containing proteins examined (ribonuclease, lysozyme and alkaline phosphatase were listed) either give a nearly quantitative yield or at least much higher yields than expected from random joining of thiols to form disulfides during oxidation of the reduced molecules, whereas the oxidation of reduced insulin was said to give the poor yield of 5–10% as expected for random joining of the thiols. Actually, it is now an established fact that the yield of insulin from the A and B chains is far better than 5–10% [20,25]. Under appropriate conditions, yields up to 70% can be achieved which was good enough for the commercial production of human insulin from its chains obtained separately by recombinant DNA techniques [26]. On the other hand, had the joining of the thiol groups to form disulfide bonds been completely random, not only insulin isomers containing one each of A and B chains, but all possible oligomeric products with different chain combinations should be considered the total number of which is immense indeed [26,63]. Consequently, the yield expected from random joining of the chains should be close to zero as has been found by oxidation of the reduced chains in 6 M guanidine HCl or in 0.1% sodium dodecyl sulfate [64]. The above was the background when it appeared to me that it was time to come back to do a little more work on this protein.

It turns out that people know of me through different aspects of my work and, as I mentioned earlier, my work on insulin is probably the best known to the general public. I was elected a Foreign Member of the Academy of Sciences of the German Democratic Republic and invited to be a Scholar-in-Residence in the Fogarty International Center of NIH probably mainly because of this work. I took full advantage of this to get better financial support both at home and abroad for my research work when I was back on insulin again.

Invited by my old friend Bert Vallee, I was in the U.S. during 1981–1982 as a visiting Professor at Harvard Medical School; I took the opportunity to travel extensively and, while discussing my work and the general situation of basic research in China during one of my visits with another old friend, Howard Mason at Portland, he remarked: 'Why don't you apply for an NIH research grant'. I discussed this matter with Bert when I was back in Boston and he suggested that continuing my earlier work on insulin would be a good idea. I gathered a few young people and started to follow my old suggestion to investigate the interaction of the insulin chains when I was back in Beijing. I applied and succeeded to get an NIH grant in 1984 with my colleague Dr. C.-C. Wang as my co-principal investigator and had it continued in 1987. This was of great help to me when hard currency was still indeed 'hard' to come by. Even a limited amount of hard currency can be very useful. Twice every year, we can place orders abroad through the Academy for reagents that are not available in China. It usually takes such a long time for these reagents to come that not infrequently when I eventually do get them, I have clean forgotten what I ordered them for!

I first set out to see whether the interaction of the insulin A and B chains can be demonstrated experimentally and was able to show by difference spectroscopy, solvent perturbation CD and NMR spectra of the separated and mixed chains that they indeed do interact in solution and that modification prevents the proper interaction of the chains [65–68]. Furthermore, Fourier transformed infrared spectra (FTIR) of insulin and its chains show that, although the separated A and B chains only partly contain the secondary structures present in insulin, a mixture of the chains has an FTIR spectrum with all the characteristic peaks for that of the intact hormone.

The failure reported in the literature to obtain native insulin from scrambled insulin by protein disulfide isomerase [30] is probably partly responsible for the proposition that the insulin structure is a metastable form among all the possible isomers. We have now shown that this can indeed be achieved under appropriate conditions with 20–30% yield. Moreover, HPLC analysis shows that native insulin is by far the predominant peak among all the products containing both chains, indicating that under suitable conditions the native structure of insulin is indeed the most stable among all possible isomers containing one each of the two chains [69]. We have recently shown with the two chains crosslinked through NαA1-NϵB29 by carbonyl-bis(*L*-methionyl *p*-nitrophenylester) that, under carefully controlled conditions, a nearly quantitative yield of crosslinked insulin with native disulfide linkages can be obtained from the *S*-sulfonate of the crosslinked molecule by treatment with protein disulfide isomerase. As it is most unlikely that the crosslinking carbonyl-bis-methionyl moiety contains any structural information to assist the correct pairing of the chains, the above results strongly suggest that the necessary structural information is provided entirely by the A and B chains themselves [70]. The yields of the various unscrambling reactions with protein disulfide isomerase are summarized in Table II.

TABLE II
Final yields of products with native disulfide bridges from unscrambling and scrambling reactions[a]

Direction of reaction	Reactant(s)	Yield, %
Unscrambling	Scrambled insulin	25
	S-sulfonates of the chains	30
	Scrambled CBM[b]-insulin	55
	CBM-insulin *S*-sulfonate (0.6 mg/ml)	75
	(0.1 mg/ml)	>90
Scrambling	Insulin	25
	CBM-insulin	75

[a] Both scrambling and unscrambling reactions were carried out with protein disulfide isomerase and dithiothreitol. The final yields were determined from the respective HPLC peaks. For insulin, measurements of biological activity gave the same results. For crosslinked insulin the identity of the product was demonstrated by the HPLC profile of the pepsin hydrolysate and by CNBr treatment to generate insulin.
[b] CBM, carbonyl-bismethionyl.

Location of the active sites of some enzymes in limited and flexible molecular regions

The effect of structural changes on the activity of enzymes has been extensively studied; however, previous authors have largely concentrated on the modification of the primary structures of the enzymes.

Although the importance of conformational integrity for enzyme activity is generally recognized and the unfolding of enzyme molecules well documented, activity changes accompanying the course of denaturation remain but little explored. It has occurred to me to see whether the inactivation parallels the conformational changes produced by denaturation which would reveal the stability of the active site relative to the molecule as a whole.

Guanidine HCl or urea at low concentrations inactivates creatine kinase without significant unfolding of the molecule and the inactivation at low guanidine HCl concentrations is not due to the dissociation of the dimeric molecule into the monomer [71–73]. With a fluorescent NAD derivative introduced at the active site of *D*-glyceraldehyde-3-phosphate dehydrogenase, it is shown that the NAD binding site is also more sensitive to guanidine HCl denaturation than the molecule as a whole [74]. Similar results have also been obtained for a small enzyme stabilized with multiple disulfide linkages such as RNase A [75] and a few other enzymes.

Furthermore, when the rates of inactivation and unfolding during denaturation of enzymes are compared, it is evident that the initial inactivation rates of some enzymes by denaturants are very rapid. The inactivation rate of creatine kinase in either guanidine HCl or urea is too fast to be measured by the conventional method of taking aliquots at different time intervals of the incubation mixture of the enzyme with the denaturant and can best be measured by the kinetic method described above with a stopped flow apparatus. In fact, the apparent inactivation rate constant in 1 M guanidine HCl was found to be 4.3 s^{-1}. For some other enzymes, the reactions are so fast that they are nearly complete within the dead time of mixing in a stopped-flow apparatus. Because of the viscosity of the concentrated guanidine HCl this is usually about 50 ms. I am fairly certain that the rapid phase of inactivation is due to denaturant induced conformational changes at the active sites of the enzyme molecules involved. However, when the relevant papers were submitted for publication, some reviewers were equally certain that it was simply a reversible inhibition of the enzyme by the denaturants (although in the relevant references there is no definitive evidence to substantiate the claim that the decrease in enzyme activity in guanidine HCl is due to a reversible inhibition). It is indeed gratifying to me to be able to show that very rapid inactivation also occurs before a detectable conformational change during heat denaturation of some enzymes in the absence of denaturants [76]. Some of the results of rate measurements are summarized in Table III.

TABLE III
Comparison of unfolding and inactivation rate constants (s^{-1})

	Guanidine M	Rate constant s^{-1}			Residual activity %	Temp. °C
		Unfolding	Dissociation	Inactivation		
Creatine kinase	0.5	0.0043		3.6	15	25
	1	0.053		4.3	0	25
	3	1.9		5.9	0	25
GAPDH	0.3	0.00016	0.0016	>50	63	25
	0.5	0.0024		>50	32	25
RNase A	1			>50	27	10
	3	0.0096			3	10

For multiphasic reactions, the rate constants of inactivation are those of the fast phase.

The results obtained are highly suggestive that enzyme-active sites are situated in relatively more flexible or mobile regions of the molecules and are hence more sensitive to heat and to denaturants. It is known from x-ray crystallographic studies that in some proteins there are exposed peptide segments, usually without ordered secondary structure, free to move about in solution and probably nonessential for the biological activity of these proteins. The mobility of the active sites could only be relative and within certain limits. It may well be dictated by the requirements for the expression of enzyme catalytic activities.

Head of a National Laboratory and years ahead

A number of National Laboratories have been established in China since 1985 attached to research institutes or to the major universities but with additional and direct financial support from the State Commissions of Planning and of Science and Technology to promote basic research and the basic aspects of applied research in different fields. These laboratories have their own independent budgets and research projects but depend on the universities or research institutes they are attached to for administrative and other supporting staffs. Facilities will be provided and research funds made available for visiting scientists from both home and foreign institutions to work for up to three years in these laboratories so as to encourage not only exchange within China but also international collaboration. As these laboratories are selected from the best in China, it is hoped to help them in this way to enter more and more into the international scene and to bring other institutions in China with less research experience up to modern standards. Apart from these, the Academy of Sciences and the Education Commission both have their own Key Laboratories attached to the research institutes and the universities, respectively.

The selection of National Laboratories from a large number of research laboratories nationwide is based on the importance of the fields of study and the past contributions of their senior scientists. For their selection, they have to be first recommended by the State Commission of Education, the Chinese Academy of Sciences or one of the Ministries to which they are attached. The senior scientists of the proposed laboratory have then to present their cases for the establishment of such a National Laboratory, including the significance of the fields of studies proposed, their past contributions and research plans for the future before a panel of scientists chosen nationwide. The State Commission of Planning approves the establishment of a National Laboratory only when the panel makes a favourable recommendation by a majority vote.

My own lab (the Laboratory of Molecular Enzymology) was made a Key Lab of the Academy in 1985. Since 1988, together with two other research groups working on x-ray crystallography and biomembranes headed by D.-C. Liang and F.-Y. Yang, respectively, was made a National Laboratory of Biomacromolecules [77] and I was appointed its first Director for the initial three years. The important thing for me is that my research projects will be adequately supported and I do not have to worry constantly about doing something which will bring quick return of money to justify myself.

It is probably strange to write about the future at 66 which, in many countries, is past the age of retirement and, for many, the time to quit research and to write books as Prof. Keilin did after his retirement. I am indeed editing a book to make Chinese biochemical research in general better known to the outside world [78]. However, being a member of the Chinese Academy of Science, I have the privilege of not having to retire from my scientific activities unless I myself wish to do so. It was the teaching of the Sage of the Chinese people, Confucius, that: 'The highest grade of this clarifying activity has no limit' [1]. I always enjoy doing science and would not think of retiring unless forced to. Now that I am adequately supported to do what I always loved to do after so much strenuous efforts throughout the years, I certainly am not going to give it all up so soon. Moreover, Confucius also said: 'At seventy, I could follow my own heart's desire without overstepping the t-square' [1] which, as I understood from my school teacher when I was a boy, means that he could do whatever he wished without doing anything wrong. I would like to see whether I shall be like him at 70, although at 66, I am absolutely nowhere near it yet!

I returned to many of the fields which interested me in the early years of my scientific life: cytochromes, enzymes, kinetics and insulin. When Bill Slater heard about this, he remarked that I must be afflicted with the 'Hopkins Syndrome'. In his later years, the first Professor of Biochemistry at the University of Cambridge, Sir Fredrick Gowland Hopkins, returned to the fields of his old love one after another. For my part, I feel strangely attracted to those subjects which had interested me when I was young. I published my first scientific papers on the cytochrome system which, introduced to me by Prof. Keilin, was my first love in biochemistry. It is interesting to note by looking up the *Science Citation Index* that these papers are still occasionally cited after 40 years. I came back recently to the cytochromes but only briefly [79]. I had my hands full with other research interests and did not have the time to keep myself up-to-date with the developments in this interesting field. In spite of the advices by a few old friends, including B. Chance and Tsoo E. King that I should remain faithful to my first love, I finally decided, reluctantly, to give it up. Nevertheless, apart from starting something new [80], I may yet return and publish another paper on cytochrome *c* before I say goodbye to active research, which I hope will be sometime after 'I could follow my own heart's desire without overstepping the t-square' [1].

REFERENCES

1 All quotations of Confucius (ca. 500 B.C.) were from the books: *Chung Yung* and *The Analects*. English translations were from *Confucius*, by Erza Pound, New Directions Books, New York, 1928.
2 C.-L. Tsou, Nature, 164 (1949) 1134.
3 C.-L. Tsou, Biochem. J., 49 (1951) 362–367.
4 C.-L. Tsou, Biochem. J., 49 (1951) 367–374.
5 C.-L. Tsou, Biochem. J., 50 (1952) 493–499.
6 D. Keilin and E. F. Hartree, Nature, 176 (1955) 200–206.
7 A. M. Pappenheimer and E. D. Hendee, J. Biol. Chem., 180 (1949) 597–609.
8 C.-L. Tsou, Biochem. J., 49 (1951) 512–520.
9 H. Wu, Chin. J. Physiol., 4 (1931) 321–344.
10 C.-L. Tsou and C.-Y. Wu, Sci. Sin. (Eng. ed.), 5 (1956) 263–270.
11 T.-Y. Wang, C.-L. Tsou and Y.-L. Wang, Sci. Sin. (Eng. ed.), 5 (1956) 73–90.
12 D. Keilin and T. E. King, Nature, 181 (1958) 1520–1522.
13 T.-Y. Wang, C.-L. Tsou and Y.-L. Wang, Sci. Sin. (Eng. ed.), 7 (1958) 65–74.
14 T. P. Singer, E. B. Kearney and N. Zastrow, Biochim. Biophys. Acta, 17 (1955) 154–155.
15 C.-Y. Wu and C.-L. Tsou, Sci. Sin. (Engl. ed.), 4 (1955) 137–155.
16 H. N. Rydon, in G. E. W. Wolstenholme and C. M. Conner (Eds.), Ciba Found. Symp. on Amino Acids and Peptides with Antimetabolic Activity, Churchill, London, 1958, pp. 192–204.
17 G. H. Dixon and A. C. Wardlaw, Nature, 188 (1960) 721–724.
18 Y.-C. Du, Y.-S. Zhang, Z.-X. Lu and C.-L. Tsou, Sci. Sin. (Engl. ed.), 10 (1961) 84–104.
19 C.-L. Tsou, Y.-C. Du and G.-J. Xu, Sci. Sin. (Engl. ed.), 14 (1961) 229–236.
20 Y.-C. Du, R.-Q. Jiang and C.-L. Tsou, Sci. Sin. (Engl. ed.), 14 (1965) 229–236.
21 C.-I. Niu, Y.-T. Kung, W.-T. Huang, L.-T. Ke, C.-C. Chen, Y.-C. Chen, Y.-C. Du, R.-Q. Jiang, C.-L. Tsou, S.-C. Hu, S.-Q. Chu and K.-Z. Wang, Sci. Sin., 15 (1966) 231–244.
22 Y. Wang, J.-Z. Hsus, W.-C. Chang, L.-L. Chen, H.-S. Li, C.-Y. Hsin, P.-T. Shi, T.-P. Loh, A.-H. Chi, C.-H. Li, Y.-H. Yieh and K.-L. Tang, Sci. Sin., 14 (1965) 1887–1890.
23 Institute of Biochemistry, Academia Sinica, Department of Chemistry, Peking University and Institute of Organic Chemistry, Academia Sinica, Kexue Tungbao, 17 (1966) 241–277.
24 H. Zahn, B. Gutte, E. F. Pfeiffer and J. Ammon, Liebig's Ann. 691 (1966) 225–231.
25 P. G. Katsoyannis, A. C. Takatellis, S. Johnson, C. Zalut and G. Schwartz, Biochemistry, 6 (1967) 2642–2655.
26 R. E. Chance, J. A. Hoffmann, E. P. Kroeff, M. G. Johnson, E. W. Schirmer, W. W. Bromer, M. J. Ross and R. Wetzed, in D. H. Rich and E. Gross (Eds.), Peptides, Synthesis, Structure and Function, Proceedings of the 7th American

Peptide Symposium, Peirce Chemical Company, Rockford, IL, 1981, pp. 721-728.
27 C. B. Anfinsen and H. A. Scheraga, Adv. Prot. Chem. 29 (1975) 205-300.
28 D. Brandenburg, H.-G. Gattner, W. Schermutzki, A. Schuttler, J. Uschkoreit, J. Weimann and A. Wollmer, in M. Friedman (Ed.), Protein Crosslinking, Part A, Plenum, New York, 1977, pp. 261-282.
29 D. A. Hillson, N. Lambert and R. B. Freedman, Methods Enzymol. 107 (1984) 281-294.
30 P. T. Varandani and M. A. Nafz, Arch. Biochem. Biophys. 141 (1970) 533-537.
31 W. J. Ray and D. E. Koshland Jr., J. Biol. Chem., 236 (1961) 1973-1979.
32 C.-L. Tsou, Sci. Sin. (Engl. ed.), 11 (1962) 1535-1558.
33 R. B. Freedman and G. K. Radda, Biochem. J., 108 (1968) 383-391.
34 A. K. Paterson and J. R. Knowles, Eur. J. Biochem., 31 (1972) 510-517.
35 R. E. Feeney, R. B. Yamasaki and K. F. Geoghegan, in R. E. Feeney and J. R. Whitaker (Eds.), Modification of Proteins. Food, Nutritional and Pharmacological Aspects, Advances in Chemistry, Vol. 198, American Chemical Society, Washington, DC, 1982, pp. 35-37.
36 K. Horiike and D. B. McCormick, J. Theoret. Biol., 79 (1979) 403-414.
37 E. T. Rakitzis, Biochem. J., 223 (1984) 259-262.
38 Z.-X. Wang, Acta Biophys. Sin., 3 (1987) 215-221.
39 M. Dixon and E. C. Webb, Enzymes, 3rd ed., Longmans, London, 1979, pp. 376-379.
40 E. Cornish-Bowden, Fundamentals of Enzyme Kinetics, Butterworth, London, 1979, 94-96.
41 R. L. Lundblad and C. M. Noyes, Chemical Reagents for Protein Modification, CRC Press, Boca Raton, FL, 1984, pp. 2-6.
42 C.-L. Tsou, Acta Biochim. Biophys. Sin., 5 (1965) 398-408.
43 C.-L. Tsou, Acta Biochim. Biophys. Sin., 5 (1965) 409-417.
44 W.-X. Tian and C.-L. Tsou, Biochemistry, 21 (1982) 1028-1032.
45 Y.-S. Ho and C.-L. Tsou, Nature, 277 (1979) 245-246.
46 C.-L. Tsou, Chin. Sci. Bull., 34 (1989) 793-799.
47 C.-L. Tsou, G.-Q. Xu, J.-M. Zhou and K.-Y. Zhao, Biochem. Soc. Trans., 11 (1983) 425-429.
48 K. J. Laidler and P. S. Bunting, The Chemical Kinetics of Enzyme Action, 2nd ed., Clarendon Press, Oxford, 1973, pp. 175-180.
49 Z.-X. Wang, B. Preiss and C.-L. Tsou, Biochemistry, 27 (1988) 5095-5100.
50 S. P. Leytus, D. L. Toledo and W. F. Mangel, Biochim. Biophys. Acta, 788 (1984) 74-86.
51 J. W. Harper, K. Hemmi and J. C. Powers, Biochemistry, 24 (1985) 1831-1843.
52 M. Silverberg, J. Longo and A. P. Kaplan, J. Biol. Chem., 261 (1986) 14965-14968.
53 R. A. Wijnands, F. Muller and A. J. W. G. Visser, Eur. J. Biochem., 163 (1987) 535-544.
54 C. Crawford, R. W. Madson, P. Wikstrom and E. Shaw, Biochem. J., 253 (1988) 751-758.

55 Wolz, R. L. and R. Zwilling, J. Inorg. Biochem., 35 (1989) 157–167.
56 W. Liu, K.-Y. Zhao and C.-L. Tsou, Eur. J. Biochem., 151 (1985) 525–529.
57 C.-L. Tsou, Adv. Enzymol., 61 (1988) 381–436.
58 J.-M. Zhou, C. Liu and C.-L. Tsou, Biochemistry, 28 (1989) 1070–1076.
59 A. L. Lehninger, Principles of Biochemistry, 3rd ed., Worth, New York, 1982, pp. 202–203.
60 T. M. Devlin, Textbook of Biochemistry with Clinical Correlations, 2nd ed., Wiley, New York, 1986, p. 756.
61 L. Stryer, Biochemistry, 3rd ed., Freeman, New York, 1988, p. 41.
62 B. Alberts, D. Bray, J. Lewis, M. Raff, K. Roberts and J. D. Watson, Molecular Biology of the Cell. 2nd ed., Garland Publishing, New York, 1989, p. 122.
63 Z.-X. Wang, M. Ju and C.-L. Tsou, J. Theoret. Biol., 124 (1987) 293-301.
64 Y.-Q. Qian and C.-L. Tsou, Biochem. Biophys. Res. Commun., 146 (1987) 437–442.
65 Q.-X. Hua, Y.-Q. Qian and C.-L. Tsou, Biochim. Biophys. Acta, 789 (1984) 234–240.
66 Q.-X. Hua, Y.-Q. Qian and C.-L. Tsou, Sci. Sin., 28B (1985) 854–862.
67 C.-C. Wang and C.-L. Tsou, Biochemistry, 25 (1986) 5336–5340.
68 Y. Tian, C.-C. Wang and C.-L. Tsou, Biol. Chem. Hoppe-Seyler, 368 (1987) 397–408.
69 J.-G. Tang, C.-C. Wang and C.-L. Tsou, Biochem. J., 255 (1988) 451–455.
70 J.-G. Tang and C.-L. Tsou, Biochem. J., 268 (1990) 429–435.
71 C.-L. Tsou, Trends Biochem. Sci., 11 (1986) 427–429.
72 Q.-Z. Yao, M. Tian and C.-L. Tsou, Biochemistry, 23 (1984) 2740–2744.
73 Q.-Z. Yao, S.-J. Liang, M. Tian and C.-L. Tsou, Sci. Sin., 28B (1985) 484–493.
74 G.-F. Xie and C.-L. Tsou, Biochim. Biophys. Acta, 911 (1987) 19–24.
75 W. Liu and C.-L. Tsou, Biochim. Biophys. Acta, 916 (1987).
76 Y.-Z. Lin, S.-J. Liang, J.-M. Zhou, C.-L. Tsou, P. Wu and Z. Zhou, Biochim. Biophys. Acta (1990), In Press.
77 C.-L. Tsou, FASEB J. 3 (1989) 2443–2444.
78 C.-L. Tsou (Ed.), Current Biochemical Research in China, Academic Press, New York, 1989.
79 D. C. Wang, S.-L. Li, and C.-L. Tsou, Sci. Sin. (Eng. ed.), 28B (1985) 942–951.
80 C.-L. Tsou, Biochemistry, 27 (1988) 1809–1812.

Name Index